New JIT, New Management Technology Principle

New JIT, New Management Technology Principle

Kakuro Amasaka

CRC Press is an imprint of the
Taylor & Francis Group, an **informa** business

CRC Press
Taylor & Francis Group
6000 Broken Sound Parkway NW, Suite 300
Boca Raton, FL 33487-2742

© 2015 by Taylor & Francis Group, LLC
CRC Press is an imprint of Taylor & Francis Group, an Informa business

No claim to original U.S. Government works

Printed on acid-free paper
Version Date: 20141003

International Standard Book Number-13: 978-1-4665-7502-8 (Hardback)

This book contains information obtained from authentic and highly regarded sources. Reasonable efforts have been made to publish reliable data and information, but the author and publisher cannot assume responsibility for the validity of all materials or the consequences of their use. The authors and publishers have attempted to trace the copyright holders of all material reproduced in this publication and apologize to copyright holders if permission to publish in this form has not been obtained. If any copyright material has not been acknowledged please write and let us know so we may rectify in any future reprint.

Except as permitted under U.S. Copyright Law, no part of this book may be reprinted, reproduced, transmitted, or utilized in any form by any electronic, mechanical, or other means, now known or hereafter invented, including photocopying, microfilming, and recording, or in any information storage or retrieval system, without written permission from the publishers.

For permission to photocopy or use material electronically from this work, please access www.copyright.com (http://www.copyright.com/) or contact the Copyright Clearance Center, Inc. (CCC), 222 Rosewood Drive, Danvers, MA 01923, 978-750-8400. CCC is a not-for-profit organization that provides licenses and registration for a variety of users. For organizations that have been granted a photocopy license by the CCC, a separate system of payment has been arranged.

Trademark Notice: Product or corporate names may be trademarks or registered trademarks, and are used only for identification and explanation without intent to infringe.

Library of Congress Cataloging-in-Publication Data

Amasaka, Kakuro, 1947-
　New JIT, new management technology principle / Kakuro Amasaka.
　　　pages cm
　Includes bibliographical references and index.
　ISBN 978-1-4665-7502-8 (hardback)
　1. Just-in-time systems. 2. Quality control. 3. Management. I. Title.

TS157.4.A43 2014
658.5'1--dc23
　　　　　　　　　　　　　　　　　　　　　　　　　　　　　　2014027311

Visit the Taylor & Francis Web site at
http://www.taylorandfrancis.com

and the CRC Press Web site at
http://www.crcpress.com

Contents

Preface xxi
Aims and Structure xxiii
Overview xxv
Author xxvii
Contributing Authors... xxix
Acknowledgments xxxi

1   Introduction 1
    References 2

2   Management Tasks of Manufacturing Companies Shifting to
    Global Production 5
    2.1   Progress of Production Control in the Manufacturing Industry 5
    2.2   Today's Management Technology Issues..7
    References 9

3   Basics of JIT: The Toyota Production System............................... 11
    3.1   Toyota Production System: The Core Concept of JIT..................... 11
    3.2   Basic Principle of Manufacturing via the Toyota Production
          System 13
          3.2.1   Simultaneous Realization of Quality and Productivity
                  via a Lean system 13
          3.2.2   Process Control and Process Improvement 14
                  3.2.2.1   Process Control and Process Improvement 15
                  3.2.2.2   Daily Control Activities at the Production Site...17
                  3.2.2.3   Daily Improvement Activities at the
                            Production Site 17
                  3.2.2.4   Innovation of the Production Process: QCD
                            Research Activities by White-Collar Staff...........18
    References 21

v

4 Innovation of Manufacturing Fundamentals Surpassing JIT 23
- 4.1 Necessity for an Advanced Production System and Advancement in Technologies and Skills 24
  - 4.1.1 What Are the Concerns of Top Management and Manager Classes? 24
  - 4.1.2 Japanese Manufacturing: Reinforcement of the Quality Technology 25
    - 4.1.2.1 Capabilities Are Needed 25
  - 4.1.3 Style of Manufacturing Required in Overseas Production Plants 27
    - 4.1.3.1 Manufacturing Products at Overseas Plants with the Same Quality as Japan at the Same Time 27
    - 4.1.3.2 From Reinforced Inspection to Enhanced Process Management 27
    - 4.1.3.3 Production System Issues at Overseas Production Plants 28
- 4.2 Advancement in Manufacturing: Reinforcement of the Production Management Function 29
- References 31

5 New JIT, New Management Technology Principle 33
- 5.1 New JIT, A New Management Technology Principle 33
  - 5.1.1 Introduction 33
  - 5.1.2 Toyota Production System and TQM: JIT 34
  - 5.1.3 Today's Management Technology Problem 34
  - 5.1.4 Proposal of New JIT for Renovating Management Technology 35
    - 5.1.4.1 The Concept of New JIT for Renovating Management Technology 35
    - 5.1.4.2 Three Subsystems of New JIT 37
  - 5.1.5 Implementation and Effect of New JIT through TQM-S 39
    - 5.1.5.1 Implementation of Vehicle Profile Design Using TDS 39
    - 5.1.5.2 Establishment of New Toyota Production System Using TPS 40
    - 5.1.5.3 Establishment of Toyota Sales Marketing System Using TMS 41
  - 5.1.6 Conclusion 41
- 5.2 Foundation of Advancing the Toyota Production System 42
  - 5.2.1 Introduction 42
  - 5.2.2 Typified Japanese Production System Called JIT 42

		5.2.3	Foundation of Advancing the Toyota Production System 43
			5.2.3.1 Demand for Advancement in Management Technology..43
			5.2.3.2 Need for Strategic QCD Studies with Affiliated and Nonaffiliated Suppliers................43
			5.2.3.3 Significance of Strategic Implementation of New JIT 44
			5.2.3.4 Cooperation between On-Site White-Collar Engineers, and Supervisors and Workers 46
		5.2.4	Application to Simultaneous Achievement of QCD through Innovation of Manufacturing Technology: Innovation of the Production Line of Automobile Rear Axle Unit for New JIT Production 46
			5.2.4.1 Outline of Production Process of Rear Axle Units ...47
			5.2.4.2 Before Process Improvement: Bottleneck-Like Manufacturing Technical Problems That Inhibit JIT..49
			5.2.4.3 After Process Improvement: Simultaneous Achievement of QCD by Innovation of Manufacturing Technology 50
		5.2.5	Solution to Bottleneck Problems Related to Processing of Rear Axle Shaft 50
			5.2.5.1 Development of Midfrequency Tempering Equipment..50
			5.2.5.2 Development of Automated Stress Relief Equipment..51
		5.2.6	Solution to Bottleneck Problems Related to Assembly Process of Rear Axle Unit..51
			5.2.6.1 Simultaneous Achievement of QCD for Corrosion Resistance Coating............................51
			5.2.6.2 Improvement of Bolt-Tightening Tools...............54
		5.2.7	Conclusion 54
	5.3	Total Quality Assurance Networking Model...................................55
		5.3.1	Introduction 55
		5.3.2	Background 55
			5.3.2.1 Manufacturing in Japan Today..........................55
			5.3.2.2 Necessity of Preventing Defects........................55
		5.3.3	Issues Faced by Quality Assurance Departments................56
		5.3.4	Creating the Total Quality Assurance Networking Model.......57
			5.3.4.1 The Total Quality Assurance Networking Model 57
			5.3.4.2 TQA-NM Approach..58

             5.3.5   Deployment of TQA-NM for Strategic Distribution 60
             5.3.6   Application Example 61
                      5.3.6.1   Technical Management 62
                      5.3.6.2   Product Management 62
                      5.3.6.3   Deploying the Total QA Network 62
             5.3.7   Conclusion 62
    5.4   Science SQC: New Quality Control Principle 62
          5.4.1   Introduction 62
          5.4.2   Need for New SQC to Improve Quality Management
                   by Manufacturing Industries 64
                   5.4.2.1   Delay in Systemization of Quality
                              Management System That Improves
                              Manufacturers' Management Technology 64
                   5.4.2.2   Necessity of New SQC as Demonstrative
                              Scientific Methodology..65
          5.4.3   Proposal and Implementation of Science SQC, New
                   Quality Control Principle 66
                   5.4.3.1   Practical Application of Scientific SQC 67
                   5.4.3.2   Establishment of SQC Technical Methods..........67
                   5.4.3.3   Construction of Integrated SQC Network TTIS....67
                   5.4.3.4   Recommendation of Management SQC.............69
          5.4.4   Demonstration Cases of Science SQC 69
                   5.4.4.1   Consistency in SQC Promotion Cycle................69
                   5.4.4.2   Systematization of Hierarchical SQC
                              Seminar and Growth of Human Resources 71
                   5.4.4.3   Quality Improvement of Drive System Unit:
                              Sealing Performance of Oil Seal for Transmission....73
          5.4.5   Conclusion 77
    Appendix 5.1 Customer Science.. 77
    Appendix 5.2 Science SQC, A New Quality Control Principle 78
    References 78

6   **Driving Force in Developing New JIT 85**
    6.1   Strategic Stratified Task Team Model 85
          6.1.1   Introduction 85
          6.1.2   Business Management and Development of Creative
                   Corporate Climate..86
          6.1.3   Reliability of Job and Importance of Team Activities 86
          6.1.4   Significance of Team Activities and Requirements for
                   Their Promotion 87
          6.1.5   Creativity and Strategy Necessary for Team Activities........88
          6.1.6   Proposal for a Strategic Stratified Task Team Model...........89

	6.1.7	Construction of the Strategic Cooperative Creation Team 91

- 6.1.7 Construction of the Strategic Cooperative Creation Team 91
- 6.1.8 Practice and Effectiveness of Strategic Stratified Task Team 92
  - 6.1.8.1 Progress of Strategic Task Team Activities 92
  - 6.1.8.2 Task Team Activity (Tasks 1 and 2) 94
  - 6.1.8.3 Task Management Team Activities (Tasks 3 and 4) 95
  - 6.1.8.4 Total Task Management Team Activity (Task 5) 96
  - 6.1.8.5 Joint Total Task Management Team Activity (Tasks 6 and 7) 97
- 6.1.9 Conclusion 98
- 6.2 Epoch-Making Innovation in the Work Environment Model 98
  - 6.2.1 Introduction 98
  - 6.2.2 Strategic Application of New JIT, New Management Technology Principle 99
    - 6.2.2.1 Establishment of Next Management Technology Principle, New JIT 99
    - 6.2.2.2 TPS, the Key to Strategic Application of New JIT 100
  - 6.2.3 Prerequisite of Strategic Global Production: Innovation in Work Quality 101
    - 6.2.3.1 Background of Concepts Regarding Consideration for Older Workers 101
    - 6.2.3.2 Consideration for Older Workers according to New JIT Strategy 102
  - 6.2.4 Definite Plan for Concept Actualization 104
    - 6.2.4.1 Formation of Project Team, AWD6P/J, and Activity Optimization 104
    - 6.2.4.2 Total Task Management Team Activity by Practical Application of Science SQC 104
  - 6.2.5 Activity Examples 106
    - 6.2.5.1 Project II: Study Work Standards to Reduce Fatigue 106
    - 6.2.5.2 Project V: Temperature Conditions Suited to Assembly Work Characteristics 109
    - 6.2.5.3 Outline of Other Projects 111
    - 6.2.5.4 Summary and Future Activities 111
  - 6.2.6 Conclusion 112
- 6.3 Partnering Performance Measurement Model 112
  - 6.3.1 Introduction 112
  - 6.3.2 Significance of Strengthening the Corporate Management of Japanese Automobile Assembly Makers and Suppliers 113

      6.3.3 Partnering Performance Measurement for Assembly Makers and Suppliers.......... 114
            6.3.3.1 Preparation of Evaluation Sheets for PPM-A, PPM-S, and PPM-AS 114
            6.3.3.2 Formulation Model of PPM-A, PPM-S, and PPM-AS 116
      6.3.4 Verification of PPM-AS 119
            6.3.4.1 Toyota Motor Corporation 120
            6.3.4.2 Nissan Motor Co., Ltd. 121
            6.3.4.3 General Motors.......... 122
      6.3.5 Conclusion 123
  6.4 Strategic Employment of the Patent Value Appraisal Method.......... 123
      6.4.1 Introduction 123
      6.4.2 Significance of Patent Acquisition as a Part of Corporate Technological Strategy.......... 124
            6.4.2.1 Current Situation of Patent Applications in Japan 124
            6.4.2.2 Corporate Reform and Patent Acquisition 124
      6.4.3 An Issue for the Intellectual Property Department, Qualitative Evaluation of Patent Value 125
      6.4.4 Strategic Patents.......... 126
            6.4.4.1 Grasping the Potential Structures of a Strategically Effective Patent.......... 126
            6.4.4.2 Grasping Engineers' Values with Regard to Strategic Patents.......... 126
      6.4.5 Proposal of Patent Value Appraisal Method.......... 127
            6.4.5.1 Redefinition of the Patent Value Appraisal Variables.......... 127
            6.4.5.2 Grasping Patent Value Appraisal and the Potential Factors 127
            6.4.5.3 The Structure of the Patent Value Appraisal Equation.......... 130
            6.4.5.4 Creating the PVAM Software.......... 132
      6.4.6 Verification of the Effectiveness of PVAM 133
      6.4.7 Conclusion 134
  References.......... 134

7 Strategic Development of New JIT 141
  7.1 Automobile Profile Design Concept Method Using Customer Science.......... 141
      7.1.1 Introduction 141
      7.1.2 Need for a New Management Technology Principle.......... 142

	7.1.3	Customer Science: Studying Consumer Value Utilizing New JIT 142
		7.1.3.1 Proposal of Customer Science 142
		7.1.3.2 Strategic Implementation of New JIT 143
	7.1.4	Constructing a Customer Science Application System: CS-CIANS 143
		7.1.4.1 New Model for Assisting the Conception of Strategic Product Development 143
		7.1.4.2 CS-CIANS: A Networking of the Customer Science Application System 144
	7.1.5	Application: Development of Customer Science, Key to Excellence Design: Lexus 147
		7.1.5.1 Lexus Design Profile Study 147
		7.1.5.2 Automobile Profile Design Concept Method for Developing Strategic Product 147
		7.1.5.3 Studies on Customers' Sense of Values Using Collages (Step 1) 148
		7.1.5.4 Studies on Customers' Design Preferences for Vehicle Appearance (Step 2) 151
		7.1.5.5 Studies on the Psychographics of the Lexus Profile Design (Step 3) 153
	7.1.6	Conclusion 155
7.2	Strategic Development Design Computer-Aided Engineering Model Employing TDS 155
	7.2.1	Introduction 155
	7.2.2	Expectations for Automotive Development Production and Simulation Technology 156
	7.2.3	Proposal of the Strategic Development Design CAE Model 157
		7.2.3.1 Total QA High Cyclization Business Process Model 157
		7.2.3.2 Stratified Intelligence CAE Development Design System 158
		7.2.3.3 Intelligence CAE Management Approach System 159
	7.2.4	Application Example: High Reliability Assurance of the Automotive Transaxle Oil Seal Leakage 161
		7.2.4.1 Oil Seal Function 161
		7.2.4.2 Understanding of the Mechanism through Visualization 161
		7.2.4.3 Fault and Factor Analyses 164

            7.2.4.4   CG Navigation and Intelligence CAE
                      Software: Oil Leakage Simulator 167
            7.2.4.5   Design Changes and Process Control for
                      Improving Reliability 169
      7.2.5 Conclusion 171
7.3 Process Layout CAE System for Intelligence TPS 171
      7.3.1 Introduction 171
      7.3.2 Proposal of the Process Layout CAE System Using
            TPS-LAS 172
            7.3.2.1   Necessity of Digital Engineering and
                      Computer Simulation in the Manufacture of
                      Motor Vehicles 172
            7.3.2.2   Concept of Process Layout CAE System
                      Using TPS-LAS 173
            7.3.2.3   Development of TPS-LAS 175
      7.3.3 Relation Model between Availability and Buffer Number 177
      7.3.4 TPS-LAS Application 179
            7.3.4.1   Result of TPS-LAS Application at Advanced
                      Car Manufacturer A 179
            7.3.4.2   Effectiveness of TPS-LAS Application 181
      7.3.5 Conclusion 182
7.4 Robot Reliability Design and Improvement Method Utilizing TPS 182
      7.4.1 Introduction 182
      7.4.2 Evolution of the TPS 183
            7.4.2.1   TPS as the Key to Success in Global Production 183
            7.4.2.2   Need for a Scientific Approach 184
      7.4.3 RRD-IM as the Key to TPS 184
            7.4.3.1   Concept and Structural Elements 185
      7.4.4 RRD-IM Application Example 189
            7.4.4.1   Practical Robot Failures 189
            7.4.4.2   Result of RRD-IM Application in Body
                      Assembly Line 190
            7.4.4.3   Evaluation on the Actual Line 192
      7.4.5 Conclusion 193
7.5 Scientific Mixed Media Model Employing TMS 194
      7.5.1 Introduction 194
      7.5.2 Need for a Marketing Strategy That Considers Market
            Trends 194
      7.5.3 Evolution of Market Creation Employing TMS 195
            7.5.3.1   Significance of TMS: The Key to Application
                      of New JIT 195
            7.5.3.2   Developing Customer Science Using Science
                      SQC 196

7.5.4 Constructing a Scientific Mixed Media Model for
Boosting Automobile Dealer Visits 196
    7.5.4.1 Publicity and Advertising as Automobile
Sales Promotion Activities 196
    7.5.4.2 Proposal of Scientific Mixed Media Model
for Boosting Automobile Dealer Visits............. 197
    7.5.4.3 VUCMIN 199
    7.5.4.4 CMP-FDM... 199
    7.5.4.5 AGTCA... 200
    7.5.4.6 PMOS-DM 201
7.5.5 Example Applications 202
    7.5.5.1 Visualizing Causal Relationships in
Customer Purchase Behavior............................202
    7.5.5.2 Application of VUCMIN 204
    7.5.5.3 Application of CMP-FDM 205
    7.5.5.4 Application of AGTCA.................................... 208
    7.5.5.5 Application of PMOS-DM 210
7.5.6 Conclusion 214
Appendix 7.1 Optimal Selection Using a Model Formula 214
A.7.1.1 Model Formula.. 215
A.7.1.2 Recipient Selection Process.. 216
References 217

8 Advanced Total Development System, Total Production System, and Total Marketing System: Key to Global Manufacturing Strategy 225

8.1 High-Linkage Model: Advanced Total Development System,
Total Production System, and Total Marketing System 225
    8.1.1 Introduction 225
    8.1.2 Need for a New Global Management Technology
Model for New JIT Strategy... 226
    8.1.3 High-Linkage Model: Advanced TDS, TPS, and TMS
for Global Manufacturing Strategy.................................227
        8.1.3.1 Advanced TDS: Total Development Design
Model 227
        8.1.3.2 Advanced TPS: Total Production
Management Model 229
        8.1.3.3 Advanced TMS: Strategic Development
Marketing Model 230
        8.1.3.4 Advanced TDS, TPS, and TMS Driven by
Science SQC..230
    8.1.4 Application: Putting into Practice and Verifying the
Validity of Advanced TDS, TPS, and TMS 231
        8.1.4.1 Effectiveness of Advanced TDS 231

			8.1.4.2	Effectiveness of Advanced TPS...................231

- 8.1.4.2 Effectiveness of Advanced TPS 231
- 8.1.4.3 Effectiveness of Advanced TMS 232
- 8.1.5 Conclusion 233
- 8.2 Automobile Exterior Color Design Development Model 233
  - 8.2.1 Introduction 233
  - 8.2.2 Background 234
  - 8.2.3 Research Method 234
    - 8.2.3.1 On-Site Investigations 234
    - 8.2.3.2 Visualizing Problem Areas and Work Processes 235
    - 8.2.3.3 Visualizing Success Factors and Elements 236
    - 8.2.3.4 Constructing the Business Approach Model 238
  - 8.2.4 Application Example 239
    - 8.2.4.1 Importance of Textual Expressions of Exterior Colors 240
    - 8.2.4.2 Proposal of the AECD-AM 243
  - 8.2.5 Verifying the Automobile Exterior Color Design Approach Model 248
  - 8.2.6 Conclusions 248
- 8.3 Highly Reliable Development Design CAE Model 248
  - 8.3.1 Introduction 248
  - 8.3.2 Need for a New Global Development Design Model 249
  - 8.3.3 Advanced TDS, Strategic Development Design Model 250
  - 8.3.4 Highly Reliable Development Design CAE Model 251
    - 8.3.4.1 Revolution in Manufacturing Development Design and the Evolution of CAE 252
    - 8.3.4.2 Highly Reliable Development Design CAE Model 254
  - 8.3.5 Application Example: Highly Reliable Numerical Simulation—Production of CAE Software for Molding Automotive Seat Pads 256
    - 8.3.5.1 Grasping the Problematic Phenomena 256
    - 8.3.5.2 Theoretical Analysis Model 257
    - 8.3.5.3 Implementation of Intelligence and High Precision 258
    - 8.3.5.4 Development of Urethane Foam Molding Simulator 259
  - 8.3.6 Conclusion 259
- 8.4 New Japanese Global Production Model 259
  - 8.4.1 Introduction 259
  - 8.4.2 Advanced TPS, Strategic Production Management Model 260
    - 8.4.2.1 Demand for New Management Technologies That Surpass JIT 260

		8.4.2.2	Basic Principle of Total Production System...... 260

                8.4.2.3   Advanced TPS: A New Japanese Global Production Model..261

                8.4.2.4   High-Cycle System for the Production Business Process Using Advanced TPS 262

                8.4.2.5   Creation of New Japanese Global Production Model Employing Advanced TPS.................... 264

        8.4.3   Example Applications 267

                8.4.3.1   TPS Layout Analysis System.............................267

                8.4.3.2   Human Intelligence-Production Operating System 268

                8.4.3.3   TPS Intelligent Production Operating System....268

                8.4.3.4   TPS Quality Assurance System 271

                8.4.3.5   Human Digital Pipeline System 271

                8.4.3.6   Virtual-Maintenance Innovated Computer System 273

        8.4.4   Conclusion 273

    8.5   Intelligent Customer Information Marketing System 274

        8.5.1   Introduction 274

        8.5.2   Need for a Marketing Strategy That Considers Market Trends 274

        8.5.3   Importance of Innovating Dealers' Sales Activities 275

        8.5.4   Advanced TMS, Strategic Development Marketing Model 276

        8.5.5   Creation of Scientific Customer Creative Model Utilizing Advanced TMS 276

        8.5.6   Application: Establishment of Intelligent Customer Information Marketing System...278

                8.5.6.1   Customer Purchasing Behavior Model of Advertisement Effect (CPBM-AE) for Automotive Sales 278

                8.5.6.2   Japanese Sales Marketing System.......................281

        8.5.7   Conclusion 282

Appendix 8.1 283

References 284

9   Innovative Deployment of an Advanced Total Development System, Total Production System, and Total Marketing System........291

    9.1   Automobile Form and Color Design Approach Model 291

        9.1.1   Introduction 291

        9.1.2   Research Background 292

                9.1.2.1   Necessity of Numerical Representation of Form Using CAD..293

		9.1.2.2	Research on Automobile Form Design Support Method 293

- 9.1.3 Creating an Automobile Form and Color Design Approach Model..294
  - 9.1.3.1 Selecting a Target (Step 1) 295
  - 9.1.3.2 Identifying the Relationship between Color and Subjective Customer Impressions (Step 2) ...295
  - 9.1.3.3 Identifying the Relationship between Form/Color and Subjective Color Customer Impressions (Step 3)..296
  - 9.1.3.4 Verification (Step 4)..298
- 9.1.4 Conclusions 301
- 9.2 Optimal Computer-Aided Engineering (CAE) Design Approach Model..301
  - 9.2.1 Introduction 301
  - 9.2.2 CAE in Product Design: Application and Issues................302
    - 9.2.2.1 CAE in the Product Design Process 302
    - 9.2.2.2 CAE Analysis: Current Status and Issues..........303
  - 9.2.3 Application of the Advanced TDS, Strategic Development Design Model..303
  - 9.2.4 Optimal CAE Design Approach Model 304
    - 9.2.4.1 Step 1: Define the Problem 304
    - 9.2.4.2 Step 2: Conduct a Visualization Experiment 306
    - 9.2.4.3 Step 3: Aligning Prototype Testing and CAE... 306
    - 9.2.4.4 Step 4: Conduct a Highly Precise CAE Analysis 307
    - 9.2.4.5 Step 5: Predict and Evaluate 307
  - 9.2.5 Application: Drivetrain Oil Seal Leaks: Cavitation Caused by Metal Particles in the Transaxle 308
    - 9.2.5.1 Oil Seal Function 308
    - 9.2.5.2 Understanding the Oil Seal Leakage Mechanism through Visualization................... 308
    - 9.2.5.3 Application of Optimal CAE Design Approach Model of Oil Simulator 310
  - 9.2.6 Conclusion 315
- 9.3 Human Digital Pipeline System for Production Preparation 316
  - 9.3.1 Introduction 316
  - 9.3.2 Necessity of Cultivating Human Resources Compatible with Digital Production 316
  - 9.3.3 Proposing HI-POS of Intelligence Production System...... 317
  - 9.3.4 Structuring "HDP" System and Its Components............. 318
    - 9.3.4.1 Hardware Configuration of HDP System.........320

                9.3.4.2 Software Configuration of HDP System 321
                9.3.4.3 Simulation Algorithm of HDP System 321
        9.3.5   Application Case: Deployment and Effect of HDP
                System 324
                9.3.5.1 Assembly Procedure Manual Using Design Data 324
                9.3.5.2 Simulation Result of HDP System.................... 325
        9.3.6   Conclusion 326
  9.4   Body Auto Fitting Model Using New Japanese Global
        Production Model 327
        9.4.1   Introduction 327
        9.4.2   Background 328
                9.4.2.1 The Current Condition and Problems of the
                         Conventional Production System....................... 328
                9.4.2.2 Ideas for Three Important Points for Global
                         Production 328
        9.4.3   Development of the New Global Production Model,
                Strategic Development of Advanced TPS 328
        9.4.4   Establishment of Body Auto Fitting Model 329
                9.4.4.1 Four Essential Techniques to Build the
                         Linkage among Production Processes in
                         BAFM for Global Production............................ 329
                9.4.4.2 Two Essential Techniques to Build the
                         Linkage to Globally Develop BAFM 330
        9.4.5   Employing the Fitting Line Utilizing BAFM.................... 331
                9.4.5.1 Necessity of the Fitting and Fastening
                         Automation 331
                9.4.5.2 Strategic Project Target..................................... 332
                9.4.5.3 Engineering Aspects 333
        9.4.6   Example Applications 333
                9.4.6.1 A Fitting Accuracy.. 333
                9.4.6.2 B Assembly Reliability: Equipment
                         Requirements to Overcome Technical
                         Hurdles... 340
                9.4.6.3 Engineering Aspects of System Requirements:
                         Development of Floor Space–Efficient System
                         Based on Review of the Transfer Equipment
                         and Installation Equipment (Robots)................. 342
                9.4.6.4 Results 343
        9.4.7   Conclusion 344
  9.5   Total Direct Mail Model to Attract Customers 344
        9.5.1   Introduction 344
        9.5.2   Necessity of Direct Mail.. 345
                9.5.2.1 Direct Mail and Customer Retention 345

xviii ■ Contents

                9.5.2.2 Direct Mail and the Attention-Interest-Desire-Action (AIDA) Model 345
                9.5.2.3 Current Research on Direct Mail Activities 346
        9.5.3 Strategically Applying a Total Direct Mail Model 346
                9.5.3.1 System for Analyzing Customer Purchase Data.....347
                9.5.3.2 System for Optimizing Direct Mail Content 349
                9.5.3.3 System for Strategically Determining Direct Mail Recipients...350
                9.5.3.4 Direct Mail Promotion System for Sales Staff...351
                9.5.3.5 Effectiveness of the Total Direct Mail Model ... 351
        9.5.4 Conclusion 352
Acknowledgments 353
References 353

10 New Manufacturing Management Employing New JIT 359
    10.1 New Software Development Model for Information Sharing 359
        10.1.1 Introduction 359
        10.1.2 Current Status and Problems in Software Development.... 360
                10.1.2.1 Current Status of Software Development 360
                10.1.2.2 Information Sharing in Software Development..... 360
        10.1.3 Constructing an NSDM to Assess the Success of Information Sharing between Customers and Vendors.....361
                10.1.3.1 Identifying Key Problem and Solution Factors in Software Development Information Sharing 361
                10.1.3.2 Collecting Data to Assess the Impact of Information-Sharing Factors on Overall Success...361
                10.1.3.3 Classifying Information-Sharing Factors 362
                10.1.3.4 Constructing an NSDM of Successful Information Sharing in Software Development...365
                10.1.3.5 Collecting Data to Assess the Impact of Success-Boosting Factors on Information-Sharing Factors... 366
                10.1.3.6 Assessing the Impact of Success-Boosting Factors on Overall Information-Sharing Success... 367
                10.1.3.7 Developing Software to Implement a Diagnostic System for Information-Sharing Success...369
                10.1.3.8 Verifying the Diagnostic System for Information-Sharing Success 370
        10.1.4 Conclusion 370
    10.2 Automotive Product Design and CAE for Bottleneck Solution 371
        10.2.1 Introduction 371

		10.2.2	Background 372

- 10.2.2 Background 372
  - 10.2.2.1 Automotive Product Design and Production 372
- 10.2.3 Proposal of Higher-Cycled Product Design CAE Model 374
- 10.2.4 Developing Highly Reliable CAE Analysis Technology Component Model 375
- 10.2.5 Application 377
  - 10.2.5.1 Automotive Bolt-Tightening Analysis Simulation Using Experiments and CAE 377
  - 10.2.5.2 Application to Similar Problems Using Higher-Cycled Product Design CAE Model.....388
- 10.2.6 Conclusion 388

10.3 Human-Integrated Assistance System for Intelligence Operators...389
- 10.3.1 Introduction 389
- 10.3.2 Need to Develop Intelligence Operators for a Highly Reliable Production System and Accelerated Training of Production Operators 390
- 10.3.3 Strategic Implementation of Advanced TPS for Intelligence Operators 391
- 10.3.4 Proposal of Human-Integrated Assistance (HIA) System: Developing Intelligence Operators for Production...391
- 10.3.5 HIA System Construction: The Key to Advanced TPS.....393
  - 10.3.5.1 Easy Creation and Modification 393
  - 10.3.5.2 Simple Know-How Accumulation and Easy Access 394
  - 10.3.5.3 Utilization of CAD and CAE Data 395
- 10.3.6 Formulating and Establishing Shortened Training Methods 396
  - 10.3.6.1 Current Issues Regarding Training Methods....396
  - 10.3.6.2 Measures for Shortened Training......................397
  - 10.3.6.3 Optimization of Training Steps 399
- 10.3.7 HIA System Application Example 400
  - 10.3.7.1 Application Case of Skill Training for Newly Employed Production Operators 400
  - 10.3.7.2 Application Case of Shortened Training for New Overseas Production Operators................402
- 10.3.8 Conclusion 404

10.4 Comparing Experienced and Inexperienced Machine Workers..... 404
- 10.4.1 Introduction 404
- 10.4.2 Skill Transfer Problem in the Manufacturing Industry.....405
- 10.4.3 Prior Research 407
- 10.4.4 Visualizing the Skills of Experienced Workers 408
  - 10.4.4.1 Comparing the Working Efficiency of Workers... 408

10.4.4.2 Comparing Experienced and Inexperienced Workers' Cognition Using Electroencephalography 410
10.4.4.3 Identifying Decision-Making Criteria of Experienced Workers Using Statistical Analysis 413
10.4.5 Evaluation of This Research 421
10.4.6 Conclusion 422
Acknowledgments 422
10.5 New Global Partnering Production Model for Expanding Overseas Strategy 422
10.5.1 Introduction 422
10.5.2 Problems with Achieving Global Production 423
10.5.3 Strategic Use of New JIT: the Key to Global Production..... 424
10.5.3.1 Current Situation of Quality Management in Japanese Manufacturing 424
10.5.3.2 Proposal of New Japanese Quality Management Model Employing New JIT 425
10.5.4 Proposal of New Global Partnering Production Model 427
10.5.4.1 Significance of Global Partnering Research 427
10.5.4.2 NGP-PM: A Proposal for Bringing Overseas Manufacturing up to Japanese Standards 427
10.5.5 Application: Proposal of the New Turkish Production System (NTPS) 429
10.5.5.1 Current Status of Japan's Leading Automobile Industry 429
10.5.5.2 Current Status of Turkey's Automobile Industry 429
10.5.5.3 Proposal of the New Turkish Production System 430
10.5.6 Conclusion 437
References 438

11 Conclusion 445

Index 447

Preface

The Japanese management technologies that made the biggest impact on the world in the second half of the twentieth century were the Toyota Production System, often also referred to as just in time (JIT), which is the most famous Japanese production system; total quality control (TQC); and total quality management (TQM).

However, as these management technologies became practiced as Lean systems around the world and were further developed and popularized, they lost their status as unique Japanese production systems, and in recent years, the superior quality of Japanese products has rapidly lost ground.

To be successful in the future, a global marketer must develop an excellent management technology that can impress consumers (customers) and continuously provide high-value products in a timely manner through corporate management for manufacturing in the twenty-first century. To realize this, the author proposes a New JIT, New Management Technology Principle that simultaneously achieves quality, cost, and delivery (QCD) and an effective management strategy.

New JIT contains hardware and software systems, as next-generation technical principles, for transforming management technology into a management strategy. The hardware system consists of the Total Development System (TDS), Total Production System (TPS), and Total Marketing System (TMS).

These are the three core elements required for establishing and transforming management technologies in the engineering, manufacturing, and marketing divisions. To improve the work process quality of all divisions concerned with development, production, and sales, the author hereby proposes TQM-S (TQM by utilizing Science Statistical Quality Control [SQC], New Quality Control Principle), called Science TQM, New Quality Management Principle, as a software system.

In addition, as a management technology strategy that enables sustainable growth, the author has proposed the Strategic Stratified Task Team Model, Eco-Making Innovation in the Work Environment Model, Partnering Performance Measurement Model, and Strategic Employment on the Patent Appraisal Method, which will become the driving force of New JIT.

Furthermore, as the key to the global manufacturing strategy of New JIT, the author believes that the effectiveness of New JIT using the high-linkage model Advanced TDS, TPS, and TMS for the innovative deployment of global management technology in advanced companies has been demonstrated, as described herein.

Aims and Structure

A future successful global marketer must develop an excellent quality management system that impresses users and continuously provides excellent, quality products in a timely manner through corporate management. To realize manufacturing that places top priority on customers with good quality, cost, and delivery (QCD) in a rapidly changing technical environment, it is essential to create a new core principle capable of changing the work process quality of all divisions for reforming super-short-term development production.

The top priority of the industrial field today is the new deployment of global marketing for surviving the era of global quality competition. To survive in the global market, the pressing management issue, particularly for Japanese manufacturers, is uniform quality worldwide and production at optimum locations, which are prerequisites for successful global production. To realize manufacturing that gives top priority to customers with a good QCD in a rapidly changing technical environment, it is essential to create a core principle capable of changing the technical development work processes of the development and design divisions.

Similarly, it is important to develop a new production technology principle and establish new process management principles to enable global production. Furthermore, new marketing activities independent from past experience are required for the sales and service divisions to achieve stronger relationships with customers. In addition, a new quality management technology principle linked with overall activities for higher work process quality in all divisions is necessary for an enterprise to survive.

One of the greatest contributions that Japan has made to the world is "just in time" (JIT). JIT is a production system that enables provision of what customers desire when they desire it. JIT has also been introduced in a number of enterprises in the United States and Europe as a key management technology. Represented by the current Toyota Production System, JIT was developed by the Toyota Motor Corporation. This production system implements total quality management (TQM) in its production process and aims to simultaneously realize quality and productivity in pursuit of maximum rationalization (optimal streamlining) while recognizing the principle of cost reduction.

This is the essential concept of JIT and has been positioned as a core part of Toyota's management technology; it is often likened to the wheels on both sides of a vehicle. However, Toyota's production system, which represents today's Japanese-style production system, has already been developed as an internationally shared system, known as a Lean system, and is no longer an exclusive technology of Toyota in Japan. In Western countries also, the importance of quality control has been recognized through studies on Japanese TQM. As a result, TQM activities have become increasingly popular. Therefore, the superiority in quality of Japanese products assured by Japanese-style quality control has been gradually undermined in recent years.

Having said this, the author conjectures that it is clearly impossible to lead the next generation by merely maintaining the two Toyota management technology principles: the Toyota Production System and TQM. To overcome this issue, it is essential not only to renovate TPS, which is the core principle of the production process, but also to establish core principles for marketing, design and development, production, and other departments.

The aim of this book is to reassess the way management technology is carried out in the manufacturing industry and establish New JIT, New Management Technology Principle. The next-generation management technology model, New JIT, which the author has proposed through theoretical and systematic analyses, is a JIT system not only for manufacturing, but also for customer relations, sales and marketing, product planning, research and development (R&D), product design, production engineering, logistics, procurement, administration, and management; for enhancing business process innovation; and for the introduction of new concepts and procedures.

New JIT contains hardware and software systems as the next-generation technical principles for accelerating the optimization (high linkage) of work process cycles of all the divisions. The first item, the hardware system, consists of the Total Development System (TDS), the Total Production System (TPS), and the Total Marketing System (TMS), which are the three core elements required for establishing new management technology principles for sales, R&D, design, engineering, and production, among others. For the second item, the strategic quality management system, the author proposes Science TQM, New Quality Management Principle, (promotion of TQM incorporating Science SQC, New Quality Control Principle as a software system for improving the business process quality).

To realize manufacturing that gives top priority to customers with good QCD in a rapidly changing technical environment, the author recommends the urgent establishment of a new global management technology model for the next generation. The author, therefore, proposes a high-linkage model, employing a structured integrated triple management technologies system—advanced TDS, advanced TPS, and advanced TMS—for expanding uniform quality worldwide and production at optimum locations. The focus of this book is thus the theory and application of strategic management technology through the application of New JIT. The effectiveness of New JIT is then demonstrated at advanced car manufacturer A (the top Japanese carmaker) and other enterprises.

Overview

In this book, the author proposes the new just in time (JIT), New Management Technology Principle, aimed at the evolution of manufacturing, and demonstrates its effectiveness.

Chapter 1 gives an overview of New JIT for the next generation of management technology. New JIT is the basis of manufacturing fundamentals arrived at by innovating the conventional JIT.

In Chapter 2, the author asserts the need for evolution in manufacturing in order to deal with the management issues that Japanese manufacturers currently face. Therefore, in this chapter, the author reconsiders the management tasks of manufacturing companies shifting to global production.

In Chapter 3, the author explains the basic JIT, the Toyota Production System, which aims to simultaneously realize quality and productivity via a Lean system and process control and process improvement.

Chapter 4 focuses on the further development of Japanese management technology, and the author explains the innovation of manufacturing fundamentals surpassing JIT as a next-generation management technology. He first addresses the shortened life cycle of products due to the diversification and sophistication of customers' needs. Second, it has been deemed essential to create a new administrative management technology for the deployment of global production.

In Chapter 5, the author proposes New JIT by composing Total Development System (TDS), Total Production System (TPS), and Total Marketing System (TMS) for manufacturing in the twenty-first century. In addition, he discusses and introduces the foundation of advancing manufacturing, Total Quality Assurance Networking Model, and "Science SQC, New Quality Control Principle for strategic development of New JIT.

In Chapter 6, the author discusses and demonstrates the effectiveness of the following as the driving forces in developing New JIT: Strategic Stratified Task Team Model, Epoch-Making Innovation in the Work Environment Model, Partnering Performance Measurement Model, and Strategic Employment of the Patent Appraisal Method.

In Chapter 7, the author discusses and demonstrates the effectiveness of the following as strategic developments in New JIT: Automobile Design Concept Method Using Customer Science, Strategic Development Design CAE Model Employing TDS, Process Layout CAE System for Intelligence TPS, Robot Reliability Design and Improvement Method Utilizing TPS, and Scientific Mixed Media Model Employing TMS.

In Chapter 8, the author discusses and demonstrates the effectiveness of the following, which are key to the global manufacturing strategy of New JIT: a high-linkage model, Advanced TDS, TPS, and TMS, for the strategic development of management technology in advanced companies that utilize the author's New JIT. In addition, the author discusses and introduces the Automobile Exterior Color Design Development Model, Highly Reliable Development Design CAE Model, New Japanese Global Production Model, and Intelligent Customer Information Marketing Model for global development of Advanced TDS, TPS, and TMS.

In Chapter 9, the author discusses and demonstrates the effectiveness of the following as innovative developments of Advanced TDS, TPS, and TMS: Automobile Form and Color Design Approach Model, Optimal CAE Design Approach Model, Human Digital Pipeline System for Production Preparation, Body Auto Fitting Model Using NJ-GPM, and Total Direct Mail Model to Attract Customers.

In Chapter 10, the author discusses and demonstrates the effectiveness of the following case studies as new manufacturing management employing New JIT: New Software Development Model for Information Sharing, Automotive Product Design and Computer-Aided Engineering for Bottleneck Solution, Human-Integrated Assistance System for Intelligence Operators, Comparison of Experienced and Inexperienced Machine Workers, and New Global Partnering Production Model for Expanding Overseas Strategy.

Finally, Chapter 11 provides an overall conclusion of the topics covered in the book.

Author

Dr. Kakuro Amasaka was born in Aomori Prefecture, Japan, on May 5, 1947. He received a bachelor of engineering degree from Hachinohe Technical College, Hachinohe, Japan, in 1968, and a doctor of engineering degree specializing in precision mechanical and system engineering, statistics, and quality control from Hiroshima University, Hiroshima, Japan, in 1997. After joining the Toyota Motor Corporation in Japan in 1968, Dr. Amasaka worked as a quality management consultant for many divisions. He was an engineer and manager of the production engineering division, quality assurance division, overseas engineering division, manufacturing division, and TQM promotion division (1968–1997) and general manager of the TQM promotion division (1998–2000).

Dr. Amasaka became a professor of the College of Science and Engineering and the Graduate School of Science and Engineering at Aoyama Gakuin University, Tokyo, Japan, in April 2000. His specialties include production engineering (just in time and Toyota Production System), statistical science, multivariate analysis, reliability engineering, and information processing engineering. Recent research conducted includes Science SQC, New Quality Control Principle; Science TQM, New Quality Management Principle; New JIT, New Management Technology Principle; customer science; *kansei* engineering; and numerical simulation (computer-aided engineering).

Dr. Amasaka is the author of a number of papers on strategic total quality management, as well as the convener of the Japanese Society for Quality Control, Japanese Operations Management and Strategy Association, Japanese Industrial Management Association, Japanese Society of Kansei Engineering, and other academic societies (e.g., Production and Operations Management Society and European Operations Management Association). He has served as the vice chairman of the Japanese Society for Production Management (2003–2007) and Japanese Operations Management and Strategy Association (2008–2010), the director of the Japanese Society for Quality Control (2001–2003), the chairman of the Japanese Operations Management and Strategy Association (2011–2012), the commissioner of the Deming Prize judging committee (2002–2013), and the director of the Japanese Operations Management and Strategy Association (2013-present).

Dr. Amasaka has acquired 72 patents concerned with production and quality control systems and engineering and measurement technology. He is a recipient of the Aichi Invention Encouragement Prize (1991), Nikkei Quality Control Literature Prize (1992, 2000, 2001, and 2010), Quality Technological Prizes (Japanese Society for Quality Control, 1993 and 1999), Statistical Quality Control Prize (Union of Japanese Scientists and Engineers, 1976), Japanese Society of Kansei Engineering Publishing Prize (2002), and others (e.g., Outstanding Paper Award, International Conference on Management and Information Systems, 2013).

Contributing Authors

Koichiro Anabiki
Aoyama Gakuin University
Tokyo, Japan

Keiichi Ebioka
Aoyama Gakuin University
Tokyo, Japan

Hisatoshi Ishiguro
Aoyama Gakuin University
Tokyo, Japan

Taku Kojima
Aoyama Gakuin University
Tokyo, Japan

Maiko Muto
Aoyama Gakuin University
Tokyo, Japan

Masahiro Nakamura
Aoyama Gakuin University
Tokyo, Japan

Yasuaki Nozawa
Aoyama Gakuin University
Tokyo, Japan

Takehiro Onodera
Aoyama Gakuin University
Tokyo, Japan

Hirohisa Sakai
Global Production Center
Toyota Motor Corporation
Toyota-shi, Japan

Mustafa Murat Sakalsiz
Aoyama Gakuin University
Tokyo, Japan

Tomoaki Sakatoku
Aoyama Gakuin University
Tokyo, Japan

Shohei Takebuchi
Aoyama Gakuin University
Tokyo, Japan

Manabu Yamaji
Aoyama Gakuin University
Tokyo, Japan

Kazuma Yanagisawa
Aoyama Gakuin University
Tokyo, Japan

Acknowledgments

This book is a combination of past studies, new scientific books, and major academic journals and explains the comprehensive theory and practices of New JIT, which has been widely used and is evolving as Japan's new management technology principle. The author would like to acknowledge the generous support of the following researchers: all those at Toyota Motor Corporation and Amasaka's New JIT Laboratory in Aoyama Gakuin University that assisted with the research, especially, Dr. H. Sakai, M. Yamaji, K. Anabuki, T. Sakatoku, K. Ebioka, T. Kojima, M. M. Sakalsiz, H. Ishiguro, S. Takebuchi, M. Muto, M. Nakamura, Y. Nozawa, T. Onodera, K. Yanagisawa, and others.

Chapter 1

Introduction

A future successful global marketer must develop an excellent management technology system that impresses users and continuously provides excellent, quality products in a timely manner through corporate management. To realize manufacturing that prioritizes customers with good quality, cost, and delivery (QCD) in a rapidly changing technical environment, it is essential to create a new core principle capable of improving the work process quality of all divisions and reforming super-short-term development production (Amasaka, 2002, 2008a).

The Japanese administrative management technology that contributed the most to the world in the latter half of the twentieth century is typified by the Japan Production System represented by the Toyota Production System. This system was kept at a high usage level by a manufacturing quality management system generally called just in time (JIT) (Ohno, 1977).

However, a close look at recent corporate management activities reveals various situations where an advanced manufacturer, which is leading the industry, is having difficulty due to unexpected quality-related problems. Some companies have slowed down their production engineering development and are thus facing a survival crisis as manufacturers.

Against this background, improvement of the Japanese administrative management technology is sorely needed at this time (Goto, 1999; Amasaka, 1999; Nihon Keizai Shinbun, 1999, 2000, 2006; Asahi Shinbun, 2005). To be successful in the future, a global marketer must develop an excellent quality management system that can impress consumers and continuously provide excellent quality products in a timely manner through corporate management for manufacturing in the twenty-first century.

In the remarkable, technologically innovative competition seen today, in order to realize manufacturing that ensures customer-first QCD, it is indispensable to first create a core technology capable of reforming the business process used for the technological development of divisions related to engineering design. It is equally important, even for production-related divisions, to develop new production technologies and establish new process management that, when combined, enable global production (Hayes and Wheelwright, 1984).

In addition, even the product promotion, sales, and service divisions are expected to carry out rationalized marketing activities that are not merely based on past experiences, so that they can strengthen ties with their customers. It is believed that the foundation of corporate survival is to establish a new quality-centered management technology that can link the management of the activities carried out by the divisions above with a view to enhance the quality of their business processes (Amasaka, 2002, 2008a). Today's Japan is not an exception when it comes to the necessity described above.

Given this context and by predicting the form of next-generation manufacturing, the author hereby establishes New JIT, New Management Technology Principle that contains a hardware system with three core elements—a Total Development System (TDS), Total Production System (TPS), and Total Marketing System (TMS)—and a software system—a Total Quality Management System (TQM-S), called Science TQM, a new quality management principle, which utilizes Science Statistical Quality Control (SQC), New Quality Control Principle—for transforming management technology into management strategy (Amasaka, 2002, 2003, 2004a,b, 2008b; Amasaka et al., 2008). New JIT is the basis of Manufacturing Fundamentals 21C, accomplished by innovating the conventional JIT system.

To realize manufacturing that gives top priority to customers with good QCD in a rapidly changing technical environment, the author recommends the urgent establishment of a "new global management technology" for the next generation. The author, therefore, proposes a high-linkage model, employing a structured integrated triple management technologies system—high-linkage model, Advanced TDS, Advanced TPS, and Advanced TMS—for expanding uniform quality worldwide and production at optimum locations.

The focus of this book is thus the theory and application of strategic management technology through the application of New JIT. The effectiveness of New JIT is then demonstrated at advanced car manufacturer A (the top maker of Japanese cars, an extremely large global enterprise) and others (Amasaka, 2007, 2009, 2010, 2011, 2013).

References

Amasaka, K. 1999. The TQM responsibilities for industrial management in Japan—The research of actual TQM activities for business management, in *Proceedings of the 10th Annual Technical Conference of the Japanese Society for Production Management*, pp. 48–54. Kyushu-Sangyo University, Fukuoka, Japan.

Amasaka, K. 2002. New JIT, a new management technology principle at Toyota, *International Journal of Production Economics*, 80: 135–144.

Amasaka, K. 2008a. Strategic QCD studies with affiliated and non-affiliated suppliers utilizing New JIT, in Putnik, G.D. and Cruz-Cunha, M.M. (eds), *Encyclopedia of Networked and Virtual Organizations*, Information Science Reference, Hershey, PA, pp. 1516–1527.

Amasaka, K. 2008b. Science TQM, a new quality management principle: The quality management strategy of Toyota, *The Journal of Management and Engineering Integration*, 1(1): 7–22.

Amasaka, K. 2009. Invitation lecture. New JIT, advanced management technology principle: The global management strategy of Toyota, in *Proceedings of the International Conference on Intelligent Manufacturing and Logistics Systems IML2009 and Symposium on Group Technology and Cellular Manufacturing*, pp. 1–16. February 17, 2009, Kitakyushu, Japan.

Amasaka, K. 2010. Keynote lecture: New manufacturing principle surpassing JIT towards the 21st century manufacturing, in *Proceedings of the 2nd Annual Conference of Japanese Operations Management and Strategy Association*, June 19, 2010, Kobe University, Japan.

Amasaka, K. 2011. Special lecture: New JIT, new management technology principle, in *The VIII Siberian Conference "Quality Management-2011"*, pp. 1–16, March 22, 2011, Krasnoyarsk, Russia.

Amasaka, K. 2013. The development of a Total Quality Management System for transforming technology into effective management strategy, *The International Journal of Management*, 30(2): 610–630.

Amasaka, K., Kurosu, S., and Morita, M. 2008. *New Manufacturing Principle: Surpassing JIT—Evolution of Just in Time*, Morikita-shuppan [in Japanese].

Asahi Shinbun. 2005. The manufacturing industry—Skill tradition feels uneasy (May 3, 2005), Tokyo, Japan (http://www.asahi-np.co.jp/).

Goto, T. 1999. *Forgotten Management Origin—Management Quality Taught by GHQ*, Seisansei-shuppan, Tokyo, Japan [in Japanese].

Hayes, R. H. and Wheelwright, S. C. 1984. *Restoring Our Competitive Edge: Competing Through Manufacturing*, Wiley, New York.

Nihon Keizai Shinbun. 1999. Corporate survey (68QCS)—Strict assessment of TQM (July 15, 1999), Tokyo, Japan.

Nihon Keizai Shinbun. 2000. Worst record: 40% increase of vehicle recalls (July 6, 2000), Tokyo, Japan.

Nihon Keizai Shinbun. 2006. Risky "quality": Apart from production increase, it is recall rapid increase (February 8, 2006), Tokyo, Japan.

Ohno, T. 1977. *Toyota Production System*, Diamond-sha, Tokyo, Japan [in Japanese].

Chapter 2

Management Tasks of Manufacturing Companies Shifting to Global Production

The top priority of the industrial field today is the new deployment of global marketing in order to survive the era of global quality competition. The pressing management issue, particularly for Japanese manufacturers, to enable survival in the global market is worldwide uniform quality and production at optimum locations, which is a prerequisite for successful global production. It is essential to realize manufacturing that gives top priority to customers with good quality, cost, and delivery (QCD) in a rapidly changing technical environment.

2.1 Progress of Production Control in the Manufacturing Industry

All over the world, including in Japan, advanced companies are shifting to global production to realize worldwide uniform quality and production at optimum locations to enable survival amid fierce competition. To attain successful global production, technical administration, production control, purchasing control, sales administration, information systems, and other administrative departments should cooperate closely with clerical and indirect departments while establishing strategic cooperative and creative business linkages with individual development, production and sales departments, and outside manufacturers (suppliers) (Amasaka, 2007).

Today, consumers have quick access to the latest information in the worldwide market thanks to the development of information technology (IT), and strategic organizational management of the production control department has become increasingly important. Simultaneous attainment of QCD requirements is the most important mission for developing highly reliable new products ahead of competitors (Amasaka, 2008). This requires the urgent establishment of an innovative production control system for the next generation (called a next-generation production control system).

With a view to assuring that future management technology is a new leap forward for Japanese "manufacturing," the "progress of production control of plants in the manufacturing industry" made so far by the manufacturing industry is summarized in Figure 2.1 (Amasaka, 2004).

In the figure, the basis of the major production control methodologies, such as industrial engineering (IE), operations research, quality control, management of administration, marketing research, production control, and IT, are plotted along the vertical axis. Along the horizontal axis, some of the key elemental technologies, management methods, scientific methodologies, and so on are mapped out in a time series.

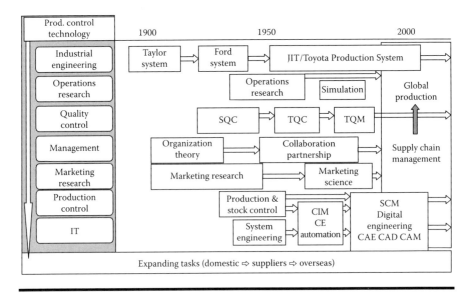

Figure 2.1 Progress of production control of plants in the manufacturing industry. SQC, statistical quality control. CIM, construction information modeling; CE, concurrent engineering; CAD, computer-aided design; CAE, computer-aided engineering; CAM, computer-aided manufacturing.

Since the beginning of the century, the operation of manufacturers has shifted from domestic production in Japan to overseas production bases, and management technology has become increasingly complicated, as depicted in Figure 2.1.

For the production control department, the key to success in global production is modeling strategic supply chain management (SCM) for domestic and overseas suppliers with a systematization of its management methods. In the implementation stage, in-depth studies of the Toyota Production System, total quality management (TQM), partnering, and digital engineering will be needed in the future (Amasaka, 2009).

2.2 Today's Management Technology Issues

For manufacturers to be successful in the future global market, they need to develop products that have a strong impression on consumers and supply these items in a timely fashion through effective corporate management. In recent years, however, the Toyota Production System representing Japanese manufacturing has been adopted as a so-called Lean system (Taylor and Brunt, 2001) and further developed in various international systems. Therefore, it is no longer exclusively Japanese (or Toyota's) technology. In the United States and European countries, the importance of quality control has been increasingly recognized through studies of Japanese TQM (or total quality control [TQC]). TQM has thus been actively promoted and recently began encroaching on the Japanese-style quality superiority, which Japanese products have previously enjoyed (Gabor, 1990; Joiner, 1994).

This shows that it is clearly impossible to continue to lead the next generation simply by adhering to and maintaining the Toyota Production System and TQM, which are the dual pillars of traditional Japanese management technology. In order to overcome these problems, it is essential not only to advance the Japanese Production System, a core technology of the production processes, but also to establish a core technology for the sales-, design-, and development-related divisions (Amasaka, 2002).

Given the above, the author (Amasaka, 2002, 2007) has conducted an awareness survey of general management personnel and executives (72 people) from 12 advanced companies belonging to the Toyota Group. Similarly, based on an awareness survey of other companies (Fuji Xerox and Daikin, among others, with 153 participants) participating in the Study Group for Manufacturing Quality Management (a.k.a. The Amasaka Forum), management technology issues have been investigated from the standpoint of corporate management.

In Figure 2.2, by incorporating Quantification Class III, the overall management technology issues have been plotted in a chart. This confirms that managers responsible for development give the highest priority to "suggestion-based new merchandise and product development" as a global merchandising strategy,

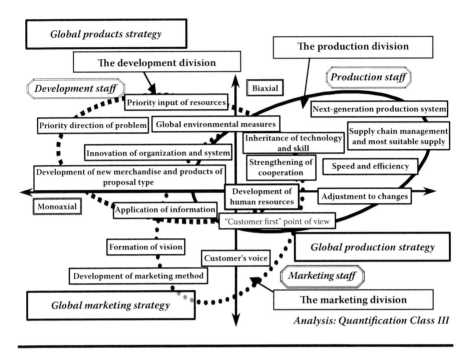

Figure 2.2 Management technology problems (positioning of opinion).

whereas production managers put efforts into establishing the "next-generation production system" in view of global production. Sales managers, on the other hand, prioritize the "development of new marketing methods" in order to be successful in global marketing. Moreover, the issue common to all was how to respond to globalization.

Therefore, in order to overcome these management issues, it will be necessary to carry out reforms, innovate the human resources cultivation system through intelligent sharing of information, and create a new management technology for closer ties among the company's divisions. The above awareness surveys and analysis clarified the core technologies necessary for the next-generation management technology principle, the basis for new management technologies, and the technological elements required for linking these core technologies.

Thus, in the future, it will be important to create management technology equipped with a new concept that enables total linkage of QCD research conducted by each of the aforementioned divisions from the standpoint of strategic corporate management, and by so doing, to create a New JIT, New Management Technology Principle. In order to promote excellent quality management capable of contributing to global manufacturing, the quality of the business processes of all divisions needs to be enhanced through clear-cut, rational JIT activities (Amasaka, 2002, 2008, 2013).

References

Amasaka, K. 2002. New JIT, a new management technology principle at Toyota, *International Journal of Production Economics*, 80: 135–144.

Amasaka, K. 2004. (Keynote lecture) The past, present, future of production management, in *The 20th Annual Technical Conference of the Japan Society for Production Management*, pp. 1–8. Nagoya Technical College, Aichi, Japan.

Amasaka, K. 2007. High linkage model "Advanced TDS, TPS & TMS"—Strategic development of "New JIT" at Toyota, *International Journal of Operations and Quantitative Management*, 13(3): 101–121.

Amasaka, K. 2008. Strategic QCD studies with affiliated and non-affiliated suppliers utilizing New JIT, in Putnik, G.D. and Cruz-Cunha, M.M. (eds), *Encyclopedia of Networked and Virtual Organizations*, Information Science Reference, Hershey, PA, pp. 1516–1527.

Amasaka, K. 2009. New JIT, advanced management technology principle: The global management strategy of Toyota, Invitation lecture at *International Conference on Intelligent Manufacturing and Logistics Systems IML2009 and Symposium on Group Technology and Cellular Manufacturing*, pp. 1–16, February 17, 2009, Kitakyushu, Japan.

Amasaka, K. 2013. The development of a Total Quality Management System for transforming technology into effective management strategy, *The International Journal of Management*, 30(2): 610–630.

Gabor, A. 1990. *The Man Who Discovered Quality; How W. Edwards Deming Brought the Quality Revolution to America—The Stories of Ford, Xerox, and GM*, Random House, New York.

Joiner, B. L. 1994. *Fourth Generation Management: The New Business Consciousness*, Joiner Associates, McGraw-Hill, Tokyo.

Taylor, D. and Brunt, D. 2001. *Manufacturing Operations and Supply Chain Management—The Lean Approach*, 1st edn., Thomson Learning, London.

Chapter 3

Basics of JIT: The Toyota Production System

In this chapter, the basics of just in time (JIT), which reformed Toyota's automobile manufacturing, are discussed. The basic principles and on-site implementation of manufacturing using the Toyota Production System, which has been adopted as a core concept of global manufacturing, will be illustrated here (Amasaka et al., 2008).

3.1 Toyota Production System: The Core Concept of JIT

JIT system, a Japanese production system typified by the Toyota Production System, is a manufacturing system that was developed by the Toyota Motor Corporation (Ohno, 1977).

The basic philosophy of the Toyota Production System (Ohno, 1977; Toyota Motor Corporation, 1987a,b) is built on the ideas of the Toyota Group founder, Sakichi Toyoda, and his business mottos: (1) be ahead of the times through endless creativity, inquisitiveness, and the pursuit of perfection; and (2) a product should never be sold unless it has been carefully manufactured and has been tested thoroughly and satisfactorily. The philosophy also reflects the ideas of Toyota Motor's founder, Kiichiro Toyoda, regarding improvement through inspection: (1) grasp the demands of consumers firsthand and reflect them in your product and (2) investigate the product quality and business operations and then improve them.

These are the basic concepts of JIT, which aims to realize quality and productivity simultaneously by effectively applying total quality control (TQC) and total quality management (TQM) to the automobile manufacturing process. It also pursues maximum efficiency (optimal streamlining, which is called a Lean system) while also being conscious of the principles of cost reduction and thereby improving the overall product quality (Ohno, 1977; Toyota Motor Corporation, 1987; Amasaka, 1988, 2002a).

In the JIT implementation stage, it is important to constantly respond to the customers' needs, promote flawless production activities, and conduct timely quality, cost, and delivery (QCD) research, as well as put it into practice (Amasaka, 2002a, 2008). Therefore, Toyota has positioned the Toyota Production System and TQM as the core management technologies for realizing *reasonable manufacturing*, and these management technologies are often likened to the wheels of an automobile (Amasaka, 1989, 1999, 2000, 2002; Hayashi and Amasaka, 1990).

In Figure 3.1, these management technologies are placed on the vertical and horizontal axes. As shown in the figure, the combination of these technologies reduces large irregularities in manufacturing to the state of tiny ripples, and average values are consistently improved in the process. This strategy is an approach used by reasonable corporate management in which the so-called *leaning process* is consistently carried out.

As indicated by the vertical and horizontal axes in the figure, when the hardware technology of the Toyota Production System and the software technology of TQM (TQC) are implemented, statistical quality control (SQC) is to be effectively incorporated to scientifically promote QCD research and achieve constant upgrading of the manufacturing quality. Another point that can be understood from the figure is that TQM and SQC are the foundations of maintaining and improving

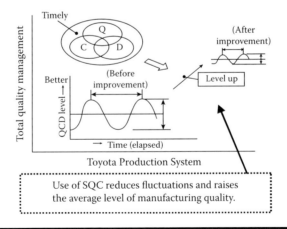

Figure 3.1 Relation between Toyota Production System and total quality management.

manufacturing quality, and have historically served as a basis for the advancement of JIT (Amasaka, 2003a, 2004a,b).

In this way, the basic concept of JIT and the Lean system approach have reformed the automobile manufacturing process used at Toyota. As a result, the effectiveness of JIT has been recognized on a worldwide scale, and it is now regarded as the core concept of manufacturing (Hayes and Wheelwright, 1984; Roos et al., 1990; Womack and Jones, 1994; Taylor and Brunt, 2001).

3.2 Basic Principle of Manufacturing via the Toyota Production System

3.2.1 Simultaneous Realization of Quality and Productivity via a Lean system

The basic principle of manufacturing via the Toyota Production System is Lean system production, as shown in Figure 3.2. In this system, manufacturing is conducted via *one-by-one* (single part) *production* and its aim is the simultaneous realization of quality and productivity (Amasaka, 1988, 2000; Hayashi and Amasaka, 1990).

The first basic principle of manufacturing is thorough quality control by means of one-by-one production. As Figure 3.2a illustrates, this system of one-by-one production on a manufacturing line using an assembly conveyor gives the assembly worker the ability to conduct a self-check on each piece. If a defective item comes to their assembly point from the previous process, they can then stop the conveyor and detect the defect without fail. Therefore, the assembly workers can provide 100% quality products to the downstream processes. It is obvious that compared to lot production, this system can considerably improve the detection of defective parts from a probabilistic viewpoint.

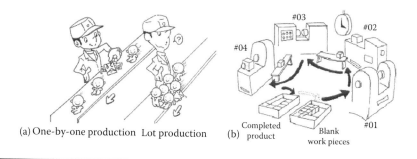

Figure 3.2 The basic principle of manufacturing via the Toyota Production System: the simultaneous realization of quality and productivity. (a) Thorough quality control by means of one-by-one production. (b) Thorough incorporation of quality into the process via one-by-one production.

In this context, one-by-one production is sometimes compared to a fine, quick, and pure flow of water upstream in a mountain stream, while lot production is likened to the broad, stagnant, and slow water found downstream. The incorporation by the Toyota Production System of this one-by-one production deserves to be recognized as Lean system production.

The second basic principle of manufacturing is the thorough incorporation of quality into the process via one-by-one production. One-by-one production in a machining process is shown in Figure 3.2b. The diagram illustrates an operation where a worker picks up a piece (work) from the parts box, conducts the machining process operation on it from Processes #01 – Processes #04 in the order shown, and finally places the completed piece into the completed parts box. Since the production operation is conducted according to a predetermined cycle time, the worker can consistently carry out the prepared standard operation in a rhythmical manner.

Similar to the previous case of operation on an assembly conveyor, a self-check can be consistently performed and the incorporation of quality is ensured so that the stabilization of production can also be promoted.

Even in the case of a production operation that involves multiple types of work pieces, the intelligent application of production engineering and process design can be used to support workers and allow them to conduct the standard operations in a rhythmical manner as shown in Figure 3.3(a). A process like this is depicted in Figure 3.3(b), where the zigzag work flow interferes with the smooth performance of the standard operations each time a work piece changes.

This causes the worker to become fatigued and triggers human errors, thus resulting in the lowering of quality and productivity.

3.2.2 Process Control and Process Improvement

The high productivity and high quality of manufacturing in Japan today are universally acknowledged (Toyota Motor Corporation, 1987; Amasaka, 2000). At manufacturing sites, the process control and process improvement, mainly promoted by

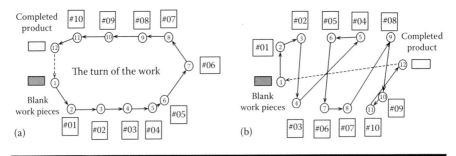

Figure 3.3 The basic principle of manufacturing via the Toyota Production System: the simultaneous realization of quality and productivity. (a) Easy process of standard work. (b) Difficult process of standard work.

production workers under the supervision of their supervisors, and QCD research, conducted by white-collar engineers, are equally valued in order to ensure the incorporation of design quality, which is demanded by customers (Amasaka, 1988, 2000, 2002b, 2009, 2008; Amasaka and Sakai, 2010).

3.2.2.1 Process Control and Process Improvement

Consideration is given here to the method of manufacturing an automobile and the resulting manufacturing cost (initial cost) and profit based on Figure 3.4. The main parts (heavy material parts) that compose an automobile are the skeletal structure, made up of the body and frame, and the parts that control the running functions, such as the engine and axles.

These parts are made of steel plates and cast forged steel. Automobile manufacturers procure these materials from steel manufacturers. Parts such as tires, engine ignition batteries, brake units, power steering units, and a wide variety of other parts such as windshields, lights, mirrors, seats, and so on are purchased from specialized parts manufacturers.

In order to produce a complete automobile, the manufacturer also purchases a variety of machining tools for processing parts, welding machines for the body and frame, and assembling robots for assembling the vehicle and other equipment from specialized machine or equipment manufacturers. Finally, the electric power needed to operate these machines and equipment is bought from power supply companies.

Generally speaking, the manufacturing costs listed above that are borne by the automobile manufacturer do not differ very much from those of other manufacturers. The labor cost (workers' wages) is also similar among manufacturers. Therefore, the manufacturing cost (initial cost) shown on the left half of the figure for automobiles of a similar design does not vary greatly from manufacturer to manufacturer.

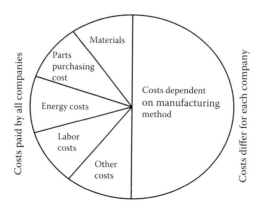

Figure 3.4 Manufacturing methods and costs.

On the other hand, the right half of the figure shows the manufacturing cost that results from the method of manufacturing the vehicle at the production site, and this can vary widely depending on how each manufacturer handles their manufacturing process control and process improvement.

The following are some poor cases of process control.

1. Due to an inconvenient process design and production process, the workers cannot carry out their operations according to the standard operation sheet, resulting in a number of irregularities in the operation. This in turn causes the assembly line to stop often, lowering the line operating rate.
2. Due to a defective standard operation sheet, the worker's operation becomes strained, wasteful, and irregular, requiring the production line to stop; therefore, the planned production volume cannot be ensured.
3. Due to insufficient maintenance and upkeep of the machines and equipment, malfunctions occur, lowering the equipment operating (availability) rate.
4. Although the production line is not operating during break times, the production equipment's power is still on, resulting in wasteful power consumption.
5. Due to inadequate instructions from the supervisor or insufficient education and training provided to new workers regarding manual (skill required) operation or handling of machines, quality defects result from operation errors or improper machine manipulation. This causes pieces with machining defects, which entails disposal or reworking of these pieces.
6. Due to a lot of waste materials (scrap for disposal) during the stamping process or excessive cutting processing that causes the blades to be damaged, the processing line has to be stopped or defective work pieces result. This increases the cost of replacing the stamping process blades.

For these and other reasons, the manufacturing cost can exceed the expected cost, and the profit rate per vehicle that was initially calculated can also decline substantially.

Cases such as these should not be neglected, and the process control at the production site needs to be thoroughly reviewed. Process improvement and preventive measures need to be devised and implemented through ingenuity, and original suggestions should be obtained via the full participation of all the on-site staff. If the efforts of the production site staff alone do not make progress, then the cooperation of the engineering staff from related departments should be obtained to thoroughly implement process improvements. The ultimate objective is to reduce the manufacturing costs and improve the profit rate.

The accumulation of a variety of process improvements at the production site can also conserve resources, resulting in lower material costs, parts purchasing costs, and energy consumption (items on the left side of Figure 3.4). In turn, this will all lead to a reduction in the number of quality defects and an improvement in the line operating rate. Furthermore, unnecessary labor costs can be eliminated by reducing excessive overtime work, and manufacturing costs can be further improved.

In addition, continuous efforts to improve process control and process improvement will enhance the skills and motivation of the production workers and contribute greatly to strengthening the work environment and work culture at the production site.

3.2.2.2 Daily Control Activities at the Production Site

Daily control activities at the production site make a large difference in the level of finished quality and productivity. For this reason, the third basic principle of manufacturing is a thorough and continuous effort to employ daily control activities through cooperation between the supervisors and workers on the production site (Amasaka, 1988, 2000, 2007).

The basic elements needed for daily control activities at the production site are the maintenance and improvement of what are called the four Ms: material, machine, man, and method (Amasaka, 2007).

In the implementation stage of daily control activities, it is vital to organize work standards and thoroughly implement standard operations by using the check sheets, such as the standard operation sheet, important quality check sheet, worker process designation list, and new worker operation sheet. These are designed to help ensure quality and productivity. Setting up reminder boards designed to remind workers about major quality defects in the past is also an effective preventive measure (Amasaka and Kamio, 1985).

In addition, an important quality control activity for ensuring manufacturing quality is the simultaneous confirmation of quality and incorporation of quality through utilization of control charts via a scientific process control method (Amasaka and Yamada, 1991).

Another factor that is vital to stably incorporate quality into the process is the control of conditions at the production facility. Condition controls take the following forms: (1) Condition controls for bolt tightening on the assembly line will indicate the operating pressure of air tools or the type of lubricating oil to be used. (2) Condition controls for the machining process line will be used to ensure the position and standard operation of quenching tempering equipment (visual control). Through daily control activities such as these, the reliability and maintainability of the production facility can be ensured, as well as the incorporation of quality (process capability [C_p] and machine capability [C_m]) and productivity by using a control chart (Amasaka 2003b).

3.2.2.3 Daily Improvement Activities at the Production Site

The fourth basic principle of manufacturing is daily improvement activities at the production site. The basis for simultaneous realization of quality and productivity via the Toyota Production System is improvement of the work environment to prevent worker fatigue, improvement of the work standards to prevent errors, improvement of control through visual confirmation, and so on.

These improvements are designed to help the workers to maintain their standard operations in a rhythmical flow. Concerted efforts are made to make small improvements that do not incur a large expense, such as safe work procedures, prioritizing quality, quickening the pace of work by one second, and so on (Amasaka and Kamio, 1985).

The main facilitators of these daily improvement activities are the workers themselves, who carry them out through small group activities, mainly in the form of QC Circles or through their own voluntary originality and ingenuity. In the implementation stage, activities such as having an improvement team support the production site or promoting greater cooperation between the supervisors and production staff are put into practice (Amasaka, 1999, 2003a).

One characteristic first example of such "daily improvement activities" is a process improvement that achieves work operations that are less tiring. Similarly, the second example is a process improvement made to an inconvenient work operation.

As demonstrated by these daily improvement examples, carrying out process improvements on a daily basis gives the workers themselves the opportunity to develop more convenient work processes, make their work operations safer, and also ensure that quality is being incorporated. At the same time, redundant work processes can be removed (Amasaka and Yamada, 1991).

3.2.2.4 Innovation of the Production Process: QCD Research Activities by White-Collar Staff

The fifth basic principle of manufacturing that is capable of quickly responding to the stringent demands of today's customers is the innovation of the production process and advancements in manufacturing that result from the QCD research activities conducted by white-collar staff (Amasaka, 2002b, 2003b, 2009, 2010; Amasaka and Sakai, 2010; Amasaka and Kamio, 1985; Amasaka and Yamada, 1991).

First, the fundamental principle of JIT production is to manufacture only what can be sold, when it can be sold, in the quantity that can be sold. To accomplish this, it is essential to establish a flexible production system that will produce and transport only what is needed, when it is needed, in the quantity that is needed, as a rational production measure.

Second, as a production mechanism that will realize the conditions listed above, it is important to incorporate production leveling, shortening of the production lead time, and a pull system into the process planning and design.

Third, with a view to reasonably carrying out the above, it is vital to always use signboards, facilitate small batch conveyance to raise precision, promote the flow of the production process, determine the takt time according to the needed volume, and then strictly adhere to these measures.

Fourth, it is imperative to reinforce the capabilities of the production site and to advance JIT production by actively developing new production technologies that will solve the bottleneck technological problems in production and substantially improve the quality and productivity of the production site.

Having said this, the following is an explanation of a small part of a QCD research activity that was promoted by the white-collar engineering staff (or engineers and managers responsible for manufacturing technology, production engineering development, process planning, process designing, and production management) that contributed to innovation of the production process.

Figure 3.5 shows an example of a process improvement for a layout that facilitates the incorporation of quality by implementing a countermeasure for the "outlying island" layout (Amasaka and Kamio, 1985).

Before the improvement, as seen in the left of Figure 3.5a, the six workers at the automatic table conveyor were separated from one another by the five parts feeders that supply the parts to the assembling machines (the five automated machines, marked A–E) located at each of the preceding processes. This process arrangement is the *outlying island* layout and when there is quality trouble or a *small stop* of the assembling machines, it is difficult for an expert worker (skilled worker) to provide instruction (assistance) to a newly assigned worker.

In addition, each worker's standard operation time differs due to the variation in experience. Therefore, an expert worker is forced to wait for work from a new worker (creating wasted time), since their operation time is shorter than that of newly assigned workers. Moreover, in the case of a production line with this outlying island layout, the takt time cannot be set according to the fluctuations (increases and decreases) in production volume because the takt time is fixed according to the operation time of the workers. Therefore, the takt time cannot be changed in a flexible manner according to the number of workers.

Given this background, the engineering staff (white-collar engineers) reviewed the production line where the workers' manual operations and automated machining processes are combined, so that they could reform the production process. As seen in Figure 3.5b, after the improvements were made, the automated machining processes were consolidated. In an effort to create a continuous processing operation from A to B to C, as well as from D to E, the set positions of the work pieces were changed and the processing jigs were altered.

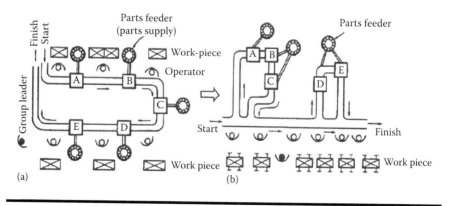

Figure 3.5 Example of a production layout that facilitates built-in quality by eliminating isolated worksites: (a) before Kaizen; (b) after Kaizen.

The parts feeders that supply the same parts were also integrated so that the number of parts feeders could be reduced from five to three. These improvement measures rearranged the automatic table conveyor into a single line, and the old work layout that had separated the workers was discarded.

The above case is a characteristic example of a process improvement that realized flexible production that could react to fluctuations in production volume; it resulted in improvements in the line operating rate and stabilized the production quality. This all effectively contributes to meeting the QCD requirements.

These improvement measures promoted (i) mutual confirmation between the workers on work progress; (ii) mutual assistance in case of any delay in work operation; (iii) observance of simultaneous confirmation of quality; (iv) improvement in the efficiency of material transport; (v) the ability to respond to the increase or decrease in the number of workers according to the changing production volume; and (vi) the ability of supervisors to respond to short stops of the automated processing line outside of their own production line, to supply parts to the parts feeders, to provide work instructions and assistance to line workers, and to create greater efficiency in the replacement work between completed work pieces and materials, and so on.

Below are a few examples of process improvements that have helped solve human errors. In recent years, it has become increasingly important to improve product quality in response to an increase in customers' sensitivity to quality.

The first example describes the efforts made to achieve high quality assurance for shock absorbers, a main part of the unit axle. Usually, the inspection to ensure the finished quality of assembled parts was dependent on the personal experience of the inspectors and their visual senses. Judgment of the Lissajous waveforms indicating the damping force characteristics (damping force and up-and-down strokes), which are displayed in rotation on the X–Y coordinates of an oscilloscope, is a difficult inspection process.

It often induced human error and misjudgments, even when performed by expert workers. In addition, it was time-consuming to find where on the work piece (product) the problem was or to conduct a disassembly investigation, fault diagnosis, and correction. This was a problem that could not be solved easily.

Then, the engineering staff and production site staff cooperated and developed an automatic diagnosis inspection device. This device simultaneously conducts a quantitative evaluation of the quality judgment and fault diagnosis. As a result, this improved the incorporation of quality into what had been a bottleneck problem in this inspection process (Amasaka and Kamio, 1985; Amasaka and Saito, 1983).

The second example is of an improvement made to help solve unforced errors such as wrong or missing parts and misassembled pieces. Generally, automobile production involves manufacturing various models and, therefore, it is necessary for the vehicle or parts assembly line workers to conduct *parts assembly* while being alert to wrong or missing parts and mistaken assembly. As a process improvement measure to reduce these sorts of errors, information technology (IT) was used to assist in accurately sorting out the work pieces on the assembly conveyor.

In this system, production instructions, electronic signboards, and radio-frequency identification tags (IC tags) are used to detect the types of work pieces. Then it lights up the assembly parts shelf that corresponds to the intended work pieces and uses sensors and visualization equipment to match the work operation with the proper work pieces (Amasaka and Kamio, 1985).

The third example is the identification of work-piece types using this IT. Until now, the parts unit signboards were visually confirmed, and the unit numbers were written down manually, but this repeatedly resulted in human errors such as mis-identification or mistaken descriptions.

After the process improvement was carried out, a bar-code signboard-reading device was developed and the unit code was automatically identified. An automatic marking device that prints out the intended code on the work-piece surface was also developed, and these devices solved the human error–based problems in this process (Amasaka and Kamio, 1985).

These examples of process improvement are the results of QCD research activities carried out through cooperation between white-collar and production site workers or supervisors in an effort to solve inconvenient work operations from the standpoint of human behavioral science.

References

Amasaka, K. 1988. Concept and progress of Toyota Production System (plenary lecture), Co-sponsorship: The Japan Society of Precision Engineering, The Japan Society for Technology of Plasticity, The Japan Society for Design and Drafting and Hachinohe Regional Advance Technology Promotion Center Foundation, Hachinohe, Aomori-ken, Japan [in Japanese].

Amasaka, K. 1989. TQC at Toyota: Actual state of quality control activities in Japan (special lecture), in *The 19th Quality Control Study Team for Europe, Union of Japanese Scientists and Engineers*, vol. 39, pp. 107–112 [in Japanese].

Amasaka, K. 1999. TQM at Toyota: Toyota's TQM activities: To create better car, in a training of trainer's course on evidence based on participatory quality improvement, international health program (TOT Course on EPQI), pp. 1–17. Touhoku University School of Medicine (WHO Collaboration Center), Sendai-city, Miyagi, Japan.

Amasaka, K. 2000. Basic thought of JIT—Concept and progress of Toyota Production System (plenary lecture), The Operations Research Society of Japan, Strategic Research Group-1, Tokyo, Japan [in Japanese].

Amasaka, K. 2002a. New JIT, a new management technology principle at Toyota, *International Journal of Production Economics*, 80: 135–144.

Amasaka, K. 2002b. Intelligence production and partnering for embodying a high quality assurance (plenary lecture), in *The Japan Society of Mechanical Engineers, The 94th Course of JSME Tokai Branch*, pp. 35–42, Nagoya, Japan.

Amasaka, K. 2003a. Proposal and implementation of the "Science SQC" quality control principle, *International Journal of Mathematical and Computer Modeling*, 38(11–13): 1125–1136.

Amasaka, K. (ed.). 2003b. *The Origin of Manufacturing: Progress of the Intelligence Quality Control Chart—Digital Engineering and High Quality Assurance*, Japanese Standards Association, Tokyo [in Japanese].

Amasaka, K. 2004a. Development of "Science TQM", a new principle of quality management: Effectiveness of strategic stratified task team at Toyota, *International Journal of Production Research*, 42(17): 3691–3706.

Amasaka, K. 2004b. *Science SQC, New Quality Control Principle: The Quality Strategy of Toyota*, Springer-Verlog, Tokyo.

Amasaka, K. 2007. Developing the high-cycled quality management system utilizing New JIT, in *Proceedings of the Seventh International Conference on Reliability and Safety*, pp. 248–253, August 23, 2007, Beijing, China.

Amasaka, K. 2008. Strategic QCD studies with affiliated and non-affiliated suppliers utilizing New JIT, in Putnik, G.D. and Cruz-Cunha, M.M. (eds), *Encyclopedia of Networked and Virtual Organizations*, Information Science Reference, Hershey, PA, pp. 1516–1527.

Amasaka, K. 2009. An intellectual development production hyper-cycle model—*New JIT* fundamentals and applications in Toyota, *International Journal of Collaborative Enterprise*, 1(1): 103–127.

Amasaka, K. and Kamio, M. 1985. Process improvement and process control. The example of axle-unit parts of the automobile, *Union of Japanese Scientist and Engineers, Statistical Quality Control*, 36(6): 38–47 [in Japanese].

Amasaka, K., Kurosu, S., and Morita, M. 2008. *New Manufacturing Principle*: Surpassing JIT—Evolution of Just in Time, Morikita-shuppan [in Japanese].

Amasaka, K. and Saito, H. 1983. Mechanization of sensory evaluation—Example of the judgment of the Lissajous waveforms indicating the damping force characteristics, in *Proceedings of the Annual Conference of Central Japan Quality Control Association 1983*, pp. 48–54, August 25, 1983, Nagoya, Japan.

Amasaka, K. and Sakai, H. 2010. Evolution of TPS fundamentals utilizing New JIT strategy—Proposal and validity of advanced TPS at Toyota, *Journal of Advanced Manufacturing Systems*, 9(2): 85–99.

Amasaka, K. and Yamada, K. 1991. Re-evaluation of the QC concept and methodology in auto-industry, *Union of Japanese Scientist and Engineers, Statistical Quality Control*, 42(4): 13–22 [in Japanese].

Hayashi, N. and Amasaka, K. 1990. Concept and progress of Toyota Production System: The physical distribution improvement in manufacture, in *Proceedings of the Annual Conference of Japan Physical Distribution Management Association 1990*, vol. 3, pp. 4-2-1–4-2-12, March 13, 1990, Tokyo, Japan [in Japanese].

Hayes, R. H. and Wheelwright, S. C. 1984. *Restoring Our Competitive Edge: Competing Through Manufacturing*, Wiley, New York.

Ohno, T. 1977. *Toyota Production System*, Diamond-sha, Tokyo, Japan [in Japanese].

Roos, D., Womack, J. P., and Jones, D. 1990. *The Machine That Changed the World—The Story of Lean Production*, Rawson/Harper Perennial, New York.

Taylor, D. and Brunt, D. 2001. *Manufacturing Operations and Supply Chain Management—The Lean Approach*, 1st edn., Thomson Learning, London.

Toyota Motor Corporation. 1987a. *The Toyota Production System and a Glossary of Special Teams at Production Job—Site at Toyota.* Toyota Motor Corporation.

Toyota Motor Corporation. 1987b. *Creation is Infinite: The History of Toyota Motor: 50 years.* Toyota Motor Corporation [in Japanese].

Womack, J. P. and Jones, D. 1994. From lean production to the lean enterprise, *Harvard Business Review*, March–April, pp. 93–103.

Chapter 4

Innovation of Manufacturing Fundamentals Surpassing JIT

Global manufacturing is currently facing two major transformations. The first is the shortened life cycle of products due to the diversification and sophistication of customers' needs. Because of this, manufacturers need to realize the shortest lead time possible in cooperation with their suppliers from the product development stage, through to manufacturing, and finally to sales, so that they are able to respond to this change. Second, it has been deemed essential to create a new administrative management technology necessary for the deployment of global production, worldwide uniform quality, and simultaneous production (production at optimal locations) and to manage this production in a systematic and organized manner (Amasaka, 2004a).

Based on this knowledge, the author deems it necessary to establish an advanced model of a new management technology toward innovation of manufacturing fundamentals surpassing just in time (JIT) for all the business processes of each department from upstream to downstream.

4.1 Necessity for an Advanced Production System and Advancement in Technologies and Skills

The increasing sophistication and diversification of customers' needs has made the development of a form of global production that acts in concert with the overseas deployment of production bases a pressing management issue. In order to succeed at global production, worldwide uniform quality and simultaneous launch (production at optimal locations) is an urgent task. The simultaneous achievement of quality, cost, and delivery (QCD) requirements that reinforce the product's appeal is required to realize this global production system (Amasaka, 2002, 2008).

In order to offer customers high-value-added products and prevail in the worldwide quality competition, it is necessary to establish an advanced production system that can intellectualize the production engineering and production management system. This will in turn produce high-performance and highly functional new products. The author believes that what determines the success of global production strategies is the advancement of technologies and skills that are capable of fully utilizing the above-mentioned advanced production system in order to realize reliable manufacturing at production sites (Amasaka, 2007b).

4.1.1 What Are the Concerns of Top Management and Manager Classes?

Amasaka (2007b) posed the question, "What are some of the issues that need to be addressed in order to prevail in the twenty-first century?" to a total of 154 respondents chosen from the top management class (board members) and managers (of divisions and departments) of the six manufacturers that participated in the Workshop on Quality Management of Manufacturers (hosted by Kakuro Amasaka, Professor of Aoyama Gakuin University from May 2004 to March 2006). The survey revealed the interests of the top management and manager classes.

Figure 4.1 is an example of the summary and analysis results of these interests (or the free opinions gathered from the survey). As shown here, their interests center

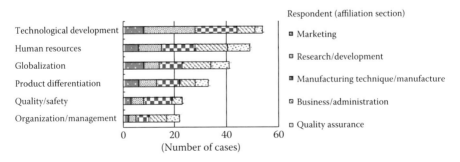

Figure 4.1 Which important issue should be tackled?

on technological development, human resources development, globalization, product differentiation, quality, safety, organization, and management. Consolidating all of these items, the management technology issue that they commonly share is the establishment of the kind of global production system that can achieve worldwide uniform quality and simultaneous launch (production at optimal locations).

Among other things, they give top priority to the realization of manufacturing with high quality assurance. This is achieved through high cyclization of the business method (or business process) related to manufacturing at overseas production plants (or production sites) in industrialized nations in the West or developing nations, not to mention in the Japanese domestic market.

In particular, they are concerned about the current situation in which the production sites are undergoing drastic changes due to the use of digital engineering and information technology (IT). The survey results confirmed that the pressing management technology issue is to overcome concerns such as "Is Japanese manufacturing reliable?", "Is it possible to carry out an overseas product launch similar to in Japan?", and "Can the same products be manufactured overseas as in Japan?" Furthermore, it was seen as critical to realize the intellectualization of the production sites so that they do not lag behind the advancements being made in technology and skills.

4.1.2 Japanese Manufacturing: Reinforcement of the Quality Technology

4.1.2.1 Capabilities Are Needed

At Japanese manufacturers that are experiencing a drastic transition in on-site manufacturing due to digital engineering, the technical capabilities of their on-site manufacturing are declining as a result of stagnation in QC Circle activities and the "hollowing out" that occurs due to the transfer of production bases overseas. Consequently, a weakening in the ability to identify and solve problems at the production site has been revealed and the incorporation of quality has become increasingly fragile (Amasaka, 2007b).

The recent increase of recall cases happening in Japanese products, which are expected to lead the world, and the remarkable quality improvement of industrially developing nations are sending out a warning message about the need to reinforce Japanese quality technology capabilities in view of global production. In the years to come, it will be urgent to establish a new and original Japanese management technology system that is suitable for the IT-adopting production sites, instead of clinging to the conventional quality management or production management methods (Amasaka, 2007a).

Launches of new products at Japanese domestic bases have been promoted through a concurrent system in which each production-related department conducts its own preparations based on its experience, while also taking into consideration a large number of design changes. However, in the case of overseas production, the experience, intuition, and know-how of the Japanese production staff are not conveyed to the local staff in each process, such as procurement, production preparation, and manufacturing.

Therefore, the practical application problems are often not solved. One countermeasure currently being taken is to launch at a production site in Japan first, solving various problems one by one, and then transferring the production as it is to an overseas site. If this current practice is continued even for a worldwide simultaneous launch, a large number of the Japanese production launch members will need to be dispatched to each of the overseas production bases.

Figure 4.2 shows conceptually what is stated above. The upper half of the figure depicts the status of production launch in Japan, while the lower half shows the status at overseas sites. As shown in the figure, in a normal production launch in Japan, planning, development, prototyping, equipment and material procurement, and production preparation overlap with one another and the project is promoted through trial and error.

As indicated in the upper half of the figure, the quality problems that occur after the production begins gradually decrease as the design and manufacturing staff take the necessary measures to solve them. A widely observed case even now is for manufacturing to be transferred to an overseas production after these quality problems have been solved.

On the other hand, in the case of a launch at an overseas plant (or local production) from the start, equipment and material procurement as well as production preparation

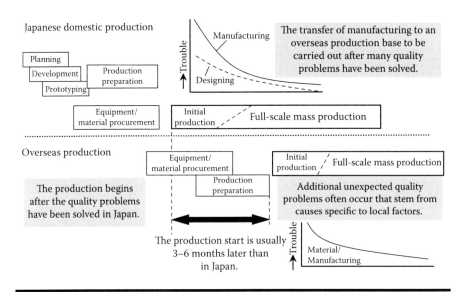

Figure 4.2 Launches of new products (Japanese domestic production and overseas production).

do not overlap with each other; rather, they are often promoted in sequence. For this reason, the production start is usually 3–6 months later than in Japan.

Even though the production begins after the quality problems have been solved in Japan, once production starts overseas, additional unexpected quality problems often occur that stem from causes peculiar to local factors (procured materials or local workers). Such additional quality problems are handled as needed on a trial and error basis similarly to in Japan, and often such troubles gradually diminish.

4.1.3 Style of Manufacturing Required in Overseas Production Plants

4.1.3.1 Manufacturing Products at Overseas Plants with the Same Quality as Japan at the Same Time

Regardless of whether it is a domestic or foreign market, the recent purchasing behavior of customers suggests that their consciousness of product value has been rapidly growing and they will only pay the price that they think matches the perceived value gained from the product. True to the saying, "the price provides temporary satisfaction, but quality provides life-long satisfaction," even in the markets of developing nations, the product specifications and functional quality of older models surely cannot gain the upper hand, even if such products are priced more inexpensively.

From now on, the basis of global production and quality assurance will be to equalize the finished quality of products regardless of where they are manufactured, either overseas or domestically. Depending on the state of the country, it will not matter whether production is carried out by the latest automated factory or by workers' manual operation. Some manufactures might decide on their production system according to the location, product specifications, or volume of the products to be manufactured. However, in order to achieve worldwide uniform quality, it is increasingly important to create a high quality assurance system that gives top priority to the customers (Amasaka, 2003a, 2007a,b).

4.1.3.2 From Reinforced Inspection to Enhanced Process Management

One of the measures that will equalize the product quality enjoyed by customers is the implementation of a thorough inspection process. It is generally difficult, however, to check in detail all the inspection items that must be assured for all products.

In addition, when the daily management based on man, machine, material, method, and environment (4M-E) is not thoroughly carried out, production that matches the design drawing will be impossible and simply result in a mountain of defective products after all the items have been inspected. This will then require a large amount of reworking or disposal and will interfere with the original production plan.

Next, the factor of logistics needs to be given similar consideration. In Japanese domestic production, many of the assembly manufacturers conduct JIT production and therefore are able to supply what is needed, when it is needed, in the amount that is needed, through cooperation with their suppliers. However, in the case where JIT production is not fully implemented at an overseas base, what will happen when the products manufactured there are exported to Japan or to other foreign countries?

The general means of transporting products is by ship and this requires a longer lead time. It also means that holding a large stock is inevitable. If a quality problem occurs under such circumstances, the damage inflicted on the manufacturer will multiply in proportion to the volume of product in stock.

Having said the above, it is not an exaggeration to say that overseas production requires a higher level of quality assurance manufacturing than Japanese domestic production. Reinforcing quality inspections is an important facet of quality assurance, but it alone is not sufficient, and attention must also be given to how quality is incorporated into the production process. In other words, in the end, it is essential to implement enhanced process management (Amasaka and Sakai, 2009).

There are a great variety of different cultures, religions, and languages in the world. When thinking about the mission of trying to realize high quality assurance from a global standpoint, while also overcoming these sorts of barriers, one cannot help but think how easy it would be if all of the company's employees at least understood a single common language for communication. Luckily, however, at the manufacturing site there is a universally common language.

It is what is called probability and statistics in statistical terminology. The majority of people agree that the scientific process management method applying statistical quality control (SQC) can be a powerful tool for maintaining and improving the production process, as well as for determining the causes of a variety of quality problems and preventing them from recurring in the future (Amasaka, 2003a, 2004b).

4.1.3.3 Production System Issues at Overseas Production Plants

In recent years, the limit of traditional Japanese manufacturing, or the so-called on-site capability, has been recognized at the overseas production sites. Up until now, the general business practice has been to bring the local production staff to Japan to educate them in the Japanese production system and then transfer them back to their overseas sites after solving various problems. These production staff members would then implement the Japanese production system and quality assurance system. The question is, "Does this achieve the overseas deployment of production that was originally intended?"

There are several issues concerning the application of the current Japanese production system at overseas production sites. These issues include correcting the declining on-site capabilities at Japanese domestic production sites, which are expected to be the model for the overseas bases, correcting the scattering of Japanese engineers to overseas bases and the subsequent decline in total technological capability, solving the concerns regarding the possibility of manufacturing products at overseas sites with the same level of quality as in Japan, establishing a new production system that will enable global production, and so on.

The author assumes that a simple extension of the current Japanese production system is not sufficient to overcome these issues and therefore has decided that the top priority issue here is the further advancement of the Japanese production system in order to realize worldwide uniform quality and simultaneous launch (production at optimal locations) as the key to success in global production (Amasaka, 2000, 2007a; Amasaka et al., 2005).

4.2 Advancement in Manufacturing: Reinforcement of the Production Management Function

As already mentioned, customers today select products that suit their lifestyle and personal values. In addition, they are strongly demanding manufacturing that will enhance customer value via the products' reliability (quality and value gained from use). For this reason, unless the manufacturer advances its manufacturing processes in such a way as to respond to these demands, while also accurately grasping the customers' preferences, it will be pushed out of the world market.

In this day and age, when customers have quick and easy access to the latest information from every corner of the world due to the permeation of IT, the production management department, which is the command center of manufacturing, needs more than ever before to have a global view and deploy strategic production management as the core of corporate management (Amasaka, 2007a).

The new mission of the production management department is the deployment of strategic cooperation between the on-site departments of engineering, production and sales, and general administration, as well as the domestic and overseas suppliers. This is done in order to realize worldwide uniform quality and simultaneous launch (production at optimal locations) ahead of competitors (Amasaka, 2000; Ebioka et al., 2007).

In other words, in order to solve the various manufacturing problems both domestically and overseas, the experienced-based implicit knowledge (such as know-how, empirical rules, intelligent information, etc.) possessed by the white-collar and on-site production workers needs to be converted to "linguistically expressed knowledge" or "explicit knowledge," as shown in Figure 4.3.

This will be done through the incorporation of the Science SQC, New Quality Control Principle that has been proven effective as a scientific problem-solving method. It is vital to deploy this advancement of manufacturing in a systematic and organizational manner by deploying the personal empirical knowledge as organizationally shared knowledge (Amasaka, 2000, 2001).

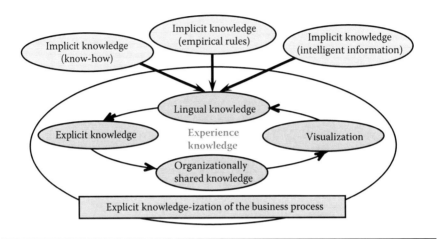

Figure 4.3 Visualization of experience-based implicit knowledge.

In the rapidly changing environment of management technology, Japanese manufacturers need to actively and courageously endeavor to offer highly reliable products of the latest model that will enhance the customer value ahead of their competitors, so that they are not eliminated from the world market.

However, the technological capability of Japanese manufacturers has become diffused due to the transfer of production to overseas bases in recent years (Amasaka, 2003b, 2007a,b). A symbol of this trend is how slowly Japanese production sites have responded to the advancements in automation and digitalization. It has also become apparent that scientific production management that ensures that design quality will be built into the process has also been weakening (Rice, 1953; Amasaka, 2003c).

On the other hand, the U.S. government has implemented the "Twenty-First Century Manufacturing Vision: New Wheel-Type Manufacturing Model" as a revival strategy for the manufacturing sector, and it has achieved a remarkable improvement in the quality level, which can be seen in recent U.S.-made cars (Nezu, 1995; J.D. Power and Associates, 2006). In contrast, the hallmark of Japanese manufacturing, namely, the utilization of process control charts, has been dying out and the process capability or machine capability needed for incorporating quality according to the design drawing has been losing its distinctive effect (Amasaka, 2003c).

In this environment, where highly precise quality control is being demanded, it is obvious that the Japanese manufacturers (1) are bearing market claim expenses that cannot be easily solved with conventional process control and (2) have been pushed into a situation of no return, where a single wrong step can lead to a recall problem and considerably damage customers' trust in the corporation (Nihon Keizai Shinbun, 1999, 2000, 2001, 2002, 2006; Nikkei Sangyo Shinbun, 2000a,b). Therefore, it can be said that the urgent task now is innovation of the Japanese production system (typified by TPS) and to not cling to past successes or conventional production systems (Amasaka, 2007a, 2011).

References

Amasaka, K. 2000. Partnering chains as the platform for quality management in Toyota, in *Proceedings of the 1st World Conference on Production and Operations Management*, pp. 1–13, August 30, 2000. Seville, Spain (CD-ROM).

Amasaka, K. 2001. Quality management in the automobile industry and the practice of the joint activities of the auto maker and the supplier, in *Japan Small Business Research Institute, Surveillance 2013—A Change in the Order System such as a Part in the Manufacturing Industry and that Correspondence: The Condition of the Supplier Continuance in the Automobile Industry*, vol. 74, No. 2, Chapter 6, pp. 135–163, Tokyo, Japan [in Japanese].

Amasaka, K. 2002. New JIT, a new management technology principle at Toyota, *International Journal of Production Economics*, 80: 135–144.

Amasaka, K. 2003a. Proposal and implementation of the "Science SQC", quality control principle, *International Journal of Mathematical and Computer Modeling*, 38(11–13): 1125–1136.

Amasaka, K. 2003b. Digital engineering and production control in a high quality assurance, at The Japan Society for Production Management (lecture), The 17th Annual Conference, Tutorial Session: The Past, the Present, and the Future of Quality Creation—Logic and Digital Engineering of Quality Creation, Gakushuin University, Tokyo, Japan [in Japanese].

Amasaka, K. (ed.). 2003c. *The Origin of Manufacturing: Progress of the Intelligence Quality Control Chart—Digital Engineering and High Quality Assurance*, Japanese Standards Association, Tokyo, Japan [in Japanese].

Amasaka, K. 2004a. Development of "Science TQM", a new principle of quality management: Effectiveness of strategic stratified task team at Toyota, *International Journal of Production Research*, 42(17): 3691–3706.

Amasaka, K. 2004b. *Science SQC, New Quality Control Principle: Quality Control Strategy at Toyota*, Springer-Verlog, Tokyo.

Amasaka, K. et al. 2005. Establishment of "Next-generation Management Technology- Japan Method", in *Proceedings of the 21th Annual Conference: 10th Anniversary Symposium of JSPM Foundation, The Japan Society for Production Management*, Separate Volume, pp. 1–16, March 12, 2005, Aoyama Gakuin University, Tokyo, Japan [in Japanese].

Amasaka, K. 2007a. New Japan production model, an advanced production management principle: Key to strategic implementation of New JIT, *The International Business and Economics Research Journal*, 6(7): 67–79.

Amasaka, K. (ed.). 2007b. *New Japan Model: Science TQM- Theory and Practice for Strategic Quality Management*, Study Group of the Ideal Situation on the Quality Management of the Manufacturing Industry, Maruzen, Tokyo, Japan [in Japanese].

Amasaka, K. 2008. Strategic QCD studies with affiliated and non-affiliated suppliers utilizing New JIT, in Putnik, G.D. and Cruz-Cunha, M.M. (eds), *Encyclopedia of Networked and Virtual Organizations*, Information Science Reference, Hershey, PA, pp. 1516–1527.

Amasaka, K. 2011. "New JIT, new management technology principle" (special lecture), in *The VIII Siberian Conference "Quality Management-2011"*, pp. 1–16. March, 22, 2011, Krasnoyarsk, Russia.

Amasaka, K. and Sakai, H. 2009. TPS-QAS, new production quality management model: Key to New JIT—Toyota's global production strategy, *International Journal of Manufacturing Technology and Management*, 18(4): 409–426.

Ebioka, K., Sakai, H., Yamaji, M., and Amasaka, K. 2007. A New Global Partnering Production Model "NGP-PM" utilizing "Advanced TPS", *Journal of Business and Economics Research*, 5(9): 1–8.

J.D. Power and Associates. 2006. http://www.jdpower.com/.

Nezu, K. 1995. *The Scenario of the U.S. Manufacturing Industry Remarkable Progress Aimed at by CALS*, Kogyo Chosakai Publishing, Tokyo, Japan [in Japanese].

Nihon Keizai Shinbun. 1999. Corporate survey (68QCS): Strict assessment TAM—development of easy-use techniques (July 15, 1999), Tokyo, Japan [in Japanese].

Nihon Keizai Shinbun. 2000. Worst record 40% increase of vehicle recalls (July 6, 2000).

Nihon Keizai Shinbun. 2001. IT innovation of manufacturing (January 1, 2001).

Nihon Keizai Shinbun. 2002. The vehicle quality survey I US—1. Toyota, 2. Honda, 3. GM (May 31, 2002), Tokyo, Japan [in Japanese].

Nihon Keizai Shinbun. 2006. Risky "quality": Apart from production increase, it is recall rapid increase (February 8, 2006), Tokyo, Japan [in Japanese].

Nikkei Sangyo Shinbun. 2000a. Improving problem—finding capabilities—reconstruction of a new system to replace QC (September 23, 2000a), Tokyo, Japan [in Japanese].

Nikkei Sangyo Shinbun. 2000b. Management viewpoints—The collapsing quality myth of japan (September 24, 2000b), Tokyo, Japan [in Japanese].

Rice, W. B. 1953. (Ishikawa, K., Translation) *Sampling at the Factory*, Maruzen, Tokyo, Japan [in Japanese].

Chapter 5

New JIT, New Management Technology Principle

5.1 New JIT, A New Management Technology Principle

5.1.1 Introduction

The production technology principle Japan contributed to the world in the latter half of the twentieth century was the Japanese-style production system typified by the Toyota Production System, enhanced by the quality management technology principle generally referred to as just in time (JIT) Today, however, improvements in the quality of Japanese-style management technology principles are strongly desired in the face of unexpected quality-related recall problems breaking out among industrial leaders, while at the same time delays in technical development cause enterprises to experience crises of existence (Goto, 1999).

To realize manufacturing of the best quality for the customer in a rapidly changing technical environment, it is essential to create a core principle capable of changing the technical development work processes of development and design divisions. Similarly, it is important for the production division to develop a new production technology principle and establish new process management principles to enable global production (Hayes and Wheelwright, 1984). Furthermore, new marketing activities independent from past experience are required for sales and service divisions to achieve firmer relationships with customers.

In addition, a new quality management technology principle linked with overall activities for higher work process quality in all divisions is necessary for an enterprise to survive. Regarding the need for these improvements, advanced car manufacturer A is no exception. For this reason, Amasaka (2000c, 2002, 2009a, 2013b), anticipating the manufacturing of the next generation, has developed a new management technology principle called New JIT as the next step in the evolution of conventional JIT. Its effectiveness is demonstrated in this section.

5.1.2 Toyota Production System and TQM: JIT

The Toyota Production System, including JIT, is the manufacturing system Toyota developed in pursuit of optimum streamlining. It aims to build quality in using total quality management (TQM) in the manufacturing process while following the principle of cost reduction. This is why Toyota often compared the Toyota Production System and TQM to the two wheels of one vehicle (Ohno, 1977; Toyota Motor Corporation, 1987). To be able to manufacture vehicles that meet customer demands without fail, timely quality, cost, and delivery (QCD) studies are important. To accomplish this, Toyota has positioned the Toyota Production System and TQM as the two figurative pillars of our management technology principle (Amasaka, 1989).

As Figure 3.1 (refer to Chapter 3) shows, these two elements combine to flatten extreme curves and constantly raise mean values by maintaining and improving QCD research with TPS as the hardware technology and TQM as the software technology. As is widely known, use of the rational technique of statistical quality control (SQC) reduces fluctuations in and raises the average level of manufacturing quality, as represented Figure 3.1. In this regard, it could perhaps be said that SQC is the root of manufacturing and the historical origin of TQM. Improvements in engineering quality are urgently needed to emerge victorious from extreme global competition in the area of quality (Amasaka, 1989).

The JIT concept and approach renovated vehicle production at Toyota (Womack et al., 1991). As a result of favorable worldwide evaluations of its effect, JIT became the core concept of manufacturing industries around the world in the twentieth century (Womack and Jones, 1994).

5.1.3 Today's Management Technology Problem

This section touches on the management technology problems that Toyota faces. The Toyota Production System has already been developed as an internationally shared system, known as a Lean system (Taylor and Brunt, 2001), and is no longer an exclusive technology of Toyota in Japan. The success of TQM principles in the United States resulting from the realization of the importance of quality management through studying Japanese TQM (also referred to as total quality control [TQC]) has diminished the superiority of Japanese product quality due to Japanese-style quality management (Gabor, 1990; Joiner, 1994).

For these reasons, we conjecture that it is clearly impossible to lead the next generation by merely maintaining the two Toyota management technology principles, the Toyota Production System and TQM. To overcome this issue, it is essential to renovate not only TPS, which is the core principle of the production process, but also to establish core principles for sales, production planning, design, development, and other departments.

In order to accomplish this, Amasaka et al. (1999e) identified technological management problems from the management's point of view based on an awareness survey of 72 executives and general management personnel from 12 Toyota group companies (Toyota, Denso, and others), as shown in Figure 2.2 (refer to Chapter 2). This figure depicts overall management technology problems by using Quantification Class III analysis (Amasaka et al., 1995a, 1999e). The development division gives priority to the development of new products, the production division to next-generation production systems, and the marketing division to the development of new marketing methods. The common problem is revision of the organization or system through judicial sharing of information for better division-to-division coordination.

Such studies and analyses have clarified the core concepts necessary for the next-generation, new management technology principle, its roots, and the technical elements required for linking core concepts. As a result, it is considered strategically necessary to create a new management technology principle. This principle will be a new version of JIT with the new concept of linking the QCD research activities of all these departments. To facilitate excellent quality management that can contribute to engineering development worldwide, it is necessary to carry on precise and reasonable JIT activities so as to enhance the quality of work processes in all divisions.

5.1.4 Proposal of New JIT for Renovating Management Technology

5.1.4.1 The Concept of New JIT for Renovating Management Technology

The creation of attractive products requires the implementation of customer science (Amasaka, 2000c, 2002, 2005) to scientifically grasp customers' tastes. To achieve this, the entire organization must be managed by each of the marketing, engineering, and production divisions (refer to Appendix 5.1 and Section 7.1). All of these are organically combined by control divisions (engineering control, production control, purchasing control, and information systems), the general administration division, and those in charge of motivating human resources and organizing the divisions as a whole. Therefore, a new organizational and systematic principle for the next-generation, new management technology principle, New JIT, for accelerating the optimization of work process cycles of all the divisions is necessary.

Amasaka (2000c, 2002, 2009a) has developed a new management technology principle, New JIT, as shown in Figure 5.1, which contains hardware and software systems as the next-generation technical principles for transforming management technology into management strategy. The first item, the hardware system, consists of the Total Development System (TDS), Total Production System (TPS), and Total Marketing System (TMS), which are the three core elements required for establishing new management technology principles for business planning, sales, R&D, design, engineering, and production, among others.

For the second item, software, Amasaka (2000a,c,d) has developed TQM-S (TQM utilizing Science SQC, New Quality Control Principle, called Science TQM, New Quality Management Principle) as the system for improving work process quality of the 13 divisions shown in Figure 5.1. In this concept, which was developed to make quality management more scientific, the quality management principle of Science SQC (Amasaka, 1997a, 2003a, 2004b) was added to TQM activities (refer to Appendix 5.2 and Section 5.4). Its advantageous effect has been demonstrated. This systemizes and organically organizes the new SQC operation by combining information technology (IT) with SQC.

Thus, divisions are linked organically, as shown in Figure 5.1. We believe that this linkage contributes to further growth and development of the three core elements of New JIT, and general solutions have to be approached by clarifying the gaps that exist in theory, testing, calculation, and actual application. For further details of Science SQC with four principles (Scientific SQC, SQC technical methods, integrated SQC network, and Management SQC), refer to Amasaka and Osaki (2003) and Amasaka (1999a, 2000a,c,d).

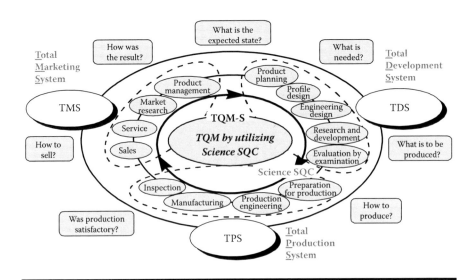

Figure 5.1   New JIT, new management technology principle.

5.1.4.2 Three Subsystems of New JIT

5.1.4.2.1 TDS: The First Principle

The expectations and role of the second principle, TDS, as shown in Figure 5.2, are the systemization of a design management method that is capable of clarifying the following:

1. Collection and analysis of updated internal and external information that emphasizes the importance of design philosophy
2. Development design process
3. Design method that incorporates enhanced design technology for obtaining general solutions
4. Design guidelines for designer development (theory, action, decision making)

The application of Science SQC to improve the process quality of design work in order to realize these criteria is called Design, SQC (Amasaka et al., 1999b). To create the latest technology in response to technological evolution, it is important to implement Design SQC so that it can contribute to the development of proprietary technology, its continuation, and its further advancement. It is important to establish general technological solutions, rather than particular solutions, by building up partial solutions. The true objective of establishing TDS is to create technologies through optimum design brought about by information sharing.

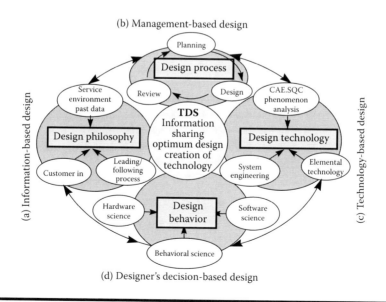

Figure 5.2 Schematic drawing of TDS.

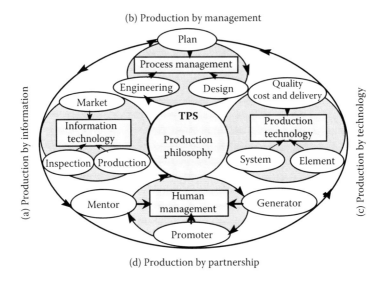

Figure 5.3 Schematic drawing of TPS.

5.1.4.2.2 TPS: The Second Principle

The expectations and new role of the third principle, TPS, as shown in Figure 5.3, comprise the following:

1. Customer-oriented production control systems that place top priority on internal and external quality information
2. Creation and management of a rational production process organization
3. QCD activities using advanced production technology
4. Creation of active workshops capable of implementing partnerships

The application of Science SQC to strengthen the overall production organization to achieve these objectives is called production SQC. One of the objectives of TPS implemented through the application of production SQC is to solve bottleneck technical problems at the production engineering (preparation) and production stages. The second objective is to establish a rational, scientific process control method for achieving a highly reliable production system (Amasaka and Sakai, 1996, 2009).

5.1.4.2.3 TMS: The Third Principle

The expectations and role of the first principle, TMS, as shown in Figure 5.4 include the following:

1. Market creation through the gathering and use of customer information
2. Improvement of product value by understanding the elements essential to raising merchandise value

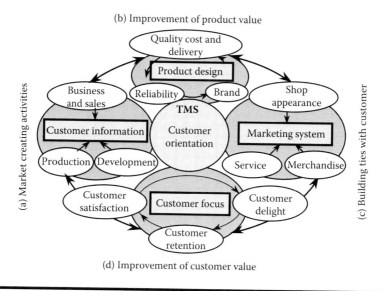

Figure 5.4 Schematic drawing of TMS.

3. Establishment of hardware and software marketing systems to form ties with customers
4. Realization of the necessary elements for adopting a corporate attitude (behavioral norm) of enhancing customer value and developing customer satisfaction (CS), customer delight (CD), customer retention (CR), and networks

The practical application of Science SQC, the effectiveness of which was demonstrated in establishing TMS in the sales division (Amasaka, 2001) as well as contributing to business through product planning departments, is termed marketing SQC. In order to scientifically conduct market surveys not confined to readily apparent sales, recognize the importance of marketing SQC, which contributes to future development, and carry out customer-oriented global marketing, the implementation of customer science is becoming more and more important, not least in terms of properly linking TMS, TDS, and TPS.

5.1.5 Implementation and Effect of New JIT through TQM-S

The following are representative examples of research, as the basis for establishing TQM-S along with TMS, TDS, and TPS, which are the core elements of New JIT.

5.1.5.1 Implementation of Vehicle Profile Design Using TDS

The main objective of establishing TDS is to provide creative products that highly impress customers ahead of the competition. Among many factors, vehicle styling is important in winning customers over. Designers have been searching for a method of supporting ideas for obtaining practical, distinctive solutions for many years. Amasaka et al. (1999e) applied TDS by using design SQC to develop the Aristo (Lexus GS400 in the United States) and successfully achieved the objective (Motor Fan, 1997).

In this instance, the author applied the technical methods of SQC to visualize vehicle images that customers might desire followed by an analysis of the association between the degree of CS and elements of vehicle appearance based on appearance assessments. Using the knowledge thus acquired, the association between vehicle image and proportion data of a vehicle was determined to help improve the design process. TDS has become the main vehicle design development method at Toyota and is contributing to the provision of high-quality and highly reliable products to customers (Amasaka and Osaki, 1999, 2003).

5.1.5.2 Establishment of New Toyota Production System Using TPS

In this section, the author discusses the establishment of a New Toyota Production System intended to meet the challenges posed by globalization using TPS. By implementing inline-online SQC, an integrated network system that utilizes IT and SQC, the next-generation, scientific process management principle New JIT can be achieved.

In specific terms, under the New Toyota Production System, information originating on the line is gathered by computer and then transferred to other related divisions. Then anyone can carry out improvements promptly as needed by accessing the quality management information of a process anytime, anywhere and eventually improve the machine capacity (shown by the machine capacity index [Cm]) and process capability (shown by the process capability index [Cp]) through two network systems that contribute to the establishment of New JIT.

The following are examples of the application of the systematic development of quality assurance technologies to test the validity of the New Toyota Production System: (1) development of the Availability and Reliability Information Administration Monitor System in the body line (ARIM-BL) (Amasaka and Sakai, 1998) and (2) typical utilization of the Quality Assurance System by using TPS in the unit-axle line (TPS-QAS) (Amasaka and Sakai, 2009).

1. ARIM-BL consists of the real-time management of trouble diagnosis of equipment using Weibull analysis, contributing to the maintenance and improvement of machine capacity.
2. TPS-QAS consists of the real-time management of process quality diagnosis using control charts, contributing to the maintenance and improvement of process capability.

With the adoption of New TPS, the expected effect can be attained (for details, see Amasaka and Sakai, 2009). In recent years, Toyota has been introducing New TPS in its domestic factories one after the other, resulting in attainment of the desired result. At present, a system applicable for Toyota group companies and overseas plants is being developed.

5.1.5.3 Establishment of Toyota Sales Marketing System Using TMS

The construction of a Toyota Sales Marketing System (Amasaka, 2001) that is capable of renewing established ties with customers is extremely important to strategic marketing, which is the basis for establishing global marketing and is one of the establishment targets of TMS. A "Netz" Toyota dealer is introduced as a model case in which vehicle sales market share was increased largely through improvements in the CR ratio at vehicle replacement.

By using a new SQC method termed categorical automatic interaction detector (CAID) analysis based on marketing SQC, we were able to demonstrate the planning measures needed to improve the CR ratio of our products. Full-scale use of this strategic system began in mid-1998, contributing to the construction and development of the Toyota Sales Marketing System at Toyota dealers (Nikkei Business, 1999).

5.1.6 Conclusion

In this paper, the author has proposed New JIT as a new management technology principle for twenty-first-century manufacturing. The proposed New JIT does not stop with the Toyota Production System and TQM, which are representative of the Japanese production system called JIT found at the Toyota Motor Corporation and elsewhere. Rather, the author has strategically developed New JIT with a concept that goes beyond production.

New JIT renovates the business process of each division, which encompasses business sales, development, and production, with the aim of customer-first quality management. It includes hardware and software systems developed according to new principles to link all activities on a company-wide basis.

The hardware system comprises three core elements: TDS, TPS, and TMS. The software system consists of the deployment of TQM-S, which is a new principle for quality management. The objectives of TQM-S are to improve the quality of work processes in all divisions and to renovate quality management activities. TQM-S utilizes Science SQC called Science TQM, which has already been demonstrated to be effective.

The author and others believe that the effectiveness of New JIT has been demonstrated, as positive results in sales, design, development, and production have been obtained from applying it within Toyota.

In the future, we will closely observe next-generation manufacturing and further develop New JIT applications for management technology from the viewpoint of management worldwide.

5.2 Foundation of Advancing the Toyota Production System

5.2.1 Introduction

The top priority of the industrial field today is the new deployment of global marketing for surviving the era of global quality competition. To realize this, manufacturing must give top priority to customers with a good QCD and in a rapidly changing technical environment.

To solve today's management technology issues, the author proposed New JIT by using new hardware and software systems, TDS, TPS, TMS, and science TQM for establishing new management technology principles for sales, R&D, design, engineering, and production, among others. The proposed New JIT does not stop with the Toyota Production System and TQM, which are representative of the Japanese production system called JIT.

In the implementation stage, the foundation for advancing the TPS by supporting TMS and TDS is based on efforts to attain simultaneous achievement of QCD through innovation of manufacturing technology utilizing New JIT. This in turn is accomplished by having cooperation between on-site white-collar engineers and supervisors and workers with affiliated and nonaffiliated suppliers (Amasaka, 2008a, 2013).

In this example, innovation of the production line of the automobile rear axle unit in advanced car manufacturer A, Factory-Motomachi, will be presented. It will demonstrate an example of solving a bottleneck manufacturing technology problem that was inhibiting New JIT production (Amasaka, 2009b).

5.2.2 Typified Japanese Production System Called JIT

The system of JIT, a Japanese production system typified by the current Toyota Production System, is a manufacturing system developed by Toyota Motor Corporation that adopts TQM in its manufacturing process. TQM recognizes the principle of cost reduction, and it pursues improvement of product quality and productivity simultaneously by pursuing maximum efficiency (optimal streamlining). TQM can also be considered the essential concept behind JIT (Amasaka, 2008a,b, 2013). For this reason, Toyota positioned this production system as the management technology of Toyota and often draws the analogy of the two systems being like the two wheels of a vehicle (Ohno, 1977; Hayes and Wheelwright, 1984).

In order to cater to customer needs and to conduct manufacturing successfully, a study into the timely simultaneous achievement of QCD is the top priority.

To accomplish this, Toyota has been viewing the two technologies above as the dual pillars of its management technology (Amasaka, 2002, 2004a). As shown in Figure 2.1 (refer to Chapter 2), through the combination of these two pillars, a rational management attitude has been consistently maintained so that deviations like large tidal waves can be reduced to small fluctuations similar to gentle ripples, which enables the constant improvement of average values.

This involves the continuation and improvement of QCD research activities while incorporating the SQC method from the standpoint of the hardware technology–based Toyota Production System and the software technology–based TQM, which are represented on the vertical and horizontal axes, respectively. As can be seen from the figure, SQC constitutes the basis for maintaining and improving product quality and the starting point of TQC (TQM) from a historical standpoint. However, in order to prevail amid strong global quality competition in the future, the technological capability of manufacturers must be urgently reinforced and reformed (Amasaka, 2008b, 2009b).

The concept of JIT and its approach have reformed automobile production at Toyota (Ohno, 1977; Amasaka, 2002, 2013). Because its effectiveness was highly praised all over the world, JIT was established as a core concept of global manufacturing in the twentieth century (Roos et al., 1990; Womack and Jones, 1994).

5.2.3 Foundation of Advancing the Toyota Production System

5.2.3.1 Demand for Advancement in Management Technology

The top priority of the industrial field today is the new deployment of global marketing for surviving the era of global quality competition (Kotler, 1999; Amasaka, 2009b). To realize manufacturing that places top priority on customers with a good QCD and in a rapidly changing technical environment, it is important to develop a new production technology principle and establish new process management principles to enable global production (Amasaka et al., 2008).

Furthermore, a new quality management technology principle linked with overall activities for higher work process quality in all divisions is necessary for an enterprise to survive (Burke and Trahant, 2000). The creation of attractive products requires each of the sales, engineering/design, and production departments to be able to carry out management that forms linkages throughout the whole organization (Seuring et al., 2003). From this point of view, the reform of Japanese-style management technology is desired once again. In this need for improvements, Toyota is no exception (Goto, 1999).

5.2.3.2 Need for Strategic QCD Studies with Affiliated and Nonaffiliated Suppliers

IT development has led to a market environment where customers can promptly acquire the latest information from around the world with ease. Today, customers select products that meet their lifestyle needs and give them a sense of value that justifies their cost. Thus, the concept of quality has expanded from being product quality–oriented to business quality–oriented and then corporate management quality–oriented. Customers demand the reliability of enterprises through the utility values (quality, reliability) of products. Advanced companies in countries all over the world, including Japan, are shifting to *global production*. The purpose of global production is to realize uniform quality worldwide and production at optimum locations in order to ensure company survival amid fierce competition.

For the manufacturing industry, the key to success in global production is systematizing its management methods when modeling strategic supply chain management (SCM) for its domestic and overseas suppliers. In-depth studies of the TPS, TQM, partnering, and digital engineering will be needed when these methods are implemented in the future. Above all, manufacturers endeavoring to become global companies are required to collaborate with not only affiliated companies but also with nonaffiliated companies to achieve harmonious coexistence among them based on cooperation and competition. In other words, a so-called federation of companies is needed (Amasaka, 2008a).

5.2.3.3 Significance of Strategic Implementation of New JIT

Having said this, it is the author's conjecture that it is clearly impossible to lead the next generation by merely maintaining the two Toyota management technology principles, Toyota Production System and TQM. To overcome this issue, it is essential to renovate not only TPS, which is the core principle of the production process, but also to establish core principles for marketing, design and development, production, and other departments (Amasaka, 2002, 2008c).

The next-generation management technology model, New JIT, which the author has proposed through theoretical and systematic analyses, as shown in Figure 5.1 (refer to Section 5.1), is the JIT system for not only manufacturing, but also customer relations, sales and marketing, product planning, R&D, design, production engineering, logistics, procurement, administration, and management, and for enhancing business process innovation and the introduction of new concepts and procedures.

New JIT contains hardware and software systems as the next-generation technical principles for accelerating the optimization (high linkage) of business process cycles of all divisions, as shown in Figure 5.5 (Amasaka, 2007d). The first item, the hardware system, consists of the TDS, TPS, and TMS, which are the three core elements required for establishing new management technology principles for sales, R&D, design, engineering, and production, among others.

The expectations and role of the second principle, TDS, are the systemization of a design management method that is capable of clarifying the following: (i) collection and analysis of updated internal and external information that emphasizes the importance of design philosophy, (ii) development design process, (iii) design method

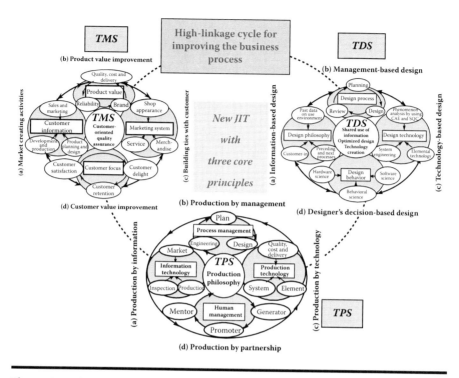

Figure 5.5   New JIT strategy, high-linkage cycle for improving business processes.

that incorporates enhanced design technology for obtaining general solutions, and (iv) design guideline for designer development (theory, action, and decision making).

The expectations and new role of the third principle, TPS, comprise the following: (i) customer-oriented production control systems that prioritize internal and external quality information, (ii) creation and management of a rational production process organization, (iii) QCD activities using advanced production technology, and (iv) creation of an active workshop capable of implementing partnership.

The expectations and role of the first principle, TMS, include the following: (i) market creation through the gathering and use of customer information, (ii) improvement of product value by understanding the elements essential to raising merchandise value, (iii) establishment of hardware and software marketing system to form ties with customers, and (iv) realization of the necessary elements for adopting a corporate attitude (behavioral norm) of enhancing customer value and developing CS, CD, CR, and networks.

For the second item, the strategic quality management system, the author is proposing a new principle of quality control, TQM-S (TQM by utilizing Science SQC) called Science TQM, in Toyota as a software system for improving the business process quality" of the 13 departments shown in the diagram as shown in Figure 5.1 (refer to Section 5.1) (Amasaka, 2002, 2003a, 2004b,c, 2008b, 2013).

5.2.3.4 Cooperation between On-Site White-Collar Engineers, and Supervisors and Workers

The foundation for advancing the TPS by supporting the TMS and TDS is based on efforts to attain simultaneous achievement of QCD through innovation of manufacturing technology utilizing New JIT. This in turn is accomplished by having on-site engineering staff carry out innovation at the manufacturing site, as well as having white-collar engineers lead the innovation and improvement of the manufacturing technology (Amasaka, 2009b; Yamaji and Amasaka, 2009). All this is done with the aim of conducting manufacturing with high quality assurance.

Some of the features of the TPS are the improvement of laborious work operations and the creation of a safe, worker-friendly manufacturing site in which the supervisors and workers take the initiative, while the manufacturing technology staff members (white-collar engineers) also cooperate. Through such a process, the improvement of daily management and maintenance can be reinforced, which realizes rhythmical and smooth work operations, thus enabling reliable manufacturing (ensuring the process capability and machine capability).

Secondly, TPS takes on the challenge of simultaneous achievement of QCD through innovation of manufacturing technology. In this case, the initiative is taken by the white-collar engineers, while the supervisors and workers on the manufacturing site cooperate. This effort is expected to drastically improve manufacturing technology (for the working environment, production equipment, production management, production technology, quality management technology, etc.), which has been a bottleneck obstacle of manufacturing sites.

In the implementation stage, the innovations mentioned above are to be carried out by a Strategic Total Task Management Team (Amasaka, 2004c) led by the on-site manufacturing technical staff and production engineering development staff, while product designing and research development staff and even suppliers (parts manufacturers, production equipment, and machinery manufacturers, etc.) also participate. By means of these activities, wide-ranging business processes from the suppliers to the vehicle manufacturers can be linked in a highly efficient manner, and therefore manufacturing can be conducted while advancing the TPS, enabling New JIT production (Amasaka, 2003b, 2007b). Section 5.2.4 contains an example of the simultaneous achievement of QCD through innovation of manufacturing technology.

5.2.4 Application to Simultaneous Achievement of QCD through Innovation of Manufacturing Technology: Innovation of the Production Line of Automobile Rear Axle Unit for New JIT Production

In this example, the innovation of the production line producing rear axle unit–related parts for automobiles will be presented (Amasaka, 2002, 2003a, 2004b,c,

2008b, 2013; Amasaka and Yamada, 1991). It will demonstrate an example of solving a bottleneck manufacturing technology problem that was inhibiting New JIT production.

5.2.4.1 Outline of Production Process of Rear Axle Units

A rear axle unit for an automobile underbody is manufactured in a production process that includes a machining process and assembly process. These are performed in order by means of the Toyota Production System. This process is generally referred to as a lean production (production of one piece at a time) system. After assembly is complete, the assembled rear axle unit is further combined with various other products via a method called the sequential parts withdrawal system and then taken to a vehicle assembly plant via the pull system production method. Figure 5.6 illustrates the machining and assembling process of a rear axle unit manufactured by a typical TPS.

The main characteristic of the TPS is to carry out production without keeping any excess parts in stock. Therefore, in this process, it is necessary to conduct assembly in an orderly manner according to the assembly sequence from the start of the conveyor driven assembly process (5) to the latter process in the vehicle assembly plant. This is done by utilizing an intraprocess instruction *kaizen* that clearly indicates the order of the processes in coordination with the downstream pull system processes.

The main component parts of this product, namely the rear axle housing and rear axle tube, come in a wide variety of types that are delivered from parts manufacturing suppliers. Both parts are subjected to machining before being supplied to the beginning of the assembly process. Similarly, the various types of rear axle shaft delivered from a designated supplier are subjected to machining and then supplied to the beginning of the assembly conveyor. For both products, what is required is synchronized production, so-called *douki-ka*, where finished processed parts of the needed types are supplied in a timely, JIT fashion to the head of the assembly conveyor, while minimizing the amount of intraprocess parts and the number of finished products kept in inventory to the greatest extent possible.

Moreover, what make this synchronized production even more difficult are the considerable differences in the operation availability of each of the processes of machining and assembly. Generally, the operation availability of the assembly process is relatively high, but the previous machining process has low operation availability due to bottleneck problems and other reasons, stated in Section 5.2.4.2.1 As discussed, the production line for the rear axle units is arranged in a one-unit-at-a-time production manner from machining to assembly. Therefore, *douki-ka* production is not easy, particularly in the case where in-process inventory is not held in each process or between processes.

48 ■ New JIT, New Management Technology Principle

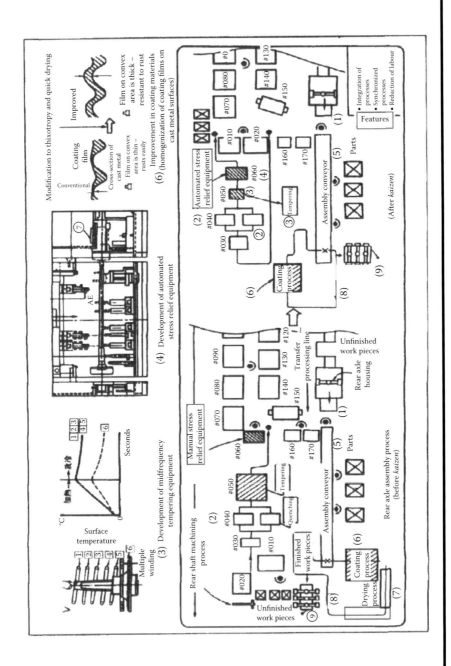

Figure 5.6 Rear axle unit processes proceeding to assembly (before and after *kaizen*).

5.2.4.2 Before Process Improvement: Bottleneck-Like Manufacturing Technical Problems That Inhibit JIT

In addition to the above, in each of the processes of machining, and assembly, bottleneck-like manufacturing technical problems that inhibit JIT exist, and these have not been solved for a long time.

5.2.4.2.1 Bottleneck Technology in Machining Process for Rear Axle Shaft

The bottleneck problems in this process are ensuring manufacturing quality and energy conservation, as well as the technical hindrances that prevent synchronized production.

As Figure 5.6 suggests, the upstream process of machining (#010–#030) and the downstream process conducted manually by production operators (#070–#170) are separated by the large, automated, high-frequency quenching equipment (#040) (2) and heat circulating–type tempering equipment (#050) (3), which consume a large amount of power and have a large amount of in-process inventory. For this reason, these processes are left like remote islands as it were. As a result, the one-operator work operations in the upstream process (#010–#030) require assistance from the supervisor in order to accommodate any extra cycle time in cases where the number of production units per day is increased. On the other hand, the operators are in standby mode when production volume per day is small.

In addition, the operation of manual stress relief equipment (marked as #060) is dependent on the empirical skills and abilities of expert production operators. The long work pieces (rear axle shafts) supplied from the upstream process (quenching and tempering equipment, #040–#050) have a stress profile of complicated spirals after heat treatment.

Also, in cases where the required heat stress adjustment is large, the number of adjustment operations (time needed for adjustment) becomes larger, resulting in extra cycle time. When a production operator is overburdened and conducts excessive adjustment operations in order to avoid the above situation, cracks can be induced inside or on the surface of these long work pieces. Moreover, if the operator has not acquired sufficient skills, this can result in human error and a supply of defective long work pieces that do not satisfy the standard stress adjustment specifications to the downstream processes (#070–#170). This will then cause a number of machining process defects.

As discussed above, many problems such as defective quality, *hanare kojima* work area arrangement, wasteful power consumption, extra cycle time, and in-process inventory have not been solved, presenting bottleneck manufacturing technical problems. Due to this situation, half-processed or finished products are stuck within the processes as inventory and synchronized production in coordination with the downstream assembly process (5) cannot be realized, particularly in the case of the welding process. This bottleneck situation has not yet been solved and this prevents the realization of simultaneous achievement of QCD.

5.2.4.2.2 Bottleneck Problems in Assembly Process for Rear Axle Units

The bottleneck problems of this process range from ensuring product quality, to energy conservation, to production response to the downstream pull system. The market requires that the quality of rust proofing of manufactured products be assured. The steel surface of a rear axle housing is flat and smooth. Therefore, the rustproof coating thickness can be ensured with (6) automated coating robots. However, the rustproof quality of the uneven areas on the cast metal surface cannot be sufficiently ensured as the coating on convex surfaces tends to be thinner due to the characteristics of the coating being used, as shown in (6) of Figure 5.6.

To avoid such uneven coating, extra coating operations are conducted to ensure sufficient thickness, and these result in a higher coating cost. Furthermore, after the coating operation indicated by (6) in the figure, the work pieces are passed through the electric heater-type drying machine (7), which requires considerable power consumption, and are then cooled down to room temperature in the cooling process area. After this, the finished products are unloaded to the pull system cart (9). For this reason, excessive inventory remains on the overhead conveyor (8), and therefore, as in the cases of the upstream welding and machining processes, this process presents a hindrance to the realization of simultaneous achievement of QCD.

5.2.4.3 After Process Improvement: Simultaneous Achievement of QCD by Innovation of Manufacturing Technology

In order to solve the above-mentioned bottleneck problems of this production line, it is necessary to innovate the manufacturing technology. Given this background, Amasaka and Yamada (1991) decided to take a systematic approach to the improvement of the fundamental process by taking into consideration the timing of production of a forthcoming new vehicle model.

Before this plan was implemented, the Amasaka (2004c, 2008b) formed a Strategic Total Task Management Team led by the technical staff on the manufacturing site, as well as the production engineering development staff. The product development staff, parts manufacturers (affiliated and nonaffiliated suppliers), and equipment machine manufacturers also participated, and the authors attempted to realize the simultaneous achievement of QCD with revolutionary approaches such as those introduced below.

5.2.5 Solution to Bottleneck Problems Related to Processing of Rear Axle Shaft

5.2.5.1 Development of Midfrequency Tempering Equipment

In place of the aforementioned, large-scale, hot air–circulating oven, panel-heater-type tempering equipment that consumes a lot of electric power (#050), a newly developed midfrequency, heat induction coil-type, tempering equipment unit was installed (Amasaka et al., 1982; Amasaka, 1983). As indicated in the upper section (3) in Figure 5.6, tempering is now conducted for one work piece at a time while concentrating on the high-frequency quenching area. This means that power consumption and in-process inventory are drastically reduced and the lead time is also substantially shortened.

5.2.5.2 Development of Automated Stress Relief Equipment

In addition to the above, the aforementioned manual stress relief equipment (#60) that depends on the empirical skills and abilities of an expert operator was replaced with a newly developed automated stress relief equipment unit (4). As indicated in the upper section (4) in Figure 5.6, the master skills commonly acquired by expert operators are simulated in order to conduct stress relief operations in an automatic and intelligent manner. This equipment is capable of properly conducting stress adjustment operations on work pieces with a variety of different shape profiles (lengths and diameters). Even in the case of a sizable spiral adjustment, stress relief can be done in a cycle time that is comparable to that of skilled operators. Also, by installing a work-piece crack detector, stable quality manufacturing was ensured and uneven cycle times were successfully eliminated.

These improvements in the facility equipment made it possible for the flow of work pieces from the high-frequency quenching equipment (3) to the midfrequency tempering equipment to the automated stress relief equipment (#040→#050→#060), as shown in Figure 5.6, to be connected in sequence along with automatic transportation of the work pieces. Consequently, these processes could be separated from the manual work processes. Also, by integrating the upstream processes (#010–#030) and downstream processes (#070–#170) as illustrated in the Figure 5.6, the problem of work operations conducted over remote island work areas was also solved.

These process improvement measures make it possible to flexibly arrange production operators according to the increase and decrease in production volume per day and therefore enhance productivity. As a result, manufactured quality was improved, the lead time was considerably shortened, the problems associated with energy conservation, labor force minimization, and skill acquisition were solved, and synchronized production was realized. This substantially contributed to the simultaneous achievement of QCD.

5.2.6 Solution to Bottleneck Problems Related to Assembly Process of Rear Axle Unit

5.2.6.1 Simultaneous Achievement of QCD for Corrosion Resistance Coating

In recent years, in an aim to improve the product value of parts related to an automobile's suspension and chassis, efforts have been made toward the simultaneous achievement of QCD in the area of corrosion-resistant coating technology by drastically improving its rust-proofing characteristics without raising the coating cost (Amasaka, 1984, 1986; Amasaka et al., 1988, 1990; Yamaji and Amasaka, 2009). In order to implement this project, the authors organized a Total Task Management Team through collaboration with the on-site staff, production engineering staff, product development and designing staff, and coating material manufacturers (Aisin Chemical Co., Ltd., one of several affiliated coating manufacturers, and also the nonaffiliated manufacturer Tokyo Paint Co., Ltd.). The manufacturing technical staff took the lead in this project.

Figure 5.7 shows the development procedure (rust-proof quality and coating cost) that aimed for the simultaneous achievement of QCD by improving the coating materials, coating equipment, and drying equipment. In addition, the authors also aimed to solve the technical problems in order to enhance the product value (VA is indicated by the rust proof quality/cost) of the rear axle units to which the coating materials manufactured by Tokyo Paint are applied (improvements (1)–(8) in Figure 5.7). The coating cost in the figure indicates not the price of coating materials, but that of the coating operation, which includes the entire cost of the coating materials, energy consumed, cleaning, and maintenance of the coating equipment per work piece.

Conventionally, styrene modified alkyd resin solvent coating was used, but in an effort to prevent initial rusting caused by the spread of antifreeze agents, mainly used overseas, improved phenol modified alkyd resin (Improvement (1)) was adopted, and this resulted in an increase in the coating cost due to the high material cost.

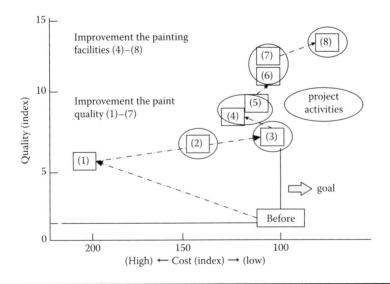

Figure 5.7 Improvement process of the paint.

Given this background, a task team was formed that set development targets in several phases, while also anticipating the required quality improvements demanded by the market, which grow year after year, and systematically promoting the project through close monitoring of the process improvement. In the early phases of improvement, mainly the coating materials were improved (Improvements (1)–(7)), and in the latter half, improvements in coating and drying equipment were promoted in parallel (Improvements (4)–(8)). More specifically, the following widely varying improvements were implemented.

1. The aim of improving the solvent coating materials was better corrosion resistance characteristics. Air-drying alkyd resin provides drying and corrosion resistance characteristics and nonpolluting rustproof colorant and filling colorant were then properly mixed with the modified alkyd resin. These were used as the base materials and improvement of the corrosion resistance was achieved.
2. Another target was to improve the quick-drying characteristics and cost, corrosion resistance, quality of the coating films, storage stability of the coating materials, and safety (prevention of spontaneous ignition by the coating mist). Improved results were obtained by using a lacquer-type phenol alkyd resin that is resistant to oxidation polymerization.
3. In order to homogenize the coating films and enhance the finished exterior of the film surface, stearin acid and organic bentonite were added to the quick-drying materials to improve thixotropy. This resulted in improvement of the finish quality, cost, and corrosion resistance characteristics.
4. To improve the coating efficiency of the airless electrostatic coating, a polar solvent and nonpolar solvent having strong polarity and high permittivity were optimally composed in such a way as to satisfy the requirements of cost, drying characteristics, solubility, workability, and storage stability. As a result, the electric resistance of the coating materials was optimally adjusted.
5. The initial material cost was reduced through an optimal choice of the coating materials, filling colorants, thinners, and various compounding ingredients.
6. In order to improve the equipment for the coating and drying processes, irregularities in the coating efficiency and films were reduced by adapting dilution-adjusted coating materials to seasonal changes, hot sprays, and so forth, so that the amount of coating materials consumed per unit was reduced by over 30%. Moreover, by implementing collection and reuse of overspray coating and optimal compounding with new agents, the initial cost was further reduced. In addition, the adoption of robots for the coating operation realized a reduction in the running cost and improved the finish quality and coating efficiency in comparison with conventional fixed multiunit airless electrostatic gun coating (Improvements (7) and (8)).

As a result, the final improved coating operation conducted 10 months later showed marked improvement. The rustproof characteristics were improved by 14 times (index based) and the visual quality was improved by 5 times (index based) by homogenizing the coating films. Furthermore, the drying facility was abolished because of the development of the quick-drying coating materials with room temperature drying characteristics. These improvements reduced the in-process inventory within the process to one-third and drastically reduced the coating cost by over 30% compared to the conventional method.

Even in the case of Aisin Chemical, similar improvement approaches have been taken and have resulted in the simultaneous achievement of QCD.

5.2.6.2 Improvement of Bolt-Tightening Tools

As shown in Figure 5.6 (on the right), the finished rear axle unit products are mounted on pull system carts located at the final process of the assembly conveyor (Amasaka, 1984, 2013a). The work hours needed for this operation have been reduced by improving the bolt-tightening tools for the brake units and disc wheel units that are assembled together with the rear axle units.

The conventional bolt-tightening operations are conducted in three steps: tightening with an impact wrench via the empirical feel of the operators, additional tightening with a torque wrench, and torque checking. These operations have been very burdensome and difficult, even for expert production operators, and tightening operations were not always uniform in quality.

Given this background, the authors developed a high-precision tightening tool with torque control (tightening torque detector), which reduced this operation to only a single step, thereby reducing the manual work time and improving and stabilizing the bolt-tightening quality at the same time. The efforts made through the task team activities of the on-site manufacturing staff and technical engineers introduced in this discussion were undertaken to realize the simultaneous achievement of QCD and in an aim to attain high quality assurance manufacturing by innovating the manufacturing technology.

These activities have been positioned as the basis for advancement of the TPS, along with daily improvement activities at the manufacturing site.

5.2.7 Conclusion

This section illustrated the foundation and effectiveness of advancing the Toyota Production System utilizing TPS through New JIT. This is based on efforts to attain simultaneous achievement of QCD through innovation of manufacturing technology by having cooperation between on-site white-collar engineers and supervisors and workers with affiliated and nonaffiliated suppliers. In this example, the author proved the effectiveness of advancing the Toyota Production System through innovation of the production line of the automobile rear axle unit.

5.3 Total Quality Assurance Networking Model

5.3.1 Introduction

In order to survive globalization and worldwide quality competition, Japanese manufacturing must work toward the simultaneous achievement of QCD in order to shorten development times, ensure high quality, and lower production costs (Amasaka, 2007c).

There are many tools, such as quality function deployment (QFD), that can be used to ensure manufacturing quality, but they do not always produce sufficient results. There are three key technological components needed to take quality assurance to the next level: (1) the experience and skill of technical personnel; (2) the combined use of methods that take a scientific approach, such as failure mode and effect analysis (FMEA) and fault tree analysis (FTA); and (3) partnerships—both internal partnerships that extend from design through production and external partnerships built with suppliers.

The Total Quality Assurance Networking Model (TQA-NM) for the New JIT strategy includes the above components and is introduced here as a new defect prevention technique (Amasaka, 2004b; Kojima and Amasaka, 2011). The TQA-NM for preventing defects is then outlined and its effectiveness verified at a leading corporation (advanced car manufacturer A).

5.3.2 Background

5.3.2.1 Manufacturing in Japan Today

Product life cycles are getting shorter as customers present more sophisticated and diverse needs. Japanese manufacturing needs to address these issues by shortening lead times throughout development, manufacturing, and sales (Amasaka, 2004b; Yamaji and Amasaka, 2008). However, recent large-scale recalls have raised social concerns, and the number of incidents as well as the number of vehicles recalled is on the rise (Ministry of Land, 2009).

These quality issues have naturally compromised user trust and caused loss in social standing for the companies themselves (Ministry of Economy, 2009). Quality issues are not limited simply to manufacturing quality issues or manufacturing reliability—it has become increasingly important to treat them as corporate organizational issues as well (Amasaka, 2004b; Amasaka et al., 2008).

5.3.2.2 Necessity of Preventing Defects

In order to grasp actual quality assurance conditions, the researchers conducted 204 interviews and surveys at the quality assurance departments of ten manufacturing companies (assembly manufacturers and suppliers), which included some major corporations. The investigation revealed that while many manufacturing

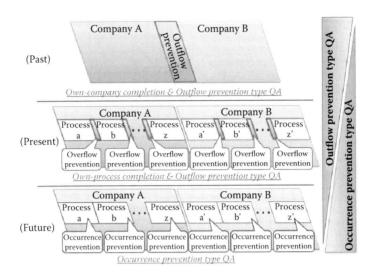

Figure 5.8 Quality assurance process transitions.

companies are looking to shorten development times, they are also spending a great deal of their defect prevention efforts on preventing outflow rather than preventing occurrence itself (Amasaka, 2007a).

It was discovered that one reason for this is that developers believe that focusing on preventing outflow contributed more to shortening development times than focusing on occurrence prevention. Occurrence prevention uses quality assurance tools such as QFD or FMEA to predict what underlying factors are causing defects. But research indicated that these tools are currently not being sufficiently implemented.

By putting effort into outflow prevention rather than occurrence prevention, fundamental solutions for reducing defects are never reached. Instead, it is critical that companies make the transition from quality assurance measures that prevent outflow to those that prevent occurrence, as shown in Figure 5.8 (Amasaka, 2007e).

5.3.3 Issues Faced by Quality Assurance Departments

A statistical analysis (type III quantification method) was conducted on the data collected in Section 5.3.2.2 (Necessity of Preventing Defects) to identify the issues faced by quality assurance departments, as shown in Figure 5.9.

The figure indicates two levels of problems: those faced by managers (i) and those faced by staff members (workers) (ii). The three problems faced by managers (i) are (1) development and deployment of quality assurance tools, (2) visualization of processes, and (3) transmission of know-how and expertise. These are the quality assurance issues from the perspective of management. The next three problems are those faced by staff members (ii). They are (4) clear quality assurance standards, (5) quality of human resources, and (6) swift response to changing market conditions.

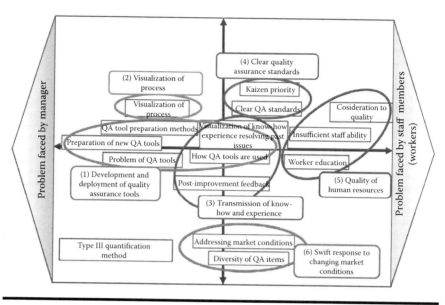

Figure 5.9 Issues faced by quality assurance.

These are the quality assurance issues from the perspective of the plant workers themselves.

As a result of the investigation and analysis, the researchers were able to clarify the critical components of achieving a high degree of quality assurance from the perspective of both managers and staff members (workers).

5.3.4 Creating the Total Quality Assurance Networking Model

5.3.4.1 The Total Quality Assurance Networking Model

Putting effort into outflow prevention is necessary to fundamentally resolve defect issues. Effective partnerships—both internal partnerships that extend from design through production and external partnerships built with suppliers—are necessary to address quality issues as organizational issues within the company (Amasaka et al., 2008). In order to establish a total quality assurance system and strategically deploy it, the authors came up with the Total Quality Assurance Networking Model (TQA-NM), as shown in Figure 5.10 (Amasaka, 2004b).

The purpose of the model is to create a new quality assurance system that can be used to comprehensively carry out quality assurance. It is called the TQA-NM to indicate the cycling total quality assurance process for achieving defect prevention (Amasaka et al., 1995b; Takahashi, 1997; Takahashi et al., 1997; Mihara et al., 2010).

The ultimate goal of the TQA-NM is to achieve CS, employee satisfaction (ES), and social satisfaction (SS) through high-quality manufacturing. This is done

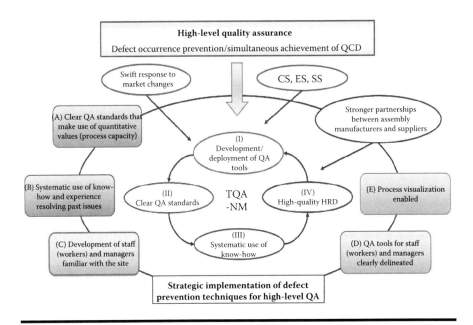

Figure 5.10 Outline of Total Quality Assurance Networking Model.

through the prevention of defect occurrence and by supporting the simultaneous achievement of QCD that come from strategically deploying a high-level quality assurance process.

One of the technological components in achieving these goals is strengthening quality assurance networks by developing and deploying quality assurance tools. A second component is establishing clear quantitative quality assurance standards (such as process capacity) that are not affected by worker experience. A third component is creating a system that can make use of worker expertise and information on past defects in an organized manner. A fourth component is the development of workers and frontline staff who are familiar with the site. The strategic implementation of these components will allow companies to prevent defect occurrence and simultaneously QCD.

5.3.4.2 TQA-NM Approach

5.3.4.2.1 Creating Total QA Networking Chart

The Total QA Networking Chart, as shown in Figure 5.11, was created in order to deploy the TQA-NM. The chart is a defect occurrence prevention technique featuring a combination of quality assurance tools. It also uses FMEA and matrix diagrams to deploy partnerships. The vertical axis of the chart identifies joint processes by suppliers and assembly manufacturers, from the arrival of goods to shipment.

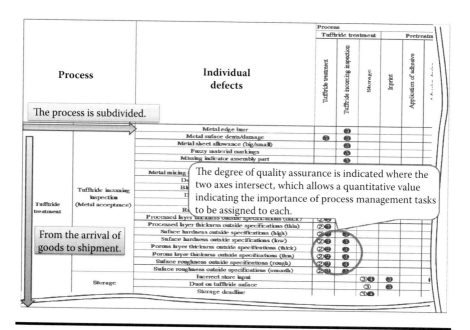

Figure 5.11 Example of the Total QA Networking Chart.

This creates a quality assurance network that eliminates gaps in quality assurance. The horizontal axis lists individual defects. The degree of quality assurance is indicated where the two axes intersect, which allows a quantitative value indicating the importance of process management tasks to be assigned to each. Each process is comprehensively evaluated on the basis of what the defect occurrence prevention tasks are for that process and how highly they are ranked in terms of the level of quality assurance.

5.3.4.2.2 Ranking Occurrence and Outflow Prevention

Table 5.1 assigns rankings to occurrence and outflow prevention using TQA Net Levels 1, 2, and 3. The importance of process management tasks is ensured based on the ranks shown for occurrence prevention and outflow prevention. This allows companies to carry out total quality assurance and establish a high-level quality assurance system.

5.3.4.2.3 Level-Based Application

Some participants in the investigation conducted in Section 5.3.2.2 (Necessity of Preventing Defects) felt that quality assurance tools did not match the scope or content of product development; they were not being used effectively, but simply being created as an empty exercise. Level-based application of the Total Quality

Table 5.1 Occurrence and Outflow Prevention Rankings

TQA Net-Level	Level 1	Level 2	Level 3
Assignment of QA level	○, ●	Subjective 1–5	According to clear standards 1–4
User	Staff (workers)	⇔	Managers
Development task	Full model change	Model change	Minor change
Level of technology at the company	Undeveloped	In transition	Advanced

Assurance Networking Chart, as shown in Figure 5.12, can resolve this issue, because it allows companies to deploy the chart in a way that matches the development of each product. Level-based application also allows quality assurance tools to be created in less time and contributes to the simultaneous achievement of QCD.

5.3.5 Deployment of TQA-NM for Strategic Distribution

Japanese companies manufacture some of the highest-quality products in the world, making successful defect occurrence prevention a top priority. The authors indicated the deployment of TQA-NM, as shown in Figure 5.13, as a way to systematize high-level quality assurance. This was done to address the importance of quality assurance through partnerships between assembly manufacturers and suppliers for strategic distribution (Amasaka et al., 1995b, 1997; Takahashi et al., 1997; Amasaka, 2000b; Ebioka et al., 2007).

The ultimate goal of the TQA-NM is to achieve high-level quality assurance through defect occurrence prevention as well as the simultaneous achievement

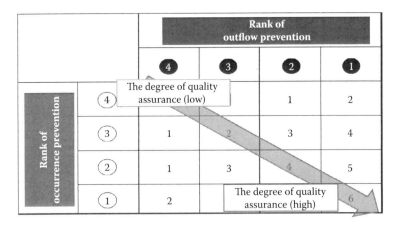

Figure 5.12 Level-based application chart.

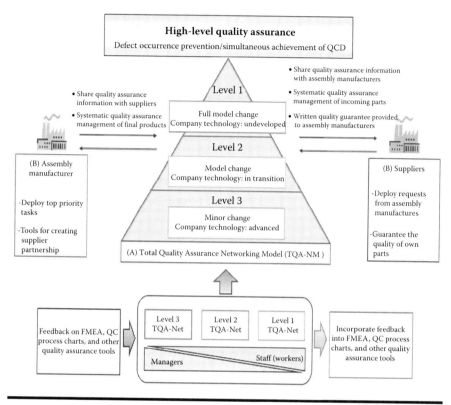

Figure 5.13 Deployment of TQA-NM.

of QCD. There are four essential guidelines that must be followed to achieve this goal.

The first is quality assurance that makes use of the TQA-NM (A in Figure 5.13). The second is further improving partnerships between assembly manufacturers and suppliers by considering the way their respective quality assurance systems are set up (B in Figure 5.13). The third requirement is establishing a quality assurance database for the TQA-NM in order to share quality assurance information. The fourth requirement is making use of quality assurance tools in subsequent product development in order to pass on know-how and experience, resolving past issues.

5.3.6 Application Example

The effectiveness of the TQA-NM was verified at a leading corporation. In this example, assembly manufacturer A (a leading automaker) and supplier B established a partnership (jointly developed activities) using the TQA-NM (Amasaka, 2004b) with the goal of improving the quality of brake pads to reduce noise.

5.3.6.1 Technical Management

Management activities took a technical approach to reducing the variations in quality that resulted in noise. The activities searched for a mechanism by conducting a variable factor analysis, which included consideration of quality evaluation techniques.

5.3.6.2 Product Management

A variable factor analysis was conducted on the brake pad products from a manufacturing perspective. The analysis sought to clarify the causal relationships leading to variations in process requirements and final products.

5.3.6.3 Deploying the Total QA Network

The researchers aimed to create more efficient business processes by taking full advantage of member expertise in technical management and in product management—and by carrying out these two activities side by side. If they are carried out completely in isolation from one another, it will be difficult for all participating members to have an awareness of shared goals. The two activities have therefore been deployed and linked using the Total QA Network Chart, as shown in Figure 5.14.

5.3.7 Conclusion

The authors focused on quality assurance activities that are integrated across different departments (from design through production) and extend to suppliers as well. The TQA-NM is constructed to establish partnerships and prevent defect occurrence, and its effectiveness is verified through application at advanced car manufacturer A (Kamio and Amasaka, 2000; Amasaka and Nagasawa, 2000).

5.4 Science SQC: New Quality Control Principle

5.4.1 Introduction

To promote quality control that contributes to the world in the future, it is necessary for us to carry out lucid and reasonable New JIT activities employing TQM-S (TQM by utilizing SQC) that will enhance the business process of all departments, as shown in Figure 5.1 (refer to Section 5.1). To do this, it is important to give thought to quality control of the manufacturing industry in the future, change the manufacturing activities accordingly, and show a good example so that a brighter future may be obtained.

Process	Individual defects	Process: Tufftride treatment: Tufftride treatment	Process: Tufftride treatment: Tufftride incoming inspection	Process: Tufftride treatment: Strage	Process: Pretreatment: Inprint	Process: Pretreatment: Application of adhesive	Process: Pretreatment: Adhesive drying	Process: Pretreatment: Starage	Present quality assurance level: Occurrence prevention: Desigh specifications: Abnormality detected Interlock function	Occurrence prevention: Desigh specifications: Abnormality detected Alarm	Occurrence prevention: Desigh specifications: Sufficient process capacity	Occurrence prevention: Standardized work: Regular equipment inspections	Occurrence prevention: Standardized work: Sufficient standardization	Occurrence prevention: Lank	Outflow prevention: Equipment specifications: Total automatic judgment Defective parts separation	Outflow prevention: Equipment specifications: Sufficient process capacity Regular sampling assessment	Outflow prevention: Standardized work: Regular gauge inspection	Outflow prevention: Standardized work: Can be determined at delivery inspection	Outflow prevention: Standardized work: Can be determined during total in-process inspections	Outflow prevention: Lank	Total Lank
Tufftride treatment / Tufftride incoming inspection (Metal acceptance)	Metal edge burr		❸								O			③				◉	O	❸	4
	Metal surface dents/damage	❸	❸																		5
	Metal sheet allowance (big/small)		❸																		3
	Fuzzy material markings		❸																		4
	Missing indicator assembly part		❸										O	③				◉	O	❸	4
	Unprocessed V-hole	❸	❸		❸								O	③				◉	O	❸	4
	Metal mixing (Different markings on the same shape)	③❸	❸		❸		❸						O	③					O	❸	4
	Defect of metal surface degree		❸								O			③		O	O			❷	3
	Blackening of the metal surface	③❸	❸		❸						O			③				◉	O	❸	4
	Dirty tufftride surface (Dust)	③❸	❸		❸								O	③				◉	O	❸	4
	Oily tufftride surface	③❸	❸		❸								O	③				◉	O	❸	4
	Rust spots on tufftride surface	③❸	❸		❸								O	③				◉	O	❸	4
	Processed layer thickness outside specifications (thick)	②❸	❸								O			②		O	O	◉		❸	2
	Processed layer thickness outside specifications (thin)	②❸	❸							O		O		②		O	O	◉		❷	2
	Surface hardness outside specifications (high)	②❸	❸							O		O		②		O	O	◉		❷	2
	Surface hardness outside specifications (low)	②❸	❸							O		O		②		O	O	◉		❷	2
	Porous layer thickness outside specifications (thick)	②❸	❸							O		O		②		O	O	◉		❷	2
	Porous layer thickness outside specifications (thin)	②❸	❸							O		O		②		O	O	◉		❷	2
	Surface roughness outside specifications (rough)	②❸	❸							O		O		②		O	O	◉		❷	2
	Surface roughness outside specifications (smooth)	②❸	❸							O		O		②		O	O	◉		❷	2
Tufftride treatment / Strage	Incorrect store input			③❸	❸		❸						O	③					O	❸	4
	Dust on tufftride surface			③	❸								O	③				O	O	❸	4
	Strage deadline			③❸									O	③				O		❸	4

Occurrence prevention

Outflow prevention

Figure 5.14 The example of the Total QA Network Chart for improving quality to reduce brake squeal.

In this regard, the author has proposed Science SQC, New Quality Control Principle as a demonstrative scientific methodology and discussed the effectiveness of this method, which improves the driving force in developing New JIT (Amasaka, 1997a,b, 1999a, 2000a) and has verified its effectiveness through development at advanced car manufacturer A and subsequent results (Amasaka, 1999b, 2003a).

5.4.2 Need for New SQC to Improve Quality Management by Manufacturing Industries

5.4.2.1 Delay in Systemization of Quality Management System That Improves Manufacturers' Management Technology

It is generally agreed that quality management activities have contributed largely to Japan's economic prosperity today. Quality management by manufacturers originated in Japan when SQC was introduced, used, and deployed by Walter A. Shewhart (1986) and W. Edwards Deming (Mary, 1988), who proposed that "quality management began and ended by control chart" (Rice, 1947), which is the basis of "quality built into process" (Kusaba et al., 1995). These activities and results were advanced by J. M. Juran (1988, 1989), systematically advancing the concept and progress method of the company-wide TQC activities. This TQC has further advanced to today's TQM activities (TQM Committee, 1998).

In the 1980s and later, U.S. companies were stimulated by introductions made by Andrea Gabor (1990), Brian L. Joiner (1994), and an MIT report (Dartouzos, 1989) to change the concept of quality from product quality to quality of customer's sense of value by learning the quality management system in Japan. There, they reviewed Japanese-style cooperative activities and the effect of SQC as a scientific approach and deployed the new quality management nationwide, while also receiving instructions from William E. Deming (1993).

In the 1980s and 1990s, however, Japanese manufacturers were too caught up in the economic boom both in Japan and overseas to sufficiently establish management technology bases to prepare for the next generation. They put too much emphasis on *jidoka* (automation), introducing large-scale equipment and taking a long period of time and large investment for completion, transferring the production system accordingly.

As a result, the production system became nonprofitable. Some companies put too much importance on automatic adjustment, with equipment having automatic adjustment functions even if man, machine, material, measuring, and method-environment (5M-E) deviated. In others, control charts simply disappeared from their manufacturing processes, reducing the scientific process management level.

As the situation continued, it became increasingly difficult to check the workmanship of products currently under production in real time. Not only for the production division, but also for the product design, production engineering, and quality assurance divisions, workmanship became difficult to confirm. It seemed that the process maintenance, management, and improvement cycle did not go well, as the skill and problem-solving capability of workshops, supervisors, and auditors that could be improved through observation of actual products and processes built into processes had been abandoned for the reasons described above (Amasaka et al., 1999a).

As alerted by T. Goto (1999b), Japanese manufacturers forgot the origin of quality management in the latter half of the 1980s toward the early 1990s, resulting in a slowdown in growth. The most important factor was the delay in systemizing next-generation-type management technology to be developed and introduced in accordance with technical advancements and changes occurring in the management environment for each manufacturer. This meant that the whole Japanese industry failed to establish a new method (a reasonable and scientific method for quality management) capable of improving management technology. This is discussed in more detail by Yoshida (1999) and Amasaka (1999).

5.4.2.2 Necessity of New SQC as Demonstrative Scientific Methodology

Successful quality control activities in manufacturing industries aimed at providing customers with attractive products consist of a reasonable way of thinking about quality control and the actual procedure to be established and followed. To be more precise, it means correctly converting customers' wishes (tacit knowledge) into engineering terms (explicit knowledge) by using the correlation technique, and so on, replacing customer's wishes with well-prepared drawings, and enhancing the process capability for early embodiment into products.

In retrospect, the transition of quality control that developed from the manufacturing industry started with application of the mathematical method of SQC. It then developed into TQC using control technology and, more recently, into TQM using various management control techniques comprehensively. And the concept of quality has expanded from being oriented to conventional product quality being oriented to business process quality, before becoming oriented to management technique quality. Along with this, the area of quality control activities expanded.

For the generator, the mentor, and the promoter, it is becoming more difficult to bundle individual or workshop techniques of the total hierarchies or departments through existing quality control activities based on past successes that depended on proprietary rule of thumb or empirical techniques. Therefore, to produce really attractive products that satisfy customers, that is, to produce customer-oriented products, common terms (management technology: quality management system) become necessary for the circular business process cycle of all departments to correctly turn from sales, service, planning, development, and design to production engineering, manufacturing, and logistics.

As far as the author knows, however, no SQC application system or demonstrative methodology for effective renovation of the business process as a language common to all departments has been seen for the new quality management for the coming age. In this regard, Amasaka (1999b, 2003a, 2004c) proposed Science SQC as a demonstrative scientific methodology for the three core techniques (TDS, TPS, and TMS) that form New JIT activities to be linked organizationally.

He also propagated and developed systematic and organized quality control. This represents the systematization of a new SQC application for creating technology by finding solutions using general solutions that can be generalized as the principles of Japanese quality management. He positions this as next-generation TQM management (TQM by utilizing SQC, so called TQM-S). He further discusses the effectiveness of this Science SQC (Amasaka et al., 1998a; Amasaka, 2004b).

5.4.3 Proposal and Implementation of Science SQC, New Quality Control Principle

For today's demonstrative research, it is necessary to implement customer science that converts customers' wishes into engineering terms (Amasaka et al., 1998b, 1999b). It is important for all departments concerned to share the objective awareness and turn tacit knowledge on business processes into explicit knowledge through coordinated activities.

The demonstrative scientific methodology for realizing this conversion is called Science SQC (Amasaka, 1997b, 2003a, 2004b), in which SQC is utilized systematically and organically under a new concept and procedures so as to allow the four cores shown in Figure 5.15 to mutually build on one another. This conceptual diagram shows the New Quality Control Principle, which forms the nucleus of TQM activities at advanced car manufacturer A (Amasaka, 1997b). For details on the demonstrative research, refer to Amasaka (1995).

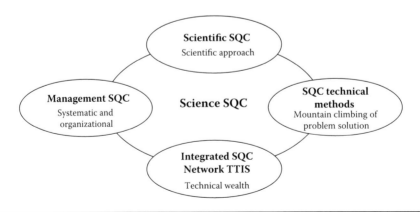

Figure 5.15 Schematic drawing of Science SQC: (a) business process for customer science; (b) scientific SQC for improving technology.

5.4.3.1 Practical Application of Scientific SQC

The primary objective of SQC as applied by the manufacturing industry is to enable all engineers and managers (hereinafter referred to as businessmen) to carry on excellent QCD research activities through insight obtained by applying SQC to scientific and inductive approaches in addition to the conventional deductive method of tackling engineering problems.

It is important to depart from mere SQC application for statistical analyses or trial and error type analysis and to scientifically use SQC in each stage from problem structuring to goal attainment by grasping the desirable form. Figure 5.16 shows a conceptual drawing of Scientific SQC (Amasaka, 1998).

5.4.3.2 Establishment of SQC Technical Methods

For solving today's engineering problems, it is possible to improve experimental and analysis designs by using the New Seven Tools for TQC (N7) and the basic SQC method based on investigation of accumulated technologies. Moreover, it is possible to solve mountain-climbing problems using a proactive combination of experimental designs, as may be required using multivariate analysis amalgamated with engineering technology.

The methodology in which the SQC method is used in combination at each stage of problem solving has spread and has been established as a SQC Technical Method (Amasaka, 1998) for efficiently improving the jobs of businesspersons. Figure 5.17 shows a conceptual diagram of the established SQC Technical Methods.

5.4.3.3 Construction of Integrated SQC Network TTIS

Various cases of successful application of SQC to actual business need to be systematized in order for them to contribute to the formation of engineering assets

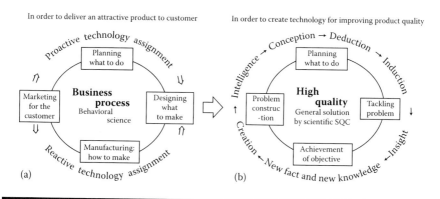

Figure 5.16 (a,b) Schematic drawing of Scientific SQC.

68 ■ *New JIT, New Management Technology Principle*

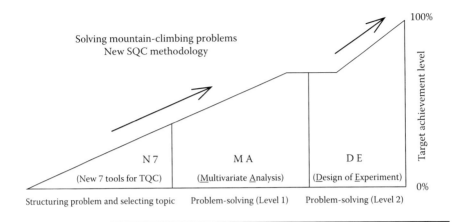

Figure 5.17 Schematic drawing of SQC Technical Methods.

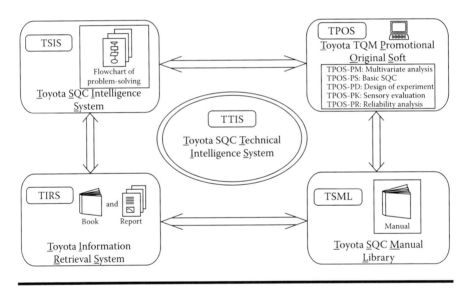

Figure 5.18 Schematic drawing of TTIS.

and to help inheritance and further development. This is a prerequisite for the development of Science SQC.

This methodology is achieved with the Total SQC Technical Intelligence System (TTIS), an integrated SQC network system that supports engineering problem solving, as shown in Figure 5.18 (Amasaka, 2000a). The TTIS is an intelligent SQC application system consisting of four main systems integrated for growth by supplementing one another. For further details, refer to Amasaka and Maki (1992).

5.4.3.4 Recommendation of Management SQC

The main objective of SQC in the manufacturing industry is to support quick solutions to deep-rooted engineering problems. Therefore, the main objective of Science SQC is to find a scientific solution to the gap generated between theories (principles and fundamental rules) and reality (events). Especially in the application of Science SQC, the differences between the principles and rules in an engineering problem should be scientifically analyzed to clarify six gaps that occur between theory, calculation, experiment, and actual result to obtain a general solution. Filling these gaps results in an organizational problem.

For problem solving, it is necessary for the planning, design, manufacturing, and marketing departments to clarify the six gaps, in other words to turn tacit knowledge on the business process into explicit knowledge for good understanding and coordination among the departments (Amasaka, 1997b).

The methodology for organizationally managing the development of Science SQC is called Management SQC. Recently, Amasaka et al. (1997) further discussed and studied Management SQC through demonstrative cases of Total Task Management Team activities at advanced car manufacturer A. Figure 5.19 shows a conceptual drawing of Management SQC for Science SQC development.

5.4.4 Demonstration Cases of Science SQC

In view of recent changes in the environment surrounding manufacturing industries, enterprises are once again required to make efforts to further enhance the technology development capabilities of engineers. To do this, the need for SQC as a behavioral science must be newly recognized. SQC must be applied from a scientific standpoint for not only solving apparent engineering problems but also foreseeing latent engineering problems, thereby contributing to the development of new technologies. The following subsections describe the demonstrative cases of Science SQC as it is applied at advanced car manufacturer A.

5.4.4.1 Consistency in SQC Promotion Cycle

Development of the SQC Promotion Cycle (implementation, practical results, education, and growth of human resources) illustrated in Figure 5.20 is essential for continuously increasing the expectations and effects of SQC applications. Amasaka (2000a) utilizes this cycle as the principle for promoting Science SQC at Toyota.

70 ■ *New JIT, New Management Technology Principle*

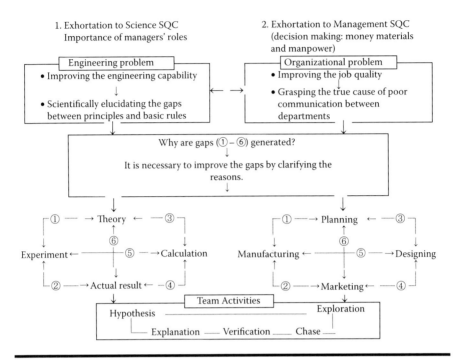

Figure 5.19 Schematic drawing of Management SQC.

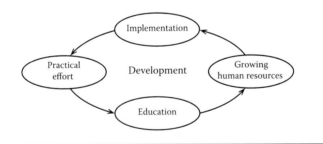

Figure 5.20 Schematic drawing of SQC Promotion Cycle.

Practically, the SQC promotion cycle means a spiral activity where SQC is used for challenging today's engineering problems to enhance proprietary and management technologies.

This will contribute to improving technologies, and subsequent results are reflected in hierarchical and practical SQC education to expand human resources, who in turn reflect the new technology in the performance of their operations. This mode of promotion constitutes the essence of Science SQC. Figure 5.20

schematically illustrates the concept of the SQC Promotion Cycle at a manufacturer (Amasaka, 1997b).

5.4.4.2 Systematization of Hierarchical SQC Seminar and Growth of Human Resources

To make the SQC Promotion Cycle spiral, as illustrated in Figure 5.20, it is important to establish a system that allows companywide development of systematic and organizational new SQC education and the growth of human resources. Conventionally, SQC seminars provided education on SQC basics, reliability analysis, and design of experiments for many years. But each of them was based on manual calculation and the curriculums were set up for individual SQC methods, dependent on analyses. As a result, a dilemma was experienced by some participants in that they were unable to put what they learned into practice.

It is important to improve the contents of seminars to make them practically applicable to SQC education. To keep abreast of the times, it is important to apply SQC personal computer software, TQM Promotional Original SQC Software (TPOS) (Amasaka and Maki, 1992), and prioritize experimental and analytical designs so as to prevent the performance of just trial and error analyses.

From the viewpoints of Sections 5.4.2 and 5.4.3, hierarchical SQC education as shown in Figure 5.21 is systematized as follows, and a step-up SQC seminar is planned and implemented for the six hierarchies of Beginner SQC, Business SQC, Intermediate SQC, Lower Advanced, Upper Advanced, and SQC Advisor courses. The seminar attendance ratio is set up as illustrated in Figure 5.21 for

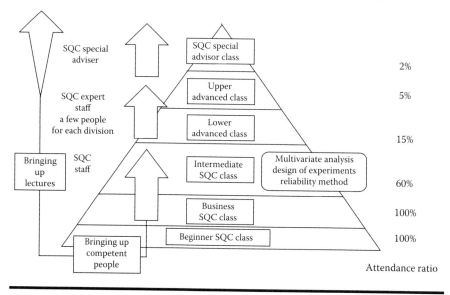

Figure 5.21 Division of SQC seminarian and human resources training.

a total of 17,000 businesspersons from all departments to ensure full growth of human resources (Amasaka and Osaki, 1999).

The beginner and business courses are for all staff and the program is designed to provide enough training for them to carry out their routine business. The Intermediate course is for professionals who can freely apply the advanced SQC methods. Figure 5.22 shows the curriculum established by the authors (Amasaka et al., 1998a) for the Advanced SQC course (for engineering staff). During the initial period, they learn the concept and procedures of Science SQC at advanced car manufacturer A (SQC Review 1–5; SR1–5) from the instructors for the SQC advisory staff (advanced course). Each participant tackles individually registered themes (problem structuring and setting) and reflects what is learnt in their business.

The second period is for experiments and analysis designs, which are important for maximizing problem solving. Lectures are given by an SQC level advisor of managerial position. The participants learn TMS (SR6–8) required for implementing Science SQC to improve New JIT activity and study demonstrative cases of Science SQC that help construct TDS (SR9,10) and TPS (SR11,12). In SR13 and SR14 in the third period, group discussion of registered themes takes place with the participation of instructors. This offers the opportunity for mutual study where the participants present their registered themes and verify the problem-solving process.

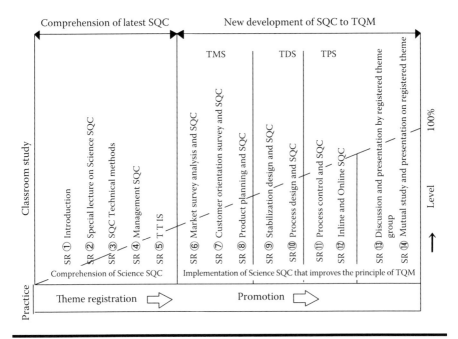

Figure 5.22 Curriculum for advanced SQC course (12-day course).

Students that have finished the advanced course and higher are qualified to be SQC special staff or SQC specialist advisors (~1100–2000 persons). Under this system, they can display their ability as promotion leaders for company-wide SQC Promotion Cycle development. Recently, Amasaka (1998a) and Amasaka et al. (1998a) set up a new 5-year course (1996 onward; for a total of 3000 engineering managers from all departments) for all managerial personnel (from directors down to divisional, departmental, and sectional managers; ~2200 persons up to 2000) in an effort to strengthen the development of Science SQC for improving New JIT activity.

5.4.4.3 Quality Improvement of Drive System Unit: Sealing Performance of Oil Seal for Transmission

This is a typical example where engineers, who had completed the new SQC training courses explained in Section 5.4.4.2, were challenged under the instruction of SQC special staff and advisors (Amasaka et al., 1998c).

5.4.4.3.1 Engineering Problem

The oil seal for the drive system works to seal lubricant inside the unit. The cause-and-effect relationship between the oil seal design parameters and sealing performance is not necessarily fully clarified. As a result, oil leakage from the oil seal is not completely eliminated, presenting a continual engineering problem.

5.4.4.3.2 Conventional Approaches to Quality Improvement

So far, oil seal quality has been improved as follows. A development designer having empirical engineering capability recovers the leaking oil seal parts from the market, analyzes the cause of leakage with proper technology, and incorporates countermeasures into the design. However, many of the parts that have recently leaked exhibit no apparent problem, and the cause of the leakage is often undetectable. This makes it difficult to map out permanent measures to eliminate the leakage.

5.4.4.3.3 Total Task Management Team Activities

It is necessary to clarify the engineering problems to be tackled by sharing the essence of the problems as a team by combining the empirical technology of each individual. Therefore, to explicate the essential engineering problems, it is necessary to improve the engineering problems generated by the lack of information among the related organizations (tacit knowledge) and their business processes. To do this, Total Task Management Team methodology, which was successfully used to improve brake performance quality, is employed (refer to Section 6.1).

74 ■ *New JIT, New Management Technology Principle*

To be more precise, a total task management team named Drive-Train Oil Seal Quality 5 Team (DOS-Q5) was organized in which the oil seal manufacturer participated. Figure 5.23 is a Relational Chart Method showing the outline of the quality assurance (QA) team's activities. The QA teams in the figure are QA1 and QA2, in charge of inquiries into the cause of oil leaks and design engineering problems, and QA3–QA5, which handle manufacturing problems relating to the driveshaft, vehicle, and transaxle. QA1 and QA2 represent collaboration-type team activities joined by the oil seal manufacturer (responsible for research, design, manufacture, and quality assurance), which carries on quality improvement of the oil seal part.

5.4.4.3.4 Implementation of Management SQC

To utilize proprietary intelligent technology and the insight of the team members (consisting of generators, mentors, and promoters) by applying Science SQC, the core technology of Management SQC is implemented. To do this, it is found that the total quality assurance activities require further upgrading. Accordingly, a Total Quality Assurance (QA) Network (Amasaka et al., 1997) is being developed based on actual records of solving this type of problem (refer to Chapter 5.3).

Moreover, to promote overall optimization by maximizing the efforts of individual teams QA1–5, problem solving is formulated using the SQC Technical Methods shown in Figure 5.24. To realize optimization of the DOS-Q5 business process, the Management SQC concept is applied in implementing information, production, and engineering management.

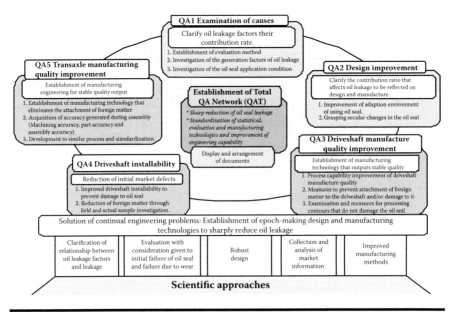

Figure 5.23 Relation chart method of activities by Total Task Management Team.

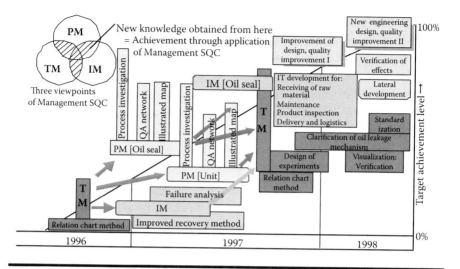

Figure 5.24 Problem solving by utilizing the SQC Technical Methods.

5.4.4.3.4.1 Improvement of Failure Analysis Process Using Information Management (IM) — The recovery method for oil leaking parts, the investigation method for oil leakage information, and other failure analysis processes are devised to explore the true cause of oil leaks. For example, nonleaking parts are recovered together with leaking parts and subjected to discriminate analysis, as shown in Figure 5.25, and other cause-and-effect analysis. As a result, new knowledge is obtained indicating that the hardness of the oil seal rubber affects oil leakage.

Moreover, Weibull analysis using the Toyota Dynamic Assurance System (DAS) for failure analysis (Sasaki, 1972) reveals a new fact, that is, a mixed model consisting of three types of oil leakage (initial, random, and wear) according to the running distance is acquired.

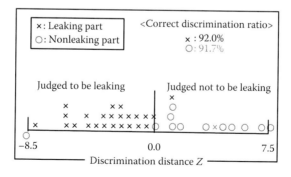

Figure 5.25 Result of discriminate analysis.

5.4.4.3.4.2 Visualization of Oil Leakage Mechanism Using Technology Management (TM)
To study the oil leakage mechanism, a device is developed to visualize the dynamic behavior of the oil leakage from the oil seal lip section. Through factorial analysis of the collected data using multivariate and experimental analyses employing a new experimental design, the oil leaking mechanism is visualized and the following causal analyses are conducted:

1. Association of running distance, lip surface roughness, and pumping volume (new fact)
2. Association of the inside diameter of the differential case, lip wear width, and roughness of the axis (new fact and quantification)
3. Association of roughness of the axis and pumping volume (new fact)
4. Association of differential case wear and driveshaft eccentricity (new fact)
5. Association of the lip tightening margin and lip wear (quantification)

As a result, the above-mentioned associations are explicated. New knowledge is acquired including the quantification of conventional empirical technology and the creation of new technology not found in the empirical technology. The oil leaking mechanism shown in Figure 5.26 is thus estimated.

5.4.4.3.4.3 Development of Total QA Network Using Production Management (PM)
For example, to improve the quality of the oil seal parts by incorporating the above-mentioned IM and TM knowledge, QA1 and QA2 analyzed and stratified the types of oil leakage and problems in the manufacturing processes, then developed illustration mapping by performing a relative analysis of these two elements. The analytical results are reflected in the QA Network Table (Amasaka et al., 1999b) used to develop process control into a science (visualization).

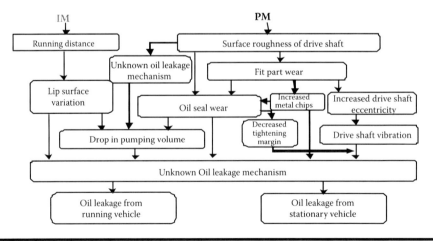

Figure 5.26 Estimated oil leakage mechanism.

The business process could be visualized from receiving to delivery inspection and logistics (distribution). Next, QA3–5 teams similarly developed the QA Network for the drive system manufacturing process. Then they established the process control science all the way from receiving to delivery. With the coordination of QA1 and QA2, they could clarify the fact (visualization of behavior) that, for example, oil leakage occurs if foreign matter the thickness of a hair (75 μm) is attached to the driveshaft.

5.4.4.3.4.4 Achievement of DOS-Q5 — Through implementation of Science SQC, the oil seal leaking mechanism whose cause was unknown was scientifically visualized, and verification was obtained using cause-and-effect analysis, with more proprietary knowledge of the technology subsequently acquired. The authors thus eliminated the oil leakage problem and achieved the target.

5.4.5 Conclusion

Demonstrative and scientific methodology is established for SQC applications aimed at improving technology. Science SQC is proposed as a new, systematic, and organizational SQC application methodology for the manufacturing industry. Advanced car manufacturer A has demonstrated that this methodology can improve the quality of work of engineers in every stage of their business process and contribute to creating products of excellent quality. In the future, Science SQC will be positioned as a quality control principle and applied to solving various practical problems. And with the accumulation of demonstrative studies, Next Generation TQM (TQM-S) designed to improve New JIT activity will hopefully be established (Amasaka, 2004c, 2008b).

Appendix 5.1 Customer Science

The author (Amasaka, 2000d) considers that the next-generation New JIT should not remain mainly for production. It is essential to establish a concept containing a new concept and core technology principle (core competence as the management technology principle) that permits corporate management based on it. Establishment of the concept of corporate management adopting the concept and method of customer science proposed by the author (Amasaka, 2000c, 2002, 2005, 2009a) and illustrated in Figure 7.1 (refer to Section 7.1) will become more and more important in the future.

Customer demands (implicit knowledge) must be developed into images (lingual knowledge) using correlation technology. These images are then accurately converted into engineering language (explicit knowledge), and reliable drawings are developed that enhance process capability for prompt merchandising. It is important for customer demands to be converted scientifically into a common language for use by all divisions.

Appendix 5.2 Science SQC, A New Quality Control Principle

It is important for all divisions to turn implicit knowledge of their business processes into explicit knowledge through integrated and collaborative activities by sharing objective awareness. To accomplish this, a new systematic and organized SQC propulsion cycle in the Toyota way, so-called Science SQC, proposed by the author (Amasaka, 1997a) was developed under a new concept using a new methodology that applied the four core principles, as shown in Figure 5.4, that enabled jobs to be scientifically performed. This conceptual diagram shows the nucleus of Toyota's and its groups' TQM activities based on a new quality principle that is the secret of success for next-generation manufacturing.

As determined from viewing Figure 5.15 (refer to Section 5.4), the four core principles in this figure are incorporated into a new SQC application system where they are closely linked to each other. The first principle of Scientific SQC refers to scientific approaches at every stage of the process ranging from determination of problem to accomplishment of objectives. The second principle of SQC Technical Methods, which use the seven new tools for TQM (N7), multivariate analysis (MA), and design of experiment (DE), refers to the mountain-climbing methodology for solving problems. The third principle of SQC, total network TTIS, represents the networking of SQC software application by using the four subcore principles. It can turn proprietary data inheritance and development into a science. The fourth principle of Management SQC is to support prompt solution of deep-rooted engineering problems.

Particularly in the practical application, the gaps between principles and rules have to be clarified scientifically as engineering problems, and general solutions have to be approached by clarifying the gaps that exist in theory, testing, calculation, and actual application. For further details on Science SQC with four principles, refer to Amasaka and Osaki (1999) and Amasaka (1998, 2000a,c,d).

References

Amasaka, K. 1983. Mechanization for operation depended on intuition and knack—Corrective method for distortion of rear axle shaft, in *The 26th Technical Conference of the Japanese Society for Quality Control*, pp. 5–10, August 29, 1983, Nagoya, Aichi, Japan [in Japanese].

Amasaka, K. 1984. Process control of the bolt tightening torque of automotive chassis parts (Part 1), *Standardization and Quality Control*, 37(9): 97–104, Japan [in Japanese].

Amasaka, K. 1986. The improvement of the corrosion resistance for the front and rear axle unit of the vehicle—QCD research activities by plant production Engineers, in *The 30th Technical Conference of the Japanese Society for Quality Control*, pp. 20–25, August 29, 1986, Nagoya, Aichi, Japan [in Japanese].

Amasaka, K. 1989. TQC at Toyota, actual state of quality control activities in Japan, in *The 19th Quality Control Study Team of Europe, Union of Japanese Scientists and Engineers*, vol. 39, pp. 107–112, September 25, 1989, Renault, France.

Amasaka, K. 1995. A construction of SQC intelligence system for quick registration and retrieval library—A visualized SQC report for technical wealth-, in *Lecture Notes in Economics and Mathematical Systems*, pp. 318–336, Springer-Verlag, Tokyo.

Amasaka, K. 1997a. A study on "Science SQC" by utilizing "Management SQC—A demonstrative study on a New SQC concept and procedure in the manufacturing industry, *Journal of Production Economics*, 60–61: 591–598.

Amasaka, K. 1997b. The development of New SQC for improving the principle of TQM in Toyota—from "SQC Renaissance" to "Science SQC", in *Proceedings of the CSQC Conference and the Asia Quality Symposium*, pp. 429–434, December 22, 1997, Tainan, Taiwan.

Amasaka, K. 1998. Application of classification and related method to the SQC renaissance in Toyota motor, in C. Hayashi et al. (eds), *Data Science, Classification and Related Methods*, pp. 684–695, Springer-verlog Tokyo.

Amasaka, K. 1999a. Proposal and implementation of the "Science SQC" quality control principle, in *Proceedings of First Western Pacific/Third Australia-Japan Work Shop on Stochastic Models*, pp. 7–16, September 24, 1999, Christchurch, New Zealand.

Amasaka, K. 1999b. A study on "Science SQC" by utilizing "Management SQC—A demonstrative study on a New SQC concept and procedure in the manufacturing industry, *International Journal of Production Economics*, 60–61: 591–598.

Amasaka, K. 1999c. The TQM responsibilities for industrial management in Japan-the research of actual TQM activities for business management lecture, in *The 10th Annual Technical Conference of the Japan Society for Production Management*, pp. 48–54, February 6, 1999, Nagoya, Japan [in Japanese].

Amasaka, K. 2000a. A demonstrative study of a New SQC concept and procedure in the manufacturing industry—Establishment of a new technical method for conducting scientific SQC, *An International Journal of Mathematical and Computer Modeling*, 31(10–12): 1–10.

Amasaka, K. 2000b. Partnering chains as the platform for quality management in Toyota, in *The 1st World Conference on Production and Operations Management*, pp. 1–13, August 20, 2000, Seville, Spain, (CD-ROM).

Amasaka, K. 2000c. New JIT, a new principle for management technology 21^C, (invitation lecture), in *Proceedings of the World Conference on Production and Operations Management*, pp. 15–27, August 19, 2000, Seville, Spain.

Amasaka, K. 2000d. "TQM-S", a new principle for TQM activities: A new demonstrative study on "Science SQC", in *Proceedings of the International Conference on Production Research*, pp. 1–6, August 3, 2000, Bangkok, Thailand, (CD-ROM).

Amasaka, K. 2001. Proposal of "Marketing SQC" for innovating dealers' sales activities—A demonstrative study on "Customer Science" by utilizing "Science SQC", in *Proceedings of 16th International Conference on Production Research ICPR16*, pp. 1–9, July 31, 2001, Prague, Czech Republic, (CD-ROM).

Amasaka, K. 2002. New JIT, a new management technology principle at Toyota, *International Journal of Production Economics*, 80: 135–144.

Amasaka, K. 2003a. Proposal and implementation of the "*Science SQC*" quality control principle, *International Journal of Mathematical and Computer Modeling*, 38(11–13): 1125–1136.

Amasaka, K. 2003b. New application of strategic quality management and SCM, in *Proceedings of Group Technology/Cellular Manufacturing World Symposium*, pp. 265–270, July 29, 2003, Columbus, Ohio.

Amasaka, K. 2004a. The past, present, future of production management (keynote lecture), in *The 20th Annual Technical Conference of the Japan Society for Production Management*, pp. 1–8, August 28, 2004, Nagoya Technical College, Aichi, Japan [in Japanese].

Amasaka, K. 2004b. *Science SQC, New Quality Control Principle*: *The Quality Strategy of Toyota*, Springer-verlog, Tokyo.

Amasaka, K. 2004c. Development of "Science TQM", a new principle of quality management: Effectiveness of strategic stratified task team at Toyota, *International Journal of Production Research*, 42(17): 3691–3706.

Amasaka, K. 2005. Constructing a customer science application system "CS-CIANS": Development of a global strategic vehicle "Lexus" utilizing New JIT, *WSEAS (World Scientific and Engineering and Society) Transactions on Business and Economics*, 3(2): 135–142.

Amasaka, K. (ed.). 2007a. *New Japan Model: Science TQM—Theory and Practice for Strategic Quality Management*, Study group of the ideal situation: the quality management of the manufacturing industry, Maruzen, Tokyo, Japan [in Japanese].

Amasaka, K. 2007b. Applying New JIT—Toyota's global production strategy: Epoch-making innovation in the work environment, *Robotics and Computer-Integrated Manufacturing*, 23(3): 285–293.

Amasaka, K. 2007c. New Japan production model, an advanced production management principle: Key to strategic implementation of New JIT, *Business and Economics Research Journal*, 6(7): 67–79.

Amasaka, K. 2007d. High linkage model "Advanced TDS, TPS & TMS"—Strategic development of New JIT at Toyota, *International Journal of Operations and Quantitative Management*, 13(3): 101–121.

Amasaka, K. 2007e. Highly reliable CAE model: The key to strategic development of Advanced TDS, *Journal of Advanced Manufacturing System*, 6(2): 159–176.

Amasaka, K. 2008a. Strategic QCD studies with affiliated and non-affiliated suppliers utilizing New JIT, in Putnik, G.D. and Cruz-Cunha, M.M. (eds), *Encyclopedia of Networked and Virtual Organizations*, volume III, PU-Z, pp. 1516–1527, Information Science Reference, Hershey, New York.

Amasaka, K. 2008b. Science TQM, a new quality management principle: The quality management strategy of Toyota, *The Journal of Management and Engineering Integration*, 1(1): 7–22.

Amasaka, K. 2008c. An intellectual development production hyper-cycle model: *New JIT* fundamentals and applications in Toyota, *The International Journal of Collaborative Enterprise*, 1(1): 103–127.

Amasaka, K. 2009a. New JIT, advanced management technology principle: The global management strategy of Toyota (tutorial talk), in *International Conference on Intelligent Manufacturing and Logistics Systems IML2009 and Symposium on Group Technology and Cellular Manufacturing* GT/CM2009, pp. 1–16, February 17, 2009, Kitakyushu, Japan.

Amasaka, K. 2009b. The foundation for advancing the Toyota Production System utilizing New JIT, *Journal of Advanced Manufacturing Systems*, 8(1): 5–26.

Amasaka, K. 2013a. *Science TQM, New Quality Management Principle, the Quality Management Strategy at Toyota*, Bentham Science Publishers.

Amasaka, K. 2013b. The development of a Total Quality Management System for transforming technology into effective management strategy, *The International Journal of Management*, 30(2): 610–630.

Amasaka, K. et al. 1995a. A study of questionnaire analysis of the free opinion—The analysis on information expressed in words using new seven QC tools and multivariate analysis together, in *The 50th Technical Conference of Journal on the Japanese Society for Quality Control*, pp. 43–46, July 28, 1995, Nagoyo, Japan [in Japanese].

Amasaka, K. et al. 1997. The development of "Total QA Network" by utilizing "Management SQC", Example of quality assurance activity for brake pad, in *The 55th Technical Conference of Journal of the Japanese Society for Quality Control*, pp. 17–20, May 31, 1997, Tokyo, Japan [in Japanese].

Amasaka, K. et al. 1998a. A study of the future quality control for bring-up of businessmen at the manufacturing industries, The significance of SQC study abroad for students in Toyota motor, in *The 60th Technical Conference of Journal of the Japanese Society for Quality Control*, pp. 25–28, July 2, 1998, Nagoyo, Japan [in Japanese].

Amasaka, K. et al. 1998b. The development of "Marketing SQC" for dealers' sales operating system—For the bond between customers and dealers, in *The 58th Technical Conference of Journal of the Japanese Society for Quality Control*, pp. 155–158, May 31, 1998, Nagoyo, Japan [in Japanese].

Amasaka, K. et al. 1998c. A proposal "TDS-D" by utilizing "Science SQC"—An improving design quality for drive-train components, in *The 60th Technical Conference of Journal of the Japanese Society for Quality Control*, pp. 29–32, July 2, 1998, Nagoyo, Japan [in Japanese].

Amasaka, K. et al. 1999a. Apply of control chart for manufacturing—A proposal of "TPS-QAS" by in line-on line SQC, in *The 29th Annual Technical Conference of Journal of the Japanese Society for Quality Control*, pp. 109–112, October 29, 1999, Tokyo, Japan [in Japanese].

Amasaka, K. et al. 1999b. Studies on "Design SQC" with the application of "Science SQC"—Improvement of business process method for automotive profile design, *Japanese Journal of Sensory Evaluation*, 3(1): 21–29.

Amasaka, K. et al. 1999c. A proposal of the New SQC education for quality management—A propose of "Science SQC" for improving the principle of TQM, *Quality, Journal of the Japanese Society for Quality Control*, 29(3): 6–13 [in Japanese].

Amasaka, K., Igarashi, M., Yamamura, N., Soeda, S., and Nomasa, H. 1995b. The QA network activity for prevent rusting of vehicle by using SQC, in *The 50th Annual Technical Conference of The Japanese Society for Quality Control*, pp. 35–38, July 28, 1995, Nagoya, Japan [in Japanese].

Amasaka, K., Kato, H., Yasudar, R., and Kosugi, T. 1999d. The development of "Science SQC" for industrial management—TQM-S in Toyota group, in *The 61st Technical Conference of Journal of the Japanese Society for Quality Control*, pp. 57–60, May 29, 1999 [in Japanese].

Amasaka, K., Kurosu, S., and Michiya, M. 2008. *New Theory of Manufacturing—Surpassing JIT: Evolution of Just-in-Time*, Morikita Publisher, Tokyo, Japan [in Japanese].

Amasaka, K. and Maki, K. 1992. Application of SQC analysis software at Toyota, *QUALITY*, 22(4): 24–30, [in Japanese].

Amasaka, K. and Nagasawa, S. 2000. *Foundation and Application of the Sensory Evaluation: For Kansei Engineering in the Car*, Japanese Standards Association, Tokyo, Japan [in Japanese].

Amasaka, K., Nagaya, A., and Shibata, W. 1999e. Studies on "Design SQC" with the application of "Science SQC", *Japanese Journal of Sensory Evaluations*, 3(1): 21–29.

Amasaka, K., Ohmi, H., and Murai, H. 1990. The improvement of the corrosion resistance for axle unit of the vehicle, *Coatings Technology*, 25(6): 230–240.

Amasaka, K., Okumura, Y., and Tamura, N. 1988. Improvement of paint quality for automotive chassis parts, *Standardization and Quality Control*, 41(2): 53–62.

Amasaka, K. and Osaki, S. 1999. The promotion of New SQC internal education in Toyota motor—A proposal of "Science SQC" for improving the principle of TQM, *The European Journal of Engineering Education on Maintenance, Reliability, Risk Analysis and Safety*, 24(3): 259–276.

Amasaka, K. and Osaki, S. 2003. Reliability of oil seal for transaxle—A Science SQC approach in Toyota, *Case Studies in Reliability and Maintenance*, pp. 571–588, Wiley, Hoboken, New Jersey.

Amasaka, K. and Sakai, H. 1996. Improving the reliability of body assembly line equipment, *International Journal of Reliability, Quality and Safety Engineering*, 3(1): 11–24.

Amasaka, K. and Sakai, H. 1998. Availability and Reliability Information Administration System "ARIM-BL" by methodology in "Inline-Online SQC", *The International Journal of Reliability and Safety Engineering*, 5(1): 55–63.

Amasaka, K. and Sakai, H. 2009. TPS-QAS, new production quality management model: Key to New JIT—Toyota's global production strategy, *International Journal of Manufacturing Technology and Management*, 18(4): 409–426.

Amasaka, K., Tsukada, H., and Ishida, M. 1982. Development of the tempering machine using mid-frequency of the rear axle shaft, in *The Annual Technical Conference of the Energy Conservation Center*, pp. 193–201, October 27, 1982, Nagoya, Aichi, Japan.

Amasaka, K. and Yamada, K. 1991. Re-evaluation of the present QC concept and methodology in auto-industry—Development of SQC renaissance at Toyota, *Quality Control*, 42(4): 13–22.

Burke, W. and Trahant, W. 2000. *Business Climate Shift: Profiles of Change Makers*, Butterworth-Heinemann, Oxford.

Dartouzos, M. L. 1989. *Made in America*, MIT Press, Cambridge, MA.

Deming, E. W. 1993. *The New Economics: For Industry, Government, Education*, MIT Press, Cambridge, MA.

Ebioka, K., Yamaji, M., and Amasaka, K. 2007. A New Global Partnering Production Model "NGP-PM" utilizing "Advanced TPS", *Journal of Business and Economics Research*, 5(9): 1–8.

Gabor, A. 1990. *The Man Who Discovered Quality: How W. Edwards Deming Brought the Quality Revolution to America*, Random House, Munich.

Goto, T. 1999. *Forgotten Management Origin—"Management Quality" Taught by GHQ*, Seisansei Shuppan, Tokyo, Japan [in Japanese].

Hayes, R. H. and Wheelwright, S. C. 1984. *Restoring Our Competitive Edge: Competing Through Manufacturing*, Wiley, New York.

Joiner, B. L. 1994. *Fourth Generation Management: The New Business Consciousness*, Joiner Associates, New York.

Juran, J. M. ed. 1988. *Juran's Quality Control Handbook*, McGraw-Hill, New York.

Juran, J. M. 1989. *Juran on Leadership for Quality—An Executive Handbook*, The Free Press.

Kamio, M. and Amasaka, K. (eds). 2000. *Science SQC, The Change of the Business Process Quality*, Nagoya QC Research Group, Japanese Standards Association, Tokyo, Japan [in Japanese].

Kojima, T. and Amasaka, K. 2011. The Total Quality Assurance Networking Model for preventing defects: Building an effective quality assurance system using a total QA network, *International Journal of Management and Information Systems*, 15(3): 1–10.

Kotler, P. 1999. *Kotler Marketing*, The Free Press, New York.

Kusaba, I. et al. 1995. *Fundamentals and Applications of Control Charts*, Japanese Standards Association, Tokyo, Japan [in Japanese].

Mary, W. 1988. *The Deming Management Method*, Dodd, Mead & Company, New York.

Mihara, R., Nakamura, M., Yamaji, M., and Amasaka, K. 2010. Study on business process navigation system "A-BPNS", *International Journal of Management and Systems*, 14(2): 51–58.

Ministry of Economy. Trade and Industry, 2009, Recovery of trust in manufacturing, Manufacturing Base Paper, Tokyo, Japan [in Japanese].
Ministry of Land. 2009. Infrastructure, transport and tourism, recall trouble information on car, http://www.mlit.go.jp/jidosha/carinf/rcl/index.html.
Motor Fan. 1997. The whole aspect of new "Aristo", a prompt report of new model car, A Separate Volume 213, pp. 24–30 [in Japanese].
Nikkei Business. 1999. Renovation of shop, product and sales method targeting young customers by "Netz", No. March 15, pp. 46–50 [in Japanese].
Ohno, T. 1977. *Toyota Production System*, Diamond-Sha, Tokyo, Japan [in Japanese].
Rice, W. B. 1947. *Control Chart in Factory Management*, Wiley, New York.
Roos, D., Womack, J. P., and Jones, D. 1990. *The Machine that Change the World—The Story of Lean Production*, Rawson/Harper Perennial, New York.
Sasaki, S. 1972. Collection and analysis of reliability information in automotive industries, in *The 2nd Reliability and Maintainability Symposium on Union of Japanese Scientists and Engineers*, pp. 385–405, Tokyo, Japan [in Japanese].
Seuring, S., Muller, M., Goldbach, M., and Schneidewind, U. (eds). 2003. *Strategy and Organization in Supply Chains*, Physica, New York.
Shewhart, W. A. 1986. *Statistical Method from the Viewpoint of Quality Control*, Edited and with a New Foreword by W. Edwards Deming, Dover Publications, New York.
Takahashi, A. 1997. *Quality Month Textbook—Meaning of TQM at Toyota*, Union of Japanese Scientists and Engineers, Tokyo, Japan [in Japanese].
Takahashi, H., Tsukeshiba, Y., Soeda, S., Hosono, H., Kosugi, T., and Kaji, Y. 1997. A study for optimum prevention rusting of vehicle (Part 2), in *The 57th Annual Technical Conference of the Japanese Society for Quality Control*, pp. 45–48, Nagoya, Japan [in Japanese].
Taylor, D. and Brunt, D. 2001. *Manufacturing Operations and Supply Chain Management: Lean Approach*, 1st edn., Thomson Learning, London.
Toyota Motor Corporation. 1987. The Toyota Production System and a glossary of special terms used at production job-sites at Toyota, Aichi, Japan [in Japanese].
TQM Committee. 1998. Overall quality management in TQM 21st century, Union of Japanese and Engineers, Tokyo, Japan [in Japanese].
Womack, J. P. and Jones, D. T. 1994. From Lean production to the Lean enterprise, *Harvard Business Review*, March–April: 93–103, Harward Business Publishing, New York.
Womack, J. P., Jones, D. T., and Doos, D. 1991. *The Machine That Changed the World—The Story of Lean Production*, Rawson/Harper Perennial, New York.
Yamaji, M. and Amasaka, K. 2008. New Japan quality management model: Implementation of New JIT for strategic management technology, *International Business and Economics Research Journal*, 7(3): 107–114.
Yamaji, M. and Amasaka, K. 2009. Strategic productivity improvement model for white-collar workers employing Science TQM, *The Journal of Japanese Operations Management and Strategy*, 1(1): 30-46.
Yoshida, Y. 1999. International strategy learned from restoring U.S.A., in *The 68th QC Symposium on Union of Japanese Scientists and Engineers*, pp. 61–65 [in Japanese].

Chapter 6
Driving Force in Developing New JIT

6.1 Strategic Stratified Task Team Model
6.1.1 Introduction

As the key to the success of globalized manufacturing in the twenty-first century lies in the enhancement of quality management, the author has proposed and discussed the effectiveness of New Just in Time (JIT), New Management Technology Principle (Amasaka, 2002a) (refer to Sections 5.1 and 5.4). The author expects that the development of New JIT with its three core principles of a Total Development System (TDS), a Total Production System (TPS), and a Total Marketing System (TMS), will contribute systematically and organically to solving quality management problems by the strategic use of the Science Statistical Quality Control (SQC), New Quality Control Principle (Amasaka, 2003a, 2004a).

As white-collar workers are prone to be obsessed with their daily work, a business process is likely to become routine. Today, it is crucial to innovate a business process through the creation of value, for which strategic team activities where related fields can collaborate autonomously and mutually are indispensable. This section investigates the significance and effectiveness of strategic joint team activities for cooperative creation, necessary for the driving force in the development of New JIT. In concrete terms, the author proposes the formation of a creative Strategic Stratified Task Team Model, and verifies its effectiveness within advanced car manufacturer A and its group (Amasaka, 2004b, 2007b).

6.1.2 Business Management and Development of Creative Corporate Climate

On viewing recent business management activities, the author has occasionally observed cases where leading enterprises are dealing with unexpected product quality problems both inside and outside their companies, or other cases where enterprises are faced with the risk of going out of business due to delays in developing technologies. On the other hand, over the past several decades, many enterprises have continued to grow vigorously through company-wide total marketing activities and management reliability enhancements.

These enterprises have consolidated their footing, fixing their gaze on world market trends and developing human resources with problem-solving capabilities. What are the backgrounds of these enterprises that make such a difference? Is management concerned with quality management prioritizing customer satisfaction (CS) and the achievement of streamlined management through the creation of a corporate climate that makes the most of individual human resources and energizes the organization?

For some time, a number of Japanese enterprises have been recruiting students with the necessary aptitudes and capabilities from universities and other institutions across the country. Therefore, it is apparent that the fate of an enterprise depends on how these young people grapple with creative work and become the forerunners of corporate innovation (Amasaka, 2002b).

6.1.3 Reliability of Job and Importance of Team Activities

This section reviews the job quality that is key for corporate management reliability and the importance of systematic and organizational collaboration across all fields. Figure 6.1 shows a typical corporate organizational chart (Amasaka, 2004a) with 13 business fields from upstream to downstream and their mission.

Table 6.1 shows six cases of job quality in terms of reliability per field (Amasaka, 2004a). In Case 1, one field has a reliability level of 99.9%, yet market claims of 1.3% are generated in terms of total reliability. Similarly, as the reliability decreases sequentially to 99.0% for Case 2, 95.0% for Case 3, and 90.0% for Case 4, total reliability rapidly drops, leading to the inevitable generation of a recall. In Case 5, a similar problem occurs merely because of a mistake made by one field. In addition, it is apparent that total reliability drops further if interfield collaboration is lost.

To upgrade the intellectual level of individuals and enhance organizational performance within a decentralized organization, the author restudies the significance of mutually collaborating team activities and touches on their effectiveness. Incorporating team activities into an organization's strategy is important from the viewpoint of key business management factors such as human resources and organization, knowledge and value, product quality and information, development and systems, and so forth. In the future, it will be necessary to develop a new methodology for achieving a higher reliability than Case 6 (99.99%).

Figure 6.1 A typical corporate organization chart.

Table 6.1 Job Reliability of 13 Departments

Case 1: 99.9%/department	$0.999^{13} = 0.987$ (1.3% problems)
Case 2: 99.0%/department	$0.990^{13} = 0.878$ (12.2% problems)
Case 3: 95.0%/department	$0.950^{13} = 0.513$ (48.7% problems)
Case 4: 90.0%/department	$0.900^{13} = 0.254$ (4.6% problems)
Case 5: When 12 departments are 99.99% but 1 department is 50.0%	$0.999^{12} \times 0.500 = 0.4999$ (50.01% problems)
Case 6: 99.99%/department	$0.9999^{13} = 99.88\%$ (0.12% problems)

6.1.4 Significance of Team Activities and Requirements for Their Promotion

When considering the association between an organization and its individuals, there are few people who could not answer the following questions: Which field in an enterprise requires no creativity? Which field does not need team activities? In corporate activities, individuals miss targets with perfect freedom. Otherwise, although individuals can have targets, everyone is scattered and individuals do not interact.

This is not to restrict freedom but to increase breakthrough productivity for management by giving direction to the whole organization. This is done by presenting a clear vision to individuals and enabling managers to guide them with a proper mission. This increases the knowledge of individuals and the organization and leads to improved human and product reliability. The author believes that team activities signify an opportunity for improving business process quality and realizing business achievement through mutual contact and enlightenment.

To obtain such effects, in the implementation stage of team activities, three factors—special knowledge, thinking capability, and the value of the job, as shown in Figure 6.2—are required (Amasaka, 2004a). To promote these factors, it is important to develop four requirements: mission, promotion cycle, performance measure, and pleasure of working. Through spiral-up cycles by meeting these requirements, the team members can steadily develop their problem-solving skills (advance from Steps (1)–(4)).

6.1.5 Creativity and Strategy Necessary for Team Activities

The business environment surrounding Japanese enterprises has been very difficult in recent times. Consequently, manufacturing requires highly reliable business management in order to cope with the globalization of manufacturing, the use of information and communications technologies, the protection of the environment, and so on. To fulfill these requirements, management has to carry out streamlined business operations through social satisfaction (SS), improvement of customer value (CS), improvement of employees' value (employee satisfaction [ES]), improvement of QCD activity, promotion of career development, and mutual trust and collaboration with the group enterprises' supply chain management (SCM) (Amasaka, 2002b).

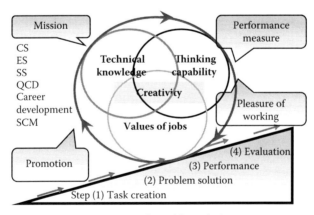

Mountaineering for problem solution

Figure 6.2 Three elements for team activities and requirements for promotion.

In the implementation stage, it will become necessary to improve human reliability through team activities that involve creativity and strategy and to consolidate merchandise power in the form of total marketing activities (Amasaka, 2004a). Team activities comprise task (force) teams as frequently observed with engineering or administration fields (knowledge-intensive type) or both, project teams, quality control (QC) Circle activities among blue-collar workers (labor-intensive type), and various other forms of activities (Amasaka, 2001a).

In particular, to survive in the abovementioned environment, today's task is to innovate the team activities of the so-called white-collars or knowledge groups. It is particularly important to innovate business processes, which are prone to become routine, as white-collars are busy with their daily work, creating value. Strategic team activities where related fields can collaborate autonomously and mutually are needed (Amasaka, 1999a).

In addition to enhancing cross-functional collaboration within their own company, strategic team activities reinforce collaboration among groups, nongroups, and overseas. To the author's knowledge, no studies have presented a Strategic Team Activity Model systematically designed from the above point of view to contribute to the advancement of management technology and verify the effectiveness of the model (Smithburg et al., 1977; Kawakita et al., 1977; Drucker, 1988; Nonaka, 1990; Doz and Hamel, 1998; Harada, 2000; Burke and Trahant, 2000; Evans and Lindsay, 2001; Gryna, 2001; Stefan et al., 2003).

6.1.6 Proposal for a Strategic Stratified Task Team Model

Figuratively speaking, business management is described as requiring 1 year for business, 3 years for financing, and 10 years for developing human resources. As a management technology strategy to enable smooth business management, the author (Amasaka, 2004a) proposes the Strategic Stratified Task Team Model. This is to move away from a job that depends on individual technique and empirical rules and toward the innovation of a business process achieved through the structured model of stratified task teams (pyramid structure from Task 1 to Task 8) as shown in Figure 6.3, where there is collaboration among other fields and vendors.

The first scenario of an innovating business process is the SQC renaissance (Amasaka, 1995) (1988–) campaign for boosting the problem-solving capabilities of technical staff who are key to New JIT activities. Here, using the inductive approach, the author developed the Total SQC Intelligence Information System (TSIS-QR) (Amasaka, 1995) and SQC Technical Methods (Amasaka, 1998).

A company-wide SQC Promotion Cycle (Amasaka, 1997) as shown in Figure 5.20 (refer to Section 5.4: implementation, achievement, education, human resource development) was developed for the technical staff to challenge the task team activities of small groups of Task 1 and of the departmental or staff level of Task 2. All of the engineering staff joined in the activities, subsequently solving chronic or bottleneck technological problems or both.

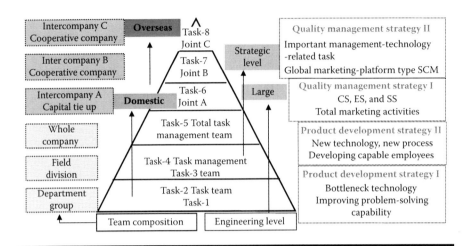

Figure 6.3 Structured model of strategic stratified task team.

The second scenario concentrated on a deductive approach to further solve latent or foreseeable engineering tasks or both. This refers to solving challenging issues through the establishment and development of the QC theory of Science SQC, New Quality Control Principle (Amasaka, 2003a) (1992–) as shown in Figure 5.15 (refer to Section 5.4). As the activities stretched across the division and its respective fields, general managers and divisional managers took the initiative to carry out the task management team activities (Task 3 and Task 4).

To create a solution that can be generalized instead of remaining an individual solution, the four core principles of scientific SQC, SQC Technical Methods, Integrated SQC Network System or so-called Total SQC Technical Intelligence System (TTIS), and mManagement SQC prove effective in developing activities in the engineering and production fields (Amasaka, 2004b).

The third scenario was project-oriented, company-wide activities, or Total Task Management Team Activities (Task 5: 1996–) that involved highly challenging tasks under initiatives taken by the top executives. Here, the author set up the process for developing New JIT activity by globalizing production quality information and innovating work.

Furthermore, the author extended the process to the business, sales, product planning, design, and administration fields to enhance the effectiveness of intercategorical collaboration. Throughout the strategic stratified task team activity, the author developed toward "Customer Science" (Amasaka, 2002a, 2005) to analyze customer's feelings from a scientific point of view, as shown in Figure 7.1 (refer to Sections 5.1 and 7.1).

Accordingly, the author employed New JIT (Amasaka, 2002d) as shown in Figure 5.1 (refer to Section 5.1). New JIT contains three hardware systems, namely, TDS, TPS, and TMS, and aims for an integrated form of next-generation management strategy.

The author used SQC Renaissance, Science SQC, and Science Total Quality Management (TQM-S) as a common language for consolidating collaboration among all fields and connecting the whole organization by establishing three core technologies through the software systems of New JIT. The fifth important quality strategy is to strengthen the quality management by utilizing the Japanese Supply System shown in Figure 6.4 (Amasaka, 2000c, 2007b).

To do this, the author adopts strategic joint total task management team activities (Tasks 6 and 7: 1997–) for working in collaboration with suppliers as the high reliability activity for strengthening our merchandise power. Here, we point ourselves toward the "platform type quality management with the supply chain." Task 6 (Joint A) is aimed at establishing a collaboration with the group suppliers with whom advanced car manufacturer A has a capital tie-up and Task 7 (Joint B) is aimed at a collaboration with suppliers that are not within its group. Task 8 (1998–), new quality strategy, is for us to grapple with and link the most important tasks to engineering creation. The mission of Task 8 (Joint C) is to strengthen cooperation with overseas suppliers as a strategic alliance.

6.1.7 Construction of the Strategic Cooperative Creation Team

In order for the strategic stratified task team proposed in Figure 6.3 (Task 1 to Task 8) to perform New JIT and solve issues of management technology, it is essential that a cooperative creation team, indicated by Figure 6.5, is formed. To empower the team, team members should collectively have the following: (1) *strategy* for systematic and organizational activities, (2) *technology* to improve core technologies, (3) *methodology* to practically identify the gap between theories and actual practice, and (4) *promotion* to fulfill the expectations and roles of the team.

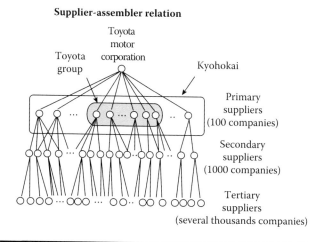

Figure 6.4 Japan supply system.

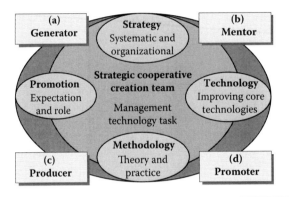

Figure 6.5 Strategic activity frame of the cooperative creation team.

If the task team tackles a strategic issue requiring high technologies, the members have to be ingenious enough as (a) *generators* and, at the same time, they have to be able to perform strategic analysis as (b) *mentors*. In addition, to infuse an effective drive force in the team activities, creativity as (c) *producers* and leadership to orchestrate all members' ideas as (d) *promoters* toward target achievement are important.

As the key to successful team activities, the team leader (administrator) should select the members who have at least one of the capabilities for (a) to (d), allocate authority and responsibility to the members, and have himself or herself concentrate on risk management. For this reason, as the leader, a person who has experience in solving business obstacles should be appointed, so that the leader is capable enough to lead the team in overcoming difficulties.

6.1.8 Practice and Effectiveness of Strategic Stratified Task Team

This section introduces evolution and the results of the strategic stratified task team activities, which are classified into eight task stages (Task 1 to Task 8) that are proposed by the author, and verifies their effectiveness at advanced car manufacturer A and its group.

6.1.8.1 Progress of Strategic Task Team Activities

As the author was in charge of promoting the strategic quality management of Toyota and the Toyota group, he organized and led task teams. The task teams tackled various engineering issues. Figure 6.6 is a chart of activity themes that the task teams undertook and accomplished (Amasaka et al., 1998). From 1988 to 2000, about 15,000 staff and 5,000 managers carried out 4,000 task team activities, part of which were presented to Japanese and overseas academic organizations as theses. The chart covers about one-fifth of those activities.

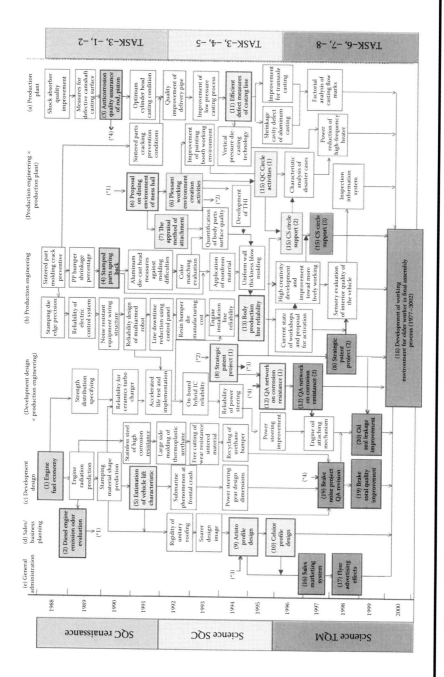

Figure 6.6 Changes in the strategic task team activities of advanced car manufacturer A.

Divisions of (a) production plant, (b) production engineering, (c) development design, (d) sales/business planning, and (e) general administration are arranged in the horizontal direction. In the chart, "×" is the symbol for cross-divisional collaboration and an arrow means that technical findings obtained in an activity are passed on to the next study (handing down of technologies). In the vertical direction, tasks are stratified as: (1) top, Task 1 (small group level: 1988–) and Task 2 (group and department level: 1990–) of task team activities in SQC renaissance (1988–); (2) middle, Task 3 (division level: 1992–) and Task 4 (field level: 1993–) of task team activities in Science SQC (1992–); and (3) bottom, Task 5 (across fields and companies: 1994–) of task team activities in TQM-S (1996–).

Furthermore, Figure 6.6 also includes (4) Task 6 (1996–) as joint total task management team activities in which Toyota collaborated with group manufacturers, (5) Task 7 (1998–) as team activities with nongroup manufacturers, and (6) Task 8 (1999–) as team activities with overseas manufacturers. The chart indicates how the technical themes organizationally and systematically evolved along with the eight strata, enhancing the level of the strategies. The evolution of these task team activities is supportive evidence of the effectiveness of the strategic stratified task team activity model that the author proposes in Figure 6.3.

6.1.8.2 Task Team Activity (Tasks 1 and 2)

As shown in Figure 6.6, SQC renaissance was triggered by (1) an analysis of the fuel economy improvement factor (Takaoka and Amasaka, 1991), which was tackled by the development design (R&D) division. The R&D members of the division formed a task team (Task 1). In Task 1, they performed a multiregression analysis of stratified curves based on existing engine fuel consumption data ($N = 22$ units) and formulated a relation model of fuel consumption and design factors.

By adjusting the identified design factors that were found to be effective for fuel saving, they succeeded in developing a fuel-saving engine in a short period. The scientific SQC approach triggered a pioneering laser engine study and a chain of new technology development shown in Figure 6.6, including (2) the development of a diesel emission odor evaluation method (Tanaka, 1990). Similarly in the production engineering division and production plant division, a chain of technological developments and production improvements occurred for production issues (1988–1989).

A framework of task teams also expanded from a small group to group and department (1990–1991). For example, in Task 2, the development of new process methods advanced, for example, (3) the studies on the rust preventive quality assurance (QA) of plated parts (Amasaka et al., 1993) and (4) an analysis of the total curvature spring back in stamped parts (Kusune et al., 1992). These remarkable achievements were underpinned by the SQC promotion cycle, as shown in Figure 6.6, and campaigns initiated by the generators and mentors for tackling upskilling themes and human resource development through stratified SQC seminars (Amasaka and Osaki, 1999).

6.1.8.3 Task Management Team Activities (Tasks 3 and 4)

After confirming the effectiveness of task team activities, the activities evolved into task management team activities within and across divisions (Task 3 and Task 4) (1992–1995). For instance, the R&D divisions unfolded Science SQC in Figure 6.6. As indicated by (5) a study on the estimation of vehicle lift characteristics (Amasaka et al., 1996a), they enhanced the speed of development by improving the accuracy of computer-aided engineering (CAE) estimation. The activities to solve these important engineering issues took place on a divisional scale (Task 3).

Furthermore, cross-divisional task teams (Task 4) were also formed and significant improvements were made. Examples include collaboration between production engineering divisions and production plants on topics of (6) the proposal of a desirable dining room environment at the employee restaurant (Uchida and Maki, 1991), the proposal to make a comfortable workplace environment (Hayashi and Kido, 1992), and (7) the proposal of an appraisal method of attachment (Komuro and Ito, 1992). One of the keys to these successful achievements was the training of managers. The managers received a Management SQC seminar (Amasaka, 1999b) as shown in Figure 6.7 (3500 managers participated over 5 years), and embodied and led Management SQC—clarifying the gap between theories and facts.

Consequently, in order to depart from work methods relying on a rule of thumb, the managers and staff applied the straightforward problem-solution approaches of SQC Technical Methods (N7/multivariate analysis/experimental design and others) and reached general solutions. In the team activities of Task 3 and Task 4, a structure in which general managers (promoters) manage managers (producers), enhancing collaboration with generators and mentors, is essential.

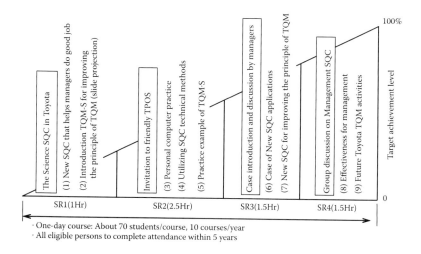

· One-day course: About 70 students/course, 10 courses/year
· All eligible persons to complete attendance within 5 years

Figure 6.7 Outline of Management SQC seminar.

6.1.8.4 Total Task Management Team Activity (Task 5)

The expansion and successful results of the team activities in Task 1 to Task 4 led company-wide activities where the management themselves assumed the leadership. A methodology to identify what customers need before they notice their needs—customer science proposed by the author—was utilized as a management technology strategy in the form of a company-wide activity (Task 5).

Here, SQC Technical Methods, which utilize combinations of seven new QC tools (N7), multivariate analysis, experimental design, reliability analysis, and so on, were applied effectively to identify and solve problems. As an example of product technology strategies, (8) a strategic patent project was jointly promoted by the development design division and production engineering division (Amasaka et al., 1996b). The project constructed a model of engineers' image concepts of good patent, enabling the evaluation of intellectual productivity. The results and approaches adopted in this project for materializing images were also utilized in (9) the LEXUS GS400 profile design (Japanese name is Aristo) (Amasaka et al., 1999d).

In the course of R&D for GS400, SQC Technical Methods—the core method of Science SQC—were applied in combination with the Total Development System-Design Technical Method (TDS-DTM) (Amasaka, 2003c) in order to identify proportion ratios that explain the universal vehicle profile of vehicles with world prestige. These methods were further applied to the development of (10) the LEXUS LS430 profile design (Japanese name is Celsior) and this model currently enjoys a worldwide reputation (J.D. Power and Associates, 2003).

Similar approaches were taken in the joint promotion by a production engineering division and a production plant division for (11) the development of a new engine casting method (Amasaka, 2000a) and (12) vehicle body rust prevention technology (Amasaka et al., 1995b). These studies led to the establishment of a QA network system. In addition, the combination of (13) a production equipment operation and maintenance system (Amasaka and Sakai, 1996a, 1998) and (14) an intelligence control chart resulted in the establishment of the Toyota Production System-Quality Assurance System (TPS-QAS; Amasaka and Sakai, 2002).

These achievements affected other production engineering development and drove company-wide (15) new QC Circle activities (Amasaka et al., 1995a). These achievements of the strategic task team activities largely contributed to the evolution of TPS and high QA in manufacturing and enhanced links with customers. Some examples include (16) the development of a Toyota sales marketing system (Amasaka, 2001b) that utilized information technology (IT) and SQC as support tools and (17) flyer advertising effects (Amasaka, 2009b) that contributed to effective advertisement. In this way, the task team activities also influenced sales and marketing activities and triggered their revision.

Furthermore, in recent years, due to the increasing number of aged and female employees, the general administration division initiated a company-wide work environment improvement activity called (18) the Aging & Work Development 6 Project (AWD6P/J) as a task team activity (Amasaka et al., 1999a). Each activity mentioned above achieved favorable results. Through these activities, the three core principles of TDS, TPS, and TMS of New JIT were built up and these techniques were gradually integrated (Amasaka, 2002d).

6.1.8.5 Joint Total Task Management Team Activity (Tasks 6 and 7)

For the survival of an automotive manufacturer in the competitive business of comprehensive assembly in the global market, partnering with parts manufacturers is essential (Amasaka, 1999a). For this reason, the author extends the task team activity structure further to Task 6 and Task 7 in which a company collaborates with group or nongroup suppliers. As Figure 6.7 shows, in Task 6, the team tackled the global quality issue of (19) brake quality improvement (Amasaka and Osaki, 1999).

The activity involved one of Toyota's group companies, Aisin, and succeeded in developing new materials and manufacturing methods that satisfied the requirements for brake noise and braking performance and durability, which were regarded as being in mutual inverse relationship.

Another example of Task 7 was (20) drive system oil seal leakage improvement (Amasaka, 2003d), for which advanced car manufacturer A worked jointly with supplier B. The achievement of this activity was the unprecedented visualization of oil leakage phenomena. Regarding the organization of the two activities, the team in the former case was AWD-6P/J, which consisted of members of Aisin divisions of R&D, production engineering, manufacturing, sales, and service, in addition to Toyota members to support the innovation of the total business process.

The project teams (TDOS-Q5 and NDOS-Q8) in the latter case invited supplier B members. They also acted as a dual total task management team, orchestrating advanced car manufacturer A and supplier B. The author led both teams and their activities are positioned as platform-based SCM for quality strategies. All the activities above utilized Science SQC to identify problem mechanisms and practiced customer science (refer to Section 7.1), achieving a 90%–95% reduction in quality problems, including field complaints and in-process defects.

Based on these achievements, since 1997, a task team activity, Task 8, which involves collaborating with overseas companies has been steadily progressing. Using approaches similar to those utilized in internal and external team activities and work, the team is seeing the achievement of their activities in human resources development using the SQC promotion cycle for problem solving with Science SQC (Amasaka et al., 1999c).

6.1.9 Conclusion

The present study has proposed a Stratified Task Team Model as one form of New JIT and has discussed the significance and effectiveness of strategic team activities that are based on cooperation and creation.

As discussed, these successes owe much firstly to the Science SQC as a scientific methodology that is adapted to the present age, secondly to the development of New JIT directed toward streamlined quality management, and thirdly to the reliability of individuals and organizations enhanced by the task teams having a stratified structure, as the author proposes.

Lastly, in order to respond to innovation by digital engineering, the author continues to promote strategic team activities with Toyota as a model by practicing New JIT for the evolution of quality management technologies (Sakai and Amasaka, 2002, 2003).

6.2 Epoch-Making Innovation in the Work Environment Model

6.2.1 Introduction

Today's challenge for business management lies in providing customers with products of excellent quality, cost, and delivery (QCD) performance in the pursuit of CS and staying ahead of competitors through market creation activities. To do this, New JIT was proposed as a new management technology principle for twenty-first-century manufacturing (Amasaka, 2002a). New JIT is configured with a hardware system that has three core elements: TDS, TPS, TMS, and the software system of Science SQC (Amasaka, 2003a) that enables the application of TQM-S (Amasaka, 2004a, 2008a, 2012).

The author believes that the key to a company's prosperity is a global production strategy that enables the supply of leading products with high QA and simultaneous global production start-up (the same quality and production at optimal locations) in both developed and developing countries. Innovation that optimizes an aging workforce is a necessary prerequisite of TPS. It is essential to identify the elements that are necessary for enhancing work value, motivation, and work energy, as well as an optimum work environment (amenities and ergonomics), through objective survey and analysis from labor science perspectives.

This section analyzes and proves the significance of a strategic driving force in developing New JIT—a global production strategy activity called Aging and Work Development 6 Project (AWD6P/J)—for epoch-making innovation of the work environment. Section 6.2 is based on the concerns of automotive manufacturers. In the Japanese automobile industry, the aging society and the expansion of overseas production are resulting in a decrease in the employment of young workers in automobile production shops. AWD6P/J has been promoted in the areas of human resources, labor, and workplace environment to innovate the workplace to respond to an increasing number of older and female employees (Amasaka, 2007c).

Under AWD6P/J, the Total Task Management Team was composed mainly of members of the production engineering and plant divisions to promote scientific approaches to (1) motivation, (2) fatigue, (3) physical strength, (4) tools and equipment, (5) temperature conditions, and (6) disease prevention.

This study selected final vehicle assembly lines as the model and investigated an attractive production line, employing a comprehensive analysis that incorporated ergonomics, physiology, and psychology. The measures for an aging workplace developed by this activity yielded practical results and are being applied to both domestic and overseas operations to improve productivity at advanced car manufacturer A.

6.2.2 Strategic Application of New JIT, New Management Technology Principle

To win the global competition, big enterprises both in Japan and overseas are actively promoting global marketing that aims to achieve the same quality and production at optimal locations (simultaneous start-up) throughout the world. Manufacturing companies, in particular, are required to understand customer needs and provide the best products to the market through global production without falling behind their competitors. Therefore, new strategic management technologies that drive a company to stay ahead of the competition have become increasingly essential on a global scale.

6.2.2.1 Establishment of Next Management Technology Principle, New JIT

The mission of enterprises is to provide customers (consumers) with products that fulfill their needs and expectations. Fulfilling this mission is the key to the continuation of a corporation. To this end, the author has recently developed the New JIT, Management Technology Principle, as shown in Figure 5.1 (refer to Section 5.1) and has shown its validity as a new management technology for twenty-first-century manufacturing.

New JIT is a next-generation management technology that innovates the business processes of each division, including sales, development, and production. New JIT includes hardware and software systems that have been developed according to new principles to link all activities throughout a company. The hardware system consists of three core elements: TDS, TPS, and TMS. Collectively, this system is called New JIT, with an excellent reputation worldwide as a *Lean system*.

The software system deploys Science SQC, which is a new QC principle for quality management from a scientific viewpoint (Amasaka, 2003a).

Here, let us consider the organizational way of proceeding with tasks under Japanese-style management and strategic development. It has demonstrated enhanced effectiveness in the divisions of engineering design, production, business sales, and others (Amasaka, 2007c,d, 2009a,b,c,d; Amasaka and Sakai, 2009). In this sense, the whole company consistently deploys total marketing (Amasaka, 2004a).

6.2.2.2 TPS, the Key to Strategic Application of New JIT

Observation of the automotive industry, which is showing an increase in global business expansion, suggests that it is representative of the general condition of various industries throughout the world. For example, while Japanese automotive companies expanded the application of digital engineering–innovated manufacturing in their workshops, the reduction of QC Circle activities and increased overseas production resulted in a decline in technical skills, problem detection, and problem-solving capabilities (Amasaka, 2007e). This ultimately lowered the workshops' ability to build in quality during each process.

Considering the recent increase in recalls with respect to Japanese vehicles and the improving quality of vehicles produced in developing countries, the position of the Japanese automotive industry as an expected leader in global production is threatened (Nihon Keizai Shimbun, 1999, 2000, 2001; Asahi Shimbun, 2005).

To overcome this situation, it is essential to eschew conventional production management and establish a new management technology principle suited to computerized workplaces. To achieve this for global production, production engineering and manufacturing divisions are expected to achieve high-level QA and productivity by using digital engineering, planning, and the implementation of (i) intelligent production systems, (ii) operations and maintenance skills, and (iii) the evolution (training and development) of manufacturing skills and training (Amasaka, 2004c).

Recently, the author (Amasaka, 2004a) referred to the effectiveness of TPS, applying Science SQC as a positive way to improve the quality of business processes in workshops, the sites where Toyota's New JIT activities take place (Refer to Section 5.1.4.2.2 and Figure 5.3). TPS uses IT and SQC in combination to produce generalizations about behavior patterns for practicing customer-oriented quality and production management, which the production workshop or production engineering department builds into the processes using core technologies. What is essential here is to circulate these four core technologies: (a) production by information—IT, (b) production by management—process management, (c) production by technology—production technology, and (d) production by partnership—human management (Refer to Section 5.1.5.2).

6.2.3 Prerequisite of Strategic Global Production: Innovation in Work Quality

Major manufacturing companies are facing the significant pressure to innovate their businesses for global production. The current and future health of manufacturing performance in Japan, as well as the possibility of simultaneously attaining the same quality level in overseas plants, remains unanswered. Depending on the situation in each country, including product specifications, production volume, and market conditions, manufacturing may be fully automated or it may require manual labor. If so, the success of global production is highly dependent on the quality of workers, indicating the necessity of work innovations. In addition, to achieve long-term growth, companies should also undertake major improvements to the work environment (Amasaka, 2004c, 2007b, 2007d).

6.2.3.1 Background of Concepts Regarding Consideration for Older Workers

Japan, famed for its big manufacturing businesses, faces industrial changes—expansion of overseas production, stagnant domestic demand, and diminished recognition of manufacturing due to the changing preferences of young people. This has resulted in reduced employment of new, young workers. In the case of Toyota, the average and oldest age of workers on vehicle assembly lines have been increasing for the past decade (1995–2005), as shown in Figure 6.8 (Amasaka, 2007d).

To cope with this trend, a more extensive consideration of worker motivation and physical conditions is essential. In other words, manufacturers should shift from work-oriented shop designs to people-oriented shop designs that focus more on the work environment. As an example of manufacturing innovation in overseas countries, various governmental actions for older workers have been taken in Scandinavian countries that have aging workforces. Though the necessity of such actions has been advocated in Japan, action has been relatively modest compared

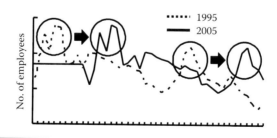

Figure 6.8 Variation of aging workers.

with that in Scandinavian countries* (Amasaka, 2000; Niemela et al., 2002; Vecchio et al., 2003).

At Toyota, improvements have been based mainly on the Toyota Verification Assembly Load (TVAL)† (Toyota Motor Corporation and Toyota Motor Kyushu Corporation, 1994) for a quantitative evaluation of workloads. At its Kyushu plant, Toyota has also implemented a worker-oriented line for the assembly process based on the new concept shown in the following list (Amasaka, 2000). A further enhancement of these activities for aging workers will be indispensable in the future.

New concept for assembly:

1. Increase worker motivation.
2. Reduce workload.
3. Provide automation that people want to work with.
4. Provide a comfortable work environment.

6.2.3.2 Consideration for Older Workers according to New JIT Strategy

In order to create a workplace that is accommodating to aged workers, four steps were planned as shown in Figure 6.9 (Suzumura et al., 1998). Step (1) was to interview middle-aged and elderly workers engaged in the assembly process. This interview brought to light both positive and negative aspects.

The firsthand information obtained from middle-aged and older workers was then classified according to their work functions. In order to ensure an accurate interpretation of the responses collected during the interviews, the study members participated in actual line operations for 5 weeks. In Step (2), objective data concerning work functions were collected through investigation and from various documents in order to clarify the implications of the interview results. Step (3) was an investigation and an evaluation of existing measures for improving conditions related to the physical attributes classified in Step 1. In Step (4), areas were identified in which countermeasures are necessary: (i) workers, (ii) car and equipment, and (iii) management.

For example, to offset insufficient muscular strength due to unnatural working postures and heavy work, assisting devices were introduced that remarkably improved the situation. However, the workshop still had a problem in terms of work speed and no measures were taken to assist work speed. In terms of endurance,

---

* Examples are the Finnish Institute of Occupational Health (Finland) and the Institute for Gerontechnology (Eindhoven University of Technology, the Netherlands) (Japan Machinery Federation and Japan Society of Industrial Machinery Manufactures, 1995a,b).
† Quantitative evaluation of assembly workloads using electromyographic values.

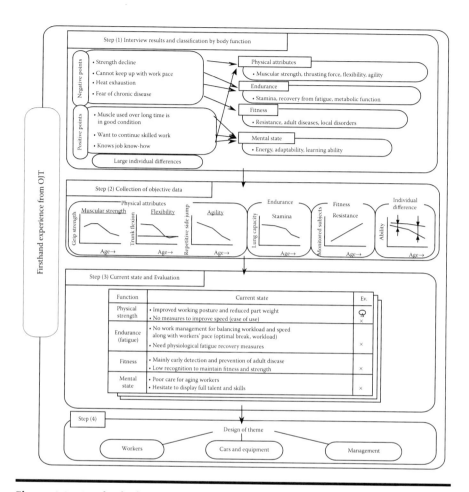

Figure 6.9 Method of setting concept of consideration for aged workers.

there was no procedure for assigning jobs according to the work speed of individual operators or for establishing effective breaks. In terms of basic physical strength, the early detection and prevention of adult diseases had been the primary consideration, while little emphasis was placed on maintaining and improving physical strength.

From the perspective of mental health, care for new employees was sufficient, but care for older employees was not sufficient. From these analyses, the following imperatives were selected—(I) boost morale, (II) study work standards to reduce fatigue, (III) build up physical strength for assembly work, (IV) alleviate heavy work by employing easy-to-use tools and devices, (V) ensure temperature conditions suited to assembly work characteristics, and (VI) reinforce preventative measures against illness and injury.

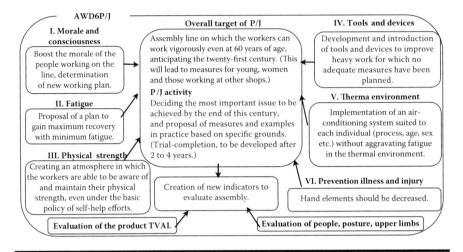

Figure 6.10 Relation diagram of overall objectives of projects and objectives of respective teams.

6.2.4 Definite Plan for Concept Actualization

6.2.4.1 Formation of Project Team, AWD6P/J, and Activity Optimization

It was concluded that tackling these six themes separately is not effective because they are strongly interrelated. As specialized investigations are necessary, a project team called AWD6P/J (Amasaka, et al., 2000; Suzumura et al., 1998) was formed within Toyota as shown in Figure 6.10, which also shows the project themes of AWD6P/J. The relation diagram in Figure 6.10 shows the interrelationship of each theme. The diagram was created to emphasize team unity during the project. The AWD6P/J team structure is also shown in Figure 6.11.

A division specializing in a theme acted as the leader, and its members, including the assembly division and related divisions, acted as the total task management team. The project was mainly promoted by the vehicle production engineering division, the safety and health promotion division, and the human resources development division. The TQM promotion division coordinated the overall project. Also, by having directors (vice president, senior managing director, and managing director) as project advisors, systematic implementation throughout the organization became possible (Eri et al., 1999).

6.2.4.2 Total Task Management Team Activity by Practical Application of Science SQC

Each team activity took the form of *total task management* (Amasaka, 2003a). By applying Science SQC, team activities by managers and staff members ensured the

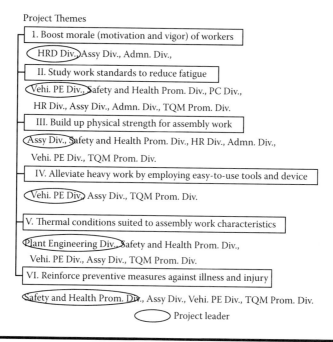

Figure 6.11 AWD6P/J structures.

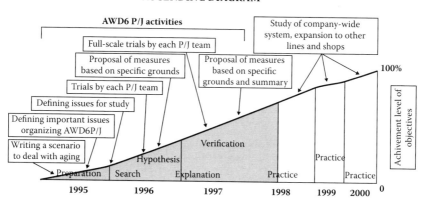

Figure 6.12 Diagram of climbing mountain of problem solving for all projects.

rotation of the plan, do, check, and action (PDCA) cycle. The relation diagram in Figure 6.10 and the mountain-climbing chart for problem solving in Figure 6.12 were made for the management of the overall activity so that all teams shared the same milestones and steps for attaining goals and recognized the interrelationship between individual teams and the direction of each activity. As the main players

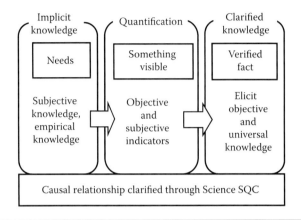

Figure 6.13 Science SQC approach.

in an assembly line are workers, a worker-oriented approach is the key to problem solving in each project team.

Figure 6.13 shows the steps for making implicit knowledge explicit via the practical application of Science SQC. First, implicit knowledge (ambiguous, subjective information) such as opinions, intuition, and worker sense should be quantified with objective and subjective indicators. These quantitative data can then be scientifically analyzed to identify causal relationships within the given phenomena. These indicators make an objective, universal evaluation possible and make implicit knowledge explicit. The next section presents examples of AWD6P/J activities.

6.2.5 Activity Examples

6.2.5.1 Project II: Study Work Standards to Reduce Fatigue

The objective of this project was to achieve work standards that minimize fatigue and maximize fatigue recovery. The following three actions were taken: (1) establishing technologies for evaluating the fatigue of assembly line workers, (2) identifying the types of fatigue experienced by aging workers, and (3) setting and testing methods for reducing fatigue (Suzumura et al., 1998; Eri et al., 1999; Amasaka, 2000, 2004a).

6.2.5.1.1 Fatigue Evaluation and Types of Fatigue Experienced by Aging Workers

Continuous assembly work results in fatigue. Fatigued workers have the desire to rest. In other words, fatigue involves changes in in both the physical and mental state. These changes first appear as symptoms such as a decline in productivity and health and an increase in operational errors as physical function decreases. Fatigue evaluation, therefore, requires an analysis of these changes. In analyzing the fatigue of workers, effective indicators have been confirmed in both subjective and objective terms. Through evaluation, it was concluded that aging workers experience chronic fatigue rather than acute fatigue.

6.2.5.1.2 Experiment by Changing the Rest Pattern and Testing the Obtained Knowledge on Model Lines

To find a way to minimize fatigue, rest patterns were studied. Two rest patterns (varying the time of continuous work and breaks) were tested experimentally to analyze differences in fatigue level. As shown in Figure 6.14, fatigue during operations gradually increases with time and decreases after each break. It was confirmed that fatigue increases as a whole with ups and downs. When the number of breaks was increased and the length of continuous working time was reduced in the afternoon (when fatigue tends to increase), the fatigue level at the end of the operation was lower, according to both subjective and objective indicators (e.g., physiological data).

To verify the abovementioned finding, two assembly lines at the Motomachi plant, where the CROWN and IPSUM car models of advanced car manufacturer A are manufactured, were selected as experimental lines for a 2-month trial. The two patterns shown in Figure 6.15 were set for the trial, with a continuous 90-min operation as the base. Pattern A adopts 5 min for the second break in the afternoon while Pattern B adopts 10 min. Pattern A follows Pattern B on the No. 1 line and Pattern B follows Pattern A on the No. 2 line. The trial was conducted with about 500 workers on the No. 1 and No. 2 lines. The effect perceived by workers and their opinions were used as subjective indicators, while the effects on productivity were used as objective indicators to provide the major basis for evaluation.

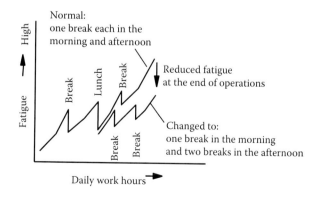

Figure 6.14 Changed break time and effect.

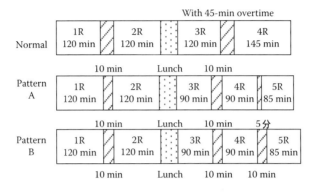

Figure 6.15 Break pattern comparison (normal and trial patterns).

6.2.5.1.3 Perceived Effect and Free Opinions (Subjective Indicators)

For both Patterns A and B, most workers found that fatigue was reduced at the end of daily operations. It was also confirmed that Pattern B produced a greater effect than Pattern A. A possible reason for this result is that workers felt psychological stress from a 5-min break, because they were used to 10-min breaks. Furthermore, answers to questions about expected retirement age based on confidence in physical strength showed an increase, indicating a reduced physical load. Free opinions on changes in physical, mental, and operating conditions during the trial were collected, classified, and summarized.

About 70% responded that breaks were "good for health" and that they felt "less load to parts of the body" in terms of their physical condition. As for the mental aspect, about 60% said that they "felt more relaxed with work time reduced to 90 min" and were able to "concentrate more on the operation." Lastly, regarding operations, about 50% declared that such changes resulted in "less operational delay" and "fewer errors." These responses indicated improved operational quality resulting from the synergy of physical and mental effects.

6.2.5.1.4 Effect on Productivity (Objective Indicators)

The line stop time data between 12:25 and 13:25 when delay is most likely to occur were sampled, as it was assumed that the line stop time is closely related to delays in operation on the No. 2 assembly line. As shown in Figure 6.16, the line stop time decreased by about 2 min on average and productivity increased by about 3% during the trial period of December and January compared with November.

Free opinions were collected from foremen who constantly watch assembly lines to check operation rates and product quality. About 50% of the foremen said that there were fewer operational delays. With regard to product quality, about

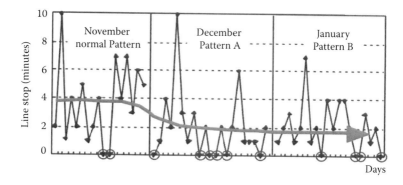

Figure 6.16 Line stop time between 12:25 and 13:25 that causes work delays.

40% of the foremen declared an improvement. These results demonstrate the effectiveness of the rest pattern change in decreasing worker fatigue and improving productivity.

6.2.5.2 Project V: Temperature Conditions Suited to Assembly Work Characteristics

The project team aimed to realize an air-conditioning system that considers individual differences in temperature preferences so that temperature conditions do not adversely affect fatigue levels. The focus of their activities was as follows: (1) clarifying the relationships between temperature and fatigue, (2) analyzing the problems of the current air-conditioning for assembly lines, and (3) developing an air-conditioning system suitable for assembly processes.

6.2.5.2.1 Suitable Temperature Conditions for Minimizing Fatigue

The project team's activities revealed that a temperature of 28°C–31°C and an airflow of 1 m/s are desirable for suppressing fatigue. When this environment was created using line-flow air-conditioning and synchronous fans, a suppression of fatigue was observed from both subjective indicators and physiological data. In addition, it was also found that nonbreathable work clothes that cause workers to sweat have an adverse affect on the body's heat control mechanism. Therefore, the development of comfortable work clothes with excellent moisture absorption and drying properties was promoted.

6.2.5.2.2 Development of Comfortable Work Clothes and Test on Model Lines

The current punch-knit work clothes made of 100% cotton offer good moisture (sweat) absorption performance but poor heat radiation, ventilation, and drying, resulting in the clothes sticking to the skin. As a result of a technical survey and development efforts, stitch-knit work clothes made of a special fiber (porous hollow-section polyester) and cotton were used as prototype, comfortable work clothes with 2.6 times the ventilation and drying capability of the current work clothes, and similar moisture absorption properties. An awareness survey of actual workers on model lines (Motomachi plant assembly lines Nos. 1 and 2) was conducted to test the following: (a) any difference in workers' comfort with and without synchronous fans (with new wide-area exposure function) and (b) differences between the comfortable work clothes and the conventional work clothes.

Since the fatigue reduction effect of line-flow air-conditioning had been made clear through past activities, we assumed line-flow air-conditioning for the model lines. Since the temperature was felt to be 1°C–2°C lower when wearing the comfortable work clothes, the air conditioner outlet temperature was raised by 1°C from that at the time of the last evaluation. Figure 6.17 shows the scatter diagram obtained by a principal component analysis of the survey results regarding the awareness of model line workers. The results show that workers felt that the comfortable work clothes were better than the conventional work clothes. Also, the installation of a synchronous fan was evaluated highly in terms of both air-conditioning and work clothes. On checking actual opinions, many workers

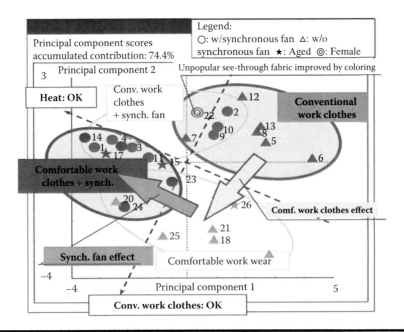

Figure 6.17 Principal component analysis (correlation matrix) of awareness survey results.

commented that the comfortable work clothes were less stuffy and sticky and allowed for easy work movement.

The comfortable work clothes were also evaluated favorably by workers involved in processes with insufficient exposure to airflow. While the initial prototype clothes (100% special fiber) were favored for processes exposed to airflow at 1 m/s or more, they were unpopular for processes exposed to less airflow because they allowed for little heat radiation (due to the heat insulation effect of the pores in the special fiber). This problem was solved by mixing the special fiber with cotton. Stitch knitting may have reduced stickiness to the skin (contributing to easy motion) because of the added surface roughness. In the actual application test, an evaluation was made by raising the air conditioner outlet temperature by 1°C from that of the previous evaluation, resulting in reduced steam consumption by the condenser of about 13%. This yielded energy savings in addition to improving the work environment.

6.2.5.3 Outline of Other Projects

1. *Project I*: Implemented a new system for line workers (new life action program) and promoted work development for skilled workers.
2. *Project III*: Verified the effectiveness of stretch exercises in fatigue recovery at a model workshop; successfully promoted stretch exercises and achieved fatigue reduction.
3. *Project IV*: Improved high-load work by providing easy-to-use tools and devices.
4. *Project VI*: Developed methods to evaluate load on fingers for disease prevention and successfully conducted disease prevention activities at a model workplace.

6.2.5.4 Summary and Future Activities

This study reported on the process of problem analysis for coping with an aging workforce in the assembly process and its results. Actual examples of activities for solving problems were also introduced. The results obtained through these project activities are being verified in model lines to further expand their application. Recently, automation in vehicle assembly lines has been reviewed, as greater emphasis has been put on workers. In the twenty-first century, as in the past, workers are indispensable to vehicle production.

Worker-oriented approaches will be a key point in future production. While worker requirements change with age, a production line should be friendly not only to older workers but also to young and female workers by responding to changes in the environment. A future study will focus on establishing a production system that enables all workers to work productively by making use of the knowledge obtained through this study.

6.2.6 Conclusion

From the viewpoint of global production, this section has proved the effectiveness of the strategic application of New JIT in AWD6P/J—a strategic management technology activity for realizing epoch-making innovation in the work environment at Toyota. This study, as a prerequisite of TPS, focused on innovation in automotive final assembly lines that have depended on young male workers, such that these workplaces will be able cope with an aging workforce and obtain successful results that help realize a comfortable workplace for older workers through an objective analysis from behavior science perspectives. This study is greatly contributing to the global production strategies of advanced car manufacturer A.

6.3 Partnering Performance Measurement Model

6.3.1 Introduction

In the midst of the rapid advancement of globalization and worldwide quality competition, Japanese manufacturers are struggling to realize the simultaneous achievement of QCD, a reduction in development periods, the assurance of high quality, and production at low cost. It can be said that they are truly facing an era of new manufacturing in the process of of realizing the simultaneous achievement of QCD (Amasaka, 2007b, 2009d; Amasaka and Sakai, 2010).

For a total assembly industry like auto manufacturing, not only is quality management for parts and units essential, but optimizing manufacturer assembly technology, as well as ensuring consistent quality from manufacturing through to sales and service, is a must. If a vehicle manufacturer or supplier wants to improve the reliability of the products manufactured in-house, quality management is crucial to jointly improving the reliability of both the manufacturers and suppliers' business processes. It is safe to say that cooperative activities requiring the ability to solve problems will remain important in the future.

In this corporate management environment, the key to success in global production is the reinforcement of product power, or the simultaneous achievement of QCD (Amasaka, 2007b, 2008b). In order to realize this, it is vital to reinforce Japanese-style partnering (Yamaji and Amasaka, 2007) or SCM between automobile manufacturers and parts suppliers. This is known as the Japan supply system (Amasaka, 2000).

Against this background, this section proposes the Partnering Performance Measurement Model for Assembly Makers and Suppliers (PPM-AS) and verifies its effectiveness (Sakatoku, 2006; Yamaji et al., 2008). The purpose of the PPM-AS is to formulate (or visualize) the actual status of Japanese partnering between automobile manufacturers and parts suppliers for a clearer evaluation and diagnosis.

This partnering has been somewhat implicitly carried out in the past, and it is viewed as the basis for deploying global SCM.

6.3.2 Significance of Strengthening the Corporate Management of Japanese Automobile Assembly Makers and Suppliers

In recent years, the management issue in the Japanese automobile industry that is playing a key role in Japanese manufacturing is worldwide uniform quality and simultaneous production launch, in response to current globalization (Amasaka, 2004a, 2007c). In this corporate management environment, the key to success in global production is the reinforcement of product power, or the simultaneous achievement of QCD (Amasaka, 2008b, 2009a). In order to realize this, it is vital to reinforce Japanese-style partnering, or partnering competition and collaboration—Japan SCM between automobile manufacturers (hereafter termed *assembly makers*) and affiliated or nonaffiliated parts manufacturers (hereafter termed *suppliers*) (Amasaka, 2008b, 2009c). This is also called the Japan supply system, as shown in Figure 6.18.

To further advance in this area, assembly makers should not only reinforce internal partnering with their own operational departments (such as engineering, production, and sales), but they must also strengthen external partnering. This means establishing cooperative relationships with other companies while advancing and establishing global partnering through strategic collaboration with both foreign and domestic suppliers. In order to implement such reinforcements, it is fundamental that a new PPM be deployed. This measurement should serve as a formulation model (a radar chart for visualization) for evaluating the actual status of Japanese partnering between assembly makers and parts suppliers, which has been somewhat implicitly carried out in the past.

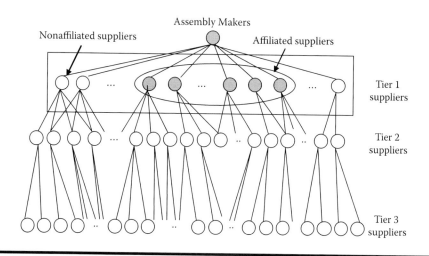

Figure 6.18 Japan supply system.

6.3.3 Partnering Performance Measurement for Assembly Makers and Suppliers

The author proposes Partnering Performance Measurement for Assembly and Suppliers (PPM-AS), which evaluates employment and Justification of "Japan Supply System" for developing global SCM and approach shown below.

According to a survey conducted by Sakatoku (2006) and Yamaji et al. (2008), in general, SCM was found to perform well at the assembly makers assessed. On the other hand, suppliers' answers indicate that they are not necessarily conducting SCM in the way that they would like to with some of their assembly makers, revealing a difference in evaluation, or a so-called disparity (deviation or gap) in evaluator awareness. These SCM evaluations are often based on the evaluators' own implicit empirical knowledge, and (as far as the author knows) evaluation scales are not usually shared (formulated as a model) between them. The author therefore attempts to formulate the evaluation causes and effects that account for the difference in implicit evaluations on both sides—the so-called gap in awareness (Amasaka, 2012). Amasaka (2004b, 2009c) then attempted to carry out a diagnosis through visualization methods by utilizing the Science SQC, New Quality Control Principle (refer to Section 5.4).

Under these circumstances, many studies on globalization (Gabor, 1990; Lagrosen, 2004; Ljungström, 2005; Manzoni and Islam, 2007) and SCM (Joiner, 1994; Taylor and Brunt, 2001; Pires and Cardoza, 2007) are consulted in the PPM-AS. The author researched the current status of partnering between assembly makers and suppliers by cooperating with the Union of Japanese Scientists and the Nihon Keizai Shinbun (Juse, 2006). It targets 528 corporations, mainly manufacturers, and does not cover the topic of partnering with suppliers. The original equipment manufacturer (OEM)-Supplier Working Relations Index (WRI) (Milo Media, 2006) defines the relationship between automobile manufacturers and their suppliers using index numbers, but specific evaluation items are not indicated.

Against this background, Sakatoku (2006) and Yamaji et al. (2008) propose the PPM-AS, including the following: (1) the Partnering Performance Measurement Model for Assembly Makers (PPM-A) and (2) the Partnering Performance Measurement Model for Suppliers (PPM-S). The authors also integrate the two and proposes (3) the PPM-AS as a comprehensive dual performance measurement to be shared by both.

6.3.3.1 Preparation of Evaluation Sheets for PPM-A, PPM-S, and PPM-AS

The author extracted evaluations of the cause-and-effect elements necessary for the preparation of the three types of evaluation sheets (PPM-A, PPM-S, and PPM-AS), based on the responses of staff members engaged in practical affairs in SCM-related departments (both assembly makers and suppliers). More specifically, evaluators from related departments on both sides were given a survey. The questions (evaluation sheet) are shown in Table 6.2. The survey comprised a total of 19 evaluation factors

Table 6.2 Evaluation Sheet

Factor	Contents
X_1	Corporate strategy
X_2	Bias in selection of suppliers
X_3	Selection of affiliated supplier
X_4	Weakening of group affiliation
X_5	Weight of suppliers' opinion
X_6	Spirit of cooperation
X_7	Appropriate evaluation
X_8	Considerate relation
X_9	Demand on discounting
X_{10}	Methods of price setting
X_{11}	Price setting considering labor
X_{12}	Parts inspection standards
X_{13}	Parts inspection items
X_{14}	Response to recalls
X_{15}	Response to contractual failure
X_{16}	Ability to improve
X_{17}	Technical capabilities
X_{18}	Total satisfaction rate
X_{19}	Gained technology and know-how

(Factors X_1–X_{19}, with evaluation contents) extracted from an affinity diagram. The diagram was constructed based on meetings with assembly makers and suppliers.

6.3.3.1.1 Evaluators of Assembly Makers

Toyota Motor Corporation (sales division 1, production preparation division 2, TQM promotion division 1), Hino Motors Ltd. (procurement division 1, TQM promotion division 1), Fuji Heavy Industries Ltd. (TQM promotion division 1, procurement division 2), Honda Motors Co. Ltd. (QA division 1), Nissan Motor Co. Ltd. (procurement division 1), General Motors (GM; procurement division 1), and Ford (TQM promotion division 1).

6.3.3.1.2 Evaluators of Suppliers

Jatco (QA division 2), JFE (QA division 2, procurement division 1), and NHK Spring Co. Ltd. (QC division 1, SQC promotion division 1).

6.3.3.2 Formulation Model of PPM-A, PPM-S, and PPM-AS

Using this evaluation sheet, the author attempted to derive a formulation model that incorporated multivariable statistical analysis. First, cause-and-effect relationships were extracted by conducting a cluster analysis (Ward method and Euclidean distance squared) and a principal component analysis. Next, using a categorical canonical correlation analysis, formulation models for PPM-A, PPM-S, and PPM-AS were derived, while the coefficients of each group's factors were calculated and a radar chart was designed for each group. One hundred (100) was set as the highest score.

6.3.3.2.1 Derivation of PPM-A

The PPM-A was derived from the results of the cluster analysis and the principal component analysis shown in Figures 6.19 and 6.20. Based on these results, evaluation factors could be categorized into five elements: (a) supplier follow-up, (b) quality inspection, (c) corporate capability, (d) supplier decisions, and (e) price setting.

Consideration was then given to PPM-A, PPM-S, and PPM-AS so that formulation models could be derived. With that in mind, a categorical canonical correlation analysis was conducted, employing optimum scaling methods in an effort to grasp the relationship between the 5 evaluation elements and the 17 evaluation factors. A calculation was performed for PPM-A using evaluation element

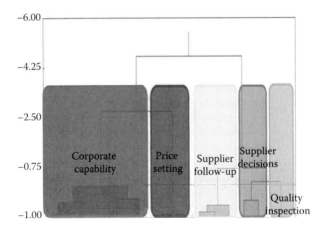

Figure 6.19 Grouping by cluster analysis.

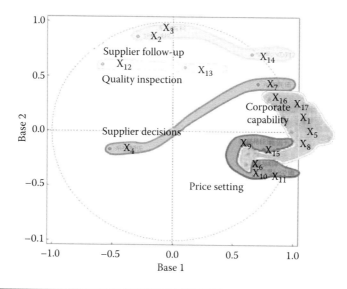

Figure 6.20 Grouping by principal component analysis.

axes (a) to (e). The weighted coefficients shown here were calculated based on the component loading values for each of the five evaluation elements, resulting in a full score of 100 points.

(a) Supplier follow-up = $5.24X_4 + 8.86X_7$
(b) Quality inspection = $2.28X_{12} + 12.01X_{13}$
(c) Corporate capability = $1.91X_1 + 2.44X_{16} + 1.85X_6 + 2.46X_{17} + 1.89X_5 + 1.81X_8 + 1.91X_{15}$
(d) Supplier decisions = $5.39X_2 + 7.23X_3 + 1.67X_{14}$
(e) Price setting = $6.80X_9 + 5.84X_{10} + 1.64X_{11}$

Using these equations, evaluation points for each of the five evaluation elements were calculated and visualized in a radar chart, as seen in Figure 6.21.

6.3.3.2.2 Derivation of PPM-S

Derivation was conducted in a similar way for the PPM-S. As a result, the relationship between the 5 evaluation elements—(f) assembly maker follow-up, (g) quality inspection, (h) corporate capability, (i) spirit of cooperation, and (j) price setting—and the 17 evaluation factors was grasped. A calculation was performed for PPM-S using the five evaluation element axes (f–j).

(f) Assembly maker follow-up = $5.83X_1 + 5.59X_4 + 2.75X_{14} + 0.12X_{15}$
(g) Quality inspection = $1.54X_{12} + 12.75X_{13}$

118 ■ *New JIT, New Management Technology Principle*

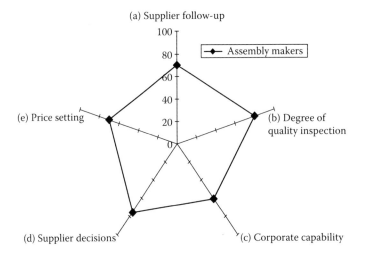

Figure 6.21 PPM-A radar chart.

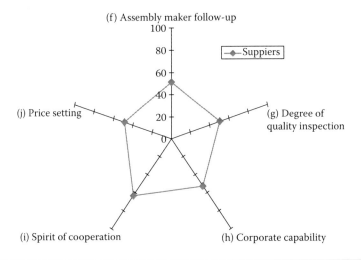

Figure 6.22 PPM-S radar chart.

(h) Corporate capability $= 5.23X_2 + 2.27X_3 + 1.34X_7 + 0.63X_{16} + 4.81X_{17}$
(i) Spirit of cooperation $= 6.46X_5 + 3.26X_6 + 4.56X_8$
(j) Price setting $= 4.17X_9 + 5.41X_{10} + 4.70X_{11}$

Using these equations, evaluation points for each of the five evaluation elements were similarly calculated and visualized by a radar chart, as shown in Figure 6.22.

6.3.3.2.3 Derivation of PPM-AS

The PPM-AS was derived in order to visualize the gap in the evaluations of both sides. This was done by closely comparing the evaluation measurements of PPM-A and PPM-S. Corporations using PPM-AS can see the gap between their evaluations and those of their suppliers at a glance. This allows conventional subjective evaluation based on each side's empirical rules to be converted into an objective analysis, while also clarifying specific problems using the component evaluation factors.

One of the features of the PPM-AS formulation model is the overlapping of the five evaluation element axes of PPM-A and PPM-S. This is because both PPM-A and PPM-S thus far consist of 5 evaluation elements and 17 evaluation factors. The three axes of quality inspection, corporate capability, and price setting are common to both, allowing these elements to be overlapped for direct comparison. In addition, the two new axes, assembly manufacturer follow-up and supplier follow-up are integrated to create a new axis of spirit of mutual support. Similarly, spirit of cooperation and supplier decisions are integrated to establish a new axis of relationship with business partners.

As a result, these evaluation elements, made up of (k) spirit of mutual support, (l) degree of quality inspection, (m) corporate capability, (n) relationship with business partners, and (o) price setting are visualized in a radar chart as shown in Figure 6.23. The figure presents example evaluations of assembly makers and suppliers.

6.3.4 Verification of PPM-AS

The author applied the proposed PPM-AS to two Japanese assembly makers (Toyota Motor Corporation and Nissan Motor Co. Ltd.), one overseas assembly manufacturer (GM), and two suppliers (Jatco and NHK Springs Co. Ltd.). Its effectiveness was thereby verified. As pointed out in Section 6.3.3, it can be generally confirmed

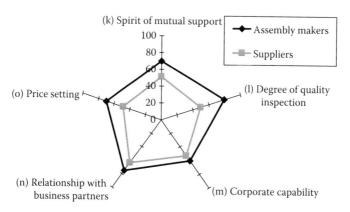

Figure 6.23 PPM-AS radar chart.

that assembly makers' evaluations are higher than those of suppliers. The suppliers' evaluations are more severe than the assembly manufacturer evaluations assume.

The following sections provide examples of PPM-AS evaluations at various corporations.

6.3.4.1 Toyota Motor Corporation

The survey results that were used for verification are shown in Table 6.3 (Toyota Motor Corporation), Table 6.4 (NHK Springs Co. Ltd. [hereafter, NHK]), and Figure 6.24 (Toyota and NHK) based on the evaluation sheet in Table 6.2. The evaluation example of Toyota Motor Corporation and NHK is further described in Figure 6.25. As the figure indicates, evaluations on both sides are generally high. There are no large differences between elements, but a slight gap can be observed in spirit of mutual support and price setting.

The following sections provide an interpretation of the obtained evaluation results. NHK's low evaluation on price setting and spirit of mutual support stems from a thorough cost price reduction carried out by Toyota, based on the company's Customer-First Principle. However, there is almost no gap in the relationship with business partners category. NHK evaluates Toyota's attitude toward problem solving and QCD improvement highly, and these activities are carried out with

Table 6.3 Evaluation Sheet of Toyota

	Excellent	Very Good	Good	Fair	Poor	Bad	Very Bad
Corporate strategy	7	6	5	4	3	2	1
Bias in selection of suppliers	7	6	5	4	3	2	1
Selection of affiliated supplier	7	6	5	4	3	2	1
Weakening of group affiliation	7	6	5	4	3	2	1
Weight of suppliers' opinion	7	6	5	4	3	2	1
Spirit of cooperation	7	6	5	4	3	2	1
Appropriate evaluation	7	6	5	4	3	2	1
Considerate relation	7	6	5	4	3	2	1

Table 6.4 Evaluation Sheet of NHK

	Excellent	Very Good	Good	Fair	Poor	Bad	Very Bad
Corporate strategy	7	6	5	4	3	2	1
Bias in selection of suppliers	7	6	5	4	3	2	1
Selection of affiliated supplier	7	6	5	4	3	2	1
Weakening of group affiliation	7	6	5	4	3	2	1
Weight of suppliers' opinion	7	6	5	4	3	2	1
Spirit of cooperation	7	6	5	4	3	2	1
Appropriate evaluation	7	6	5	4	3	2	1

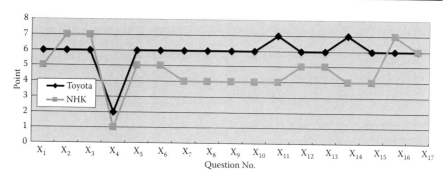

Figure 6.24 Questionnaire result of Toyota and NHK.

mutual support. Therefore, NHK names Toyota as the corporation that they would most like to do business with.

6.3.4.2 Nissan Motor Co., Ltd.

The evaluation example of Nissan and Jatco is shown in Figure 6.26. One outstanding feature is a generally lower evaluation than in the case of Toyota. Gaps can be observed in both price setting and corporate capability. These results can be interpreted as showing partnering activity that is not necessarily based on mutual trust, thus generating gaps in price setting and corporate capability.

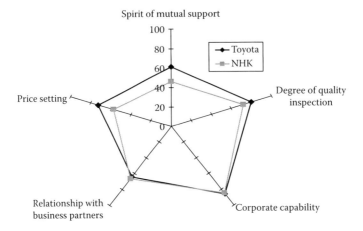

Figure 6.25 Radar chart of Toyota and NHK.

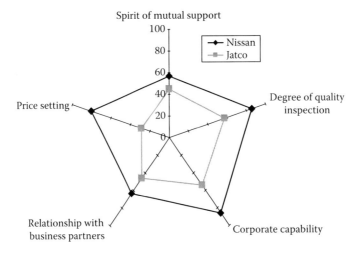

Figure 6.26 Radar chart of Nissan and Jatco.

6.3.4.3 General Motors

Figure 6.27 shows the examples of GM and Jatco. The characteristic of these results is a large gap in the relationship between price setting and relationship with business partners. This is because Jatco cannot fully respond to GM's severe, one-sided demands regarding QCD. This leads to a gap in awareness regarding good business partnerships based on mutual trust. Similarly, an evaluation was implemented in Fuji Heavy Industries Ltd. and Honda Motors Co. Ltd., and the expected results were obtained with regard to its effectiveness.

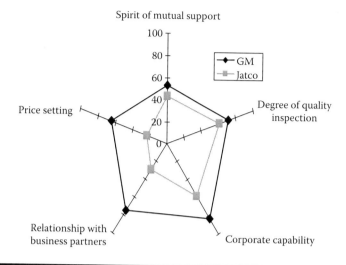

Figure 6.27 Radar chart of GM and Jatco.

6.3.5 Conclusion

This study proposed the PPM-AS. The purpose of the PPM-AS is to formulate (or visualize) the actual status of Japanese partnering between automobile manufacturers and parts suppliers for a clearer evaluation and diagnosis. This partnering has been somewhat implicitly carried out in the past, and it is viewed as the basis for deploying global SCM. This measurement method allowed the visualization of gaps in assembly manufacturer and supplier evaluations of partnering among affiliated companies. The evaluation was implemented in advanced companies, and the expected results were obtained with regard to its effectiveness.

6.4 Strategic Employment of the Patent Value Appraisal Method

6.4.1 Introduction

Looking to attain a high-performance business model for corporate management technology in Japanese companies, the author (Amasaka, 2002a) proposes and verifies the validity of the next-generation management technology principle, New JIT, New Management Technology Principle, employing the three core technologies of TDS, TPS, and TMS (refer to Section 5.1). Among these, the intellectual property divisions are the think tank of corporate strategy and play a key role in acquiring patent rights and enhancing the intellectual productivity of the company through collaborative activities.

This section proposes the application of the Patent Value Appraisal Model (PVAM) to corporate strategy (Amasaka, 2009a; Anabuki et al., 2011). It is generally stated that a good patent refers to an invention or a right or both that is useful for the entity that owns it for retaining its main business, while affecting the business of others. What is available in this field, however, is mere analysis of simple metrical statistics that are available from patent information. In other words, no qualitative analysis is available of the engineers' level of consciousness about the depth of "what is a good patent?" for them, the patent inventors.

Therefore, the author undertook the task of formulating measures to define the concept of the PVAM. He defined it by grasping individual engineers' recognition of the present status of their patents using verbal information and understanding it quantitatively using Science SQC (Amasaka, 2003a, see Section 6.3). An improvement of patent value signifies engineers' value creation at work (invention). The author established a PVAM that consists of several elements, each for a strategic patent. In recent years, the author has continued to apply the PVAM successfully and has proved the validity of the proposal with advanced car manufacturer A and other leading corporations (Amasaka, 2009a; Anabuki et al., 2011).

This method provides new insight that contributes to the quality patent creation process for innovating the intellectual property function and enhancing engineering strategy.

6.4.2 Significance of Patent Acquisition as a Part of Corporate Technological Strategy

6.4.2.1 Current Situation of Patent Applications in Japan

In recent years, the number of patent applications in Japan is about 400,000 (Japan Patent Office, 2008). Of these, about 30% are actually implemented and the remaining 70% are said to be dormant patents. This situation has been brought about because defensive patents account for 40% or more of the patent applications, which are not intended for commercial applications (Nikkei, 2002).

Recently, the number of patent applications has been declining. All over Japan, there has been a shift from a phase of numerous applications to that of selective applications. However, a specific methodology for selecting a highly effective patent is not available at this time, and this is sorely needed in different industrial fields (Amasaka, 2003b).

6.4.2.2 Corporate Reform and Patent Acquisition

One of today's most relevant issues concerning corporate technological strategy is to conduct cutting-edge and innovative technological development from both a mid- and long-term standpoint with a view to improving the quality of patents so that exclusive and competitively superior patents can be acquired (Amasaka et al., 1996b). Patents need to be new and progressive; therefore, the significance of patent acquisition is for engineers to create and realize strategic technologies with proactive ideas (ingenuity).

The activities of such engineers in creating a patent while searching for new technologies also promote self-growth and contribute greatly to the innovation and development of corporations (Amasaka, 2002c). In this way, the intellectual productivity of white-collar workforces can be improved. Moreover, the act of invention and patent registration, while being conscious of corporate strategy, can also be an essential basis of corporate strategy (Amasaka, 2003a, 2004a).

6.4.3 An Issue for the Intellectual Property Department, Qualitative Evaluation of Patent Value

Based on the idea that a good patent is a product of good development, in order to acquire a strong and wide-ranging patent, the conventional issues involved have been (1) the patent management system (from quantity to quality), (2) the reinforcement of patent acquisition support for the development departments, and (3) an improvement in the capabilities of the intellectual property department staff. However, at the core of these fundamental issues is the fact that there have been no systematic or effective patent evaluation methods emphasizing strategic competitiveness with regard to the elements necessary for developing technological patents for commercialization.

These elements are (i) product power, (ii) practicality, (iii) strength of right, and (iv) technological development capabilities. The main reason for such a situation is that the methodology for (a) engineers (inventors) who create patents, and their superiors; (b) the intellectual property department (patent examiners) who decide whether or not to apply for the patents; and (c) chartered patent agents (outsourced experts) who prepare the patent application forms, to qualitatively assess and evaluate the patent value based on objective evaluation standards has been insufficient (Amasaka et al., 1996b).

In the conventional quality evaluation of patents, which is dependent on the empirical rules of the three evaluating groups (a), (b), and (c) above, the intention of development engineers concerning the four elements of patent application cannot be sufficiently appreciated by their superiors, the patent management staff, the patent examiners, or the chartered patent agents. As a result, the rejection of patent examination filings, the mismatch between the content of the invention and the scope of the patent claims, and the deterioration in the quality of the patent due to restrictions have been observed from time to time.

Such problems in the process of patent creation have not been effectively solved. Therefore, technological development, which is supposed to strengthen corporate strategy, and the resulting continual creation of strategic patents are being delayed. This also interferes with the improvement of intellectual productivity.

Generally, a strategically effective patent is said to be the kind of invention or right that develops the company's business operations, influences competitors, and places the company in a superior position (Amasaka et al., 1996b; Amasaka, 2002c, 2003b, 2009a). However, as far as the author is aware, no studies on the qualitative evaluation of patent value have been conducted (Umezawa, 1999; Kevin and David, 2001). For example, many of the studies on patents are limited to statistical analyses, mainly for the stratified categorization of patent development trends or the contents of inventions developed by corporations and inventors (Amasaka et al., 1996b).

Some considerations of the value (cash flow) of intellectual property have been done in the past, but mainly for calculating the monetary value of a patent, and not for promoting the growth of technology or improving the quality of patents (Taketomi et al., 1997). According to the interviews that the author conducted with 5 manufacturers, 12 chartered patent agents, Japanese patent attorneys, and the Japan Patent Office, disparities were observed in their subjective evaluations since these were based on their own experiences. Against this background, the establishment of a consistent qualitative appraisal of patent value is eagerly anticipated from a variety of relevant industrial circles.

6.4.4 Strategic Patents

6.4.4.1 Grasping the Potential Structures of a Strategically Effective Patent

The author conducted a questionnaire survey of engineers working for advanced manufacturer A (97 engineers from 7 departments: 2 from R&D, 1 from designing, 2 from production engineering, 1 from machinery and tools, and 2 from management) whose development or inventions have been patented, for the purpose of finding a strategically effective patent for corporations. Based on the analysis of the collected linguistic information (by means of cluster analysis and factor analysis), potential structures to be considered were clarified.

These structures included (1) the realistic group (emphasizing profit), (2) the preemptive group (opting for originality), and (3) the innovative group (trying to be advanced) (Amasaka, 2007a). Furthermore, new knowledge was obtained that suggests that the values (recognition) of engineers concerning what makes an effective patent differ according to their assigned department (R&D, production engineering, manufacturing, etc.) or according to their research history (practicality, superiority to competitors, initiative minded, etc.). Through this analysis, the author was able to establish the PVAM.

6.4.4.2 Grasping Engineers' Values with Regard to Strategic Patents

Similarly, in connection with the previously mentioned dormant patents, the author carried out an investigation at Toyota on the change in the engineers' consciousness (values) toward simply good patents that are contemplated by engineers on a daily basis for acquiring patents, and strategically effective patents that are designed to strengthen the corporate technological strategy.

As a result, a new understanding was gained through factor analyses and other methods about the aforementioned four elements of patent application. The results indicated that the simply good patents generally tend to focus on (i) product power and (ii) practicality, whereas the emphasis of strategically effective patents shifts to focus on (iii) authoritativeness of rights and (iv) technological development capabilities (Amasaka et al., 1996b; Amasaka, 2003b).

6.4.5 Proposal of Patent Value Appraisal Method

Based on the knowledge acquired above, the author hereby proposes the PVAM, which supports the creation of strategic patents as a basis of corporate strategy (Amasaka, 2009a; Anabuki et al., 2011).

6.4.5.1 Redefinition of the Patent Value Appraisal Variables

The relevant factors (explanation variables X_1–X_{11}) from the linguistic information gathered from advanced car manufacturer A were extracted and summarized (Amasaka, 2003b, 2007a).

Moreover, based on the interviews conducted by the author, the idea was formulated that, for an evaluation of patents, an international viewpoint as well as a time-series viewpoint might be necessary in addition to the 11 explanation variables. Fourteen total variables (X_1–X_{14}) were defined as explaining strategic patents as shown in Table 6.5.

6.4.5.2 Grasping Patent Value Appraisal and the Potential Factors

Through the interviews conducted by the author, it has been established that engineers, the intellectual property department, and chartered patent agents are separately evaluating patents based on their own empirical rules. Here, the disagreement in the opinions of these three groups is one of the causes of the declining quality of patents.

For the purpose of investigating what is given priority when these three groups appraise strategic patents, and also to clarify the importance of and the relationships between the 14 evaluation indices, the author prepared a seven-point scoring questionnaire (7: extremely important, 6: fairly important, 5: important, 4: neither important nor unimportant, 3: unnecessary, 2: fairly unnecessary, and 1: extremely unnecessary) and conducted a survey targeting 69 participants (40 engineers, 17 intellectual property department staff, and 12 chartered patent agents).

Table 6.5 Explanation Variables of "Strategic Patents"

Explanation Variables	Evaluation Indices
X_1	Technological lead ahead of competitors
X_2	International technological lead
X_3	Technological superiority to own company's alternative technology
X_4	Product appeal
X_5	Applicability
X_6	Profitability
X_7	Innovativeness
X_8	Fundamental idea
X_9	Abundance of cases
X_{10}	Patent right–based influence on competitors
X_{11}	International patent right–based influence
X_{12}	Exhaustiveness of the patent's scope
X_{13}	Trendiness
X_{14}	Patent's consideration of the future of the company

As a result of a cluster analysis based on the survey, all three groups surveyed—engineers, intellectual property department staff, and chartered patent agents—could be clustered into five groups. However, there was a difference in the way that the explanation variables were divided among the three groups. In other words, these three groups emphasize different points when evaluating a patent. Having said that, assuming that these five separate groups are potential factors for patent evaluation, attempts were then made to grasp the implications of these results.

6.4.5.2.1 Grasping the Potential Factors of Chartered Patent Agents

Figure 6.28 shows the cluster analysis results of the questionnaire survey for the chartered patent agents. The first group of chartered patent agents herein is comprised of X_1, X_2, X_6, and X_{12}. This group can be interpreted as concerned with technological ability that puts emphasis on profitability and the exhaustive coverage of the patent's scope. Therefore, it can be considered as a potential factor related to (A1: technological capability). Similarly, the group made up of X_7, X_8, X_9,

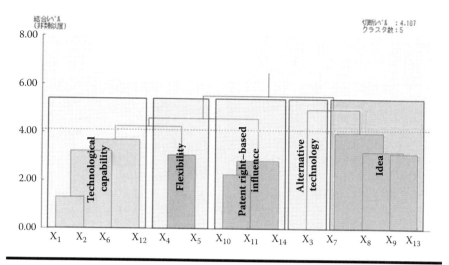

Figure 6.28 Cluster analysis results for the chartered patent agents.

and X_{13} focuses on innovative ideas while considering the current trends, and thus emphasis is given to (A2: idea).

The group made up of X_{10}, X_{11}, and X_{14} is concerned with the influence of the patent rights on the company's future, giving emphasis to (A3: patent right–based influence), and the group focused on X_4 and X_5 is concerned about the product appeal as well as the development and application possibilities of patents. Therefore, this group can be categorized as (A4: flexibility), while the X_3 group highly values (A7: alternative technology).

6.4.5.2.2 Grasping the Potential Factors of Development Engineers

The potential factors were clarified for these groups in the same way as for the chartered patent agents. The first group from the engineers, which opted for X_1, X_2, and X_3, is considered to be a factor that gives emphasis to (A1: technological capability). The group choosing X_7, X_8, and X_{14} indicate that they put emphasis on (A2: idea). The group highly valuing X_6, X_{10}, X_{11}, and X_{12} puts emphasis on (A3: patent right-based influence). The group choosing X_4 and X_5 puts emphasis on (A4: flexibility), and the group choosing X_9 and X_{13} puts most emphasis on (A5: trendiness).

6.4.5.2.3 Grasping the Potential Factors of the Intellectual Property Department Staff

The potential factors were clarified for this group in the same way as for the development engineers. The group opting for X_1, X_2, and X_7 is considered to be a factor that gives emphasis to (A1: technological capability). The group choosing X_3,

X_4, and X_8 is putting emphasis on (A2: idea). The group choosing X_6, X_{10}, and X_{11} is most concerned about (A3: patent right-based influence). The group opting for X_5, X_{13}, and X_{14} puts emphasis on (A4: flexibility), and the group choosing X_9 and X_{12} highly values the (A6: exhaustive coverage of the patent's scope).

The evaluation indices from A1 to A4 are factors that are common to all three groups, but the fifth factor proved to be different. The thinking behind it is something common to all types of specialties. For example, the main job of chartered patent agents is to prepare descriptions of patents, but what they are most concerned about is whether or not the same technology already exists.

The potential factor that was revealed for chartered patent agents is a reflection of such thinking. In other words, the fifth factor is a point that deserves attention from each group or department, and thus can be considered to be an important index.

6.4.5.3 The Structure of the Patent Value Appraisal Equation

As shown in the analysis results in Section 6.4.5.2, patent value appraisal based on the 14 indices has been proven to have its basis in the 5 potential factors of each appraising group. It was thought that by investigating the relationships between the 14 appraisal indices and the 5 potential factors, as well as the relationships between the potential factors and the patent strength, that a PVAM might be constructed.

Consequently, a covariance structure analysis was used by considering the five potential factors as primary potential factors and comprehensive patent strength as a secondary potential factor in order to investigate their relationship. Using the partial regression coefficient calculated from the covariance structure, patent value appraisal equations were established as shown below (Amasaka, 2007a).

1. Development engineers

$$\text{Comprehensive patent strength} = 0.24 * A1 + 0.66 * A2 + 0.59 * A3$$
$$+ 0.54 * A4 + 0.24 * A5$$
$$A1 = 0.70 * X_1 + 0.64 * X_2 + 0.68 * X_3$$
$$A2 = 0.42 * X_7 + 0.59 * X_8 + 0.59 * X_{14}$$
$$A3 = 0.36 * X_6 + 0.65 * X_{10} + 0.74 * X_{11}$$
$$+ 0.39 * X_{12}$$
$$A4 = 0.48 * X_4 + 0.63 * X_5$$
$$A5 = 0.74 * X_9 + 0.39 * X_{13} \tag{6.1}$$

2. Intellectual property department staff

Comprehensive patent strength = $0.36 * A1 + 0.69 * A2 + 0.45 * A3$
$+ 0.42 * A4 + 0.56 * A6$

$A1 = 0.81 * X_1 + 0.87 * X_2 + 0.76 * X_7$

$A2 = 0.42 * X_3 + 0.49 * X_4 + 0.39 * X_8$

$A3 = 0.45 * X_6 + 0.50 * X_{10} + 1.33 * X_{11}$

$A4 = 0.47 * X_5 + 0.96 * X_{13} + 0.6 * X_{14}$

$A6 = 0.39 * X_9 + 1.78 * X_{12}$ (6.2)

3. Chartered patent agents

Comprehensive patent strength = $0.21 * A1 + 0.35 * A2 + 0.1.99 * A3$
$+ 0.08 * A4 + 0.22 * A7$

$A1 = 1.19 * X_1 + 0.74 * X_2 + 0.30 * X_6$
$+ 0.31 * X_{12}$

$A2 = 0.46 * X_6 + 0.74 * X_8 + 0.92 * X_9$
$+ 0.70 * X_{13}$

$A3 = 0.97 * X_{10} + 0.83 * X_{11} + 0.78 * X_{14}$

$A4 = 2.2 * X_4 + 0.34 * X_5$

$A7 = X_3$ (6.3)

By means of the above procedure, the larger picture of how development engineers, intellectual property department staff, and chartered patent agents appraise patents has been clarified.

By preparing a radar chart using the above patent appraisal equations, the appraisal of a patent's value can be strategically visualized. A detailed evaluation is visualized in Figure 6.29, and the potential factors are graphically presented in Figure 6.30.

132 ■ *New JIT, New Management Technology Principle*

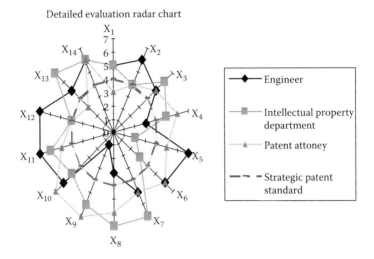

Figure 6.29 Radar chart of patent power at advanced car manufacturer A.

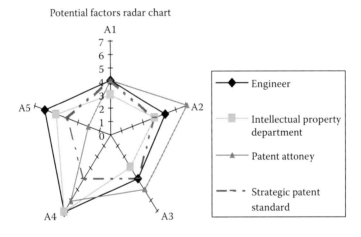

Figure 6.30 Radar chart of potential factors at advanced car manufacturer A.

6.4.5.4 Creating the PVAM Software

In an effort to make the patent evaluation method generally applicable and convenient to use, it was standardized as PVAM software (Ishigaki and Niihara, 2002; Anabuki et al., 2007). Its structure and characteristics can be seen in Figure 6.31 (Example 1, Amasaka Laboratory's PVAM). When appraising a patent and scoring the abovementioned explanation variables X_1–X_{14} from 1 to 7 points, the appraisal results can be visualized in a radar chart of the five potential factors. Moreover, a function is provided for conducting the evaluation so that the difference in

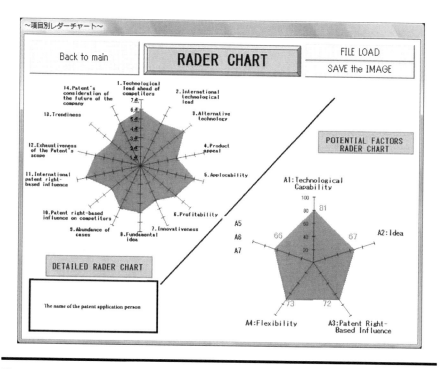

Figure 6.31 Structure and characteristics of PVAM software.

the appraisals by the three groups involved in patent acquisition (development engineers, intellectual property department staff, and chartered patent agents) is made clear.

By using these two radar charts, the way that these three groups think about the appraised patents, as well as their possible strong and weak points, can be clearly seen at a glance. Furthermore, the conventional subjective appraisal based on the three groups' empirical rules has been converted into an objective appraisal, and this facilitates opinion exchange among the groups as well. PVAM therefore automates the visualization process, consisting of the appraisal calculation and radar charts, and supports the convenient use of this method for those who use this software.

The evaluation of patents using PVAM can easily present their weak points, so that more strategic patents can be created.

6.4.6 Verification of the Effectiveness of PVAM

The author applied the established PVAM at a company to verify its effectiveness. Figure 6.32 shows an application case at a leading company B (Ishigaki and and Niihara, 2002; Anabuki et al., 2007; Yamaji et al., 2008). Here, PVAM was modified to fit the characteristics of the company.

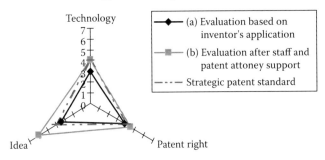

Figure 6.32 An application example of PVAM.

As the figure shows, the evaluation of application patents made by (a) the development engineers in the research department was later supported by the (b) patent management staff from the intellectual property department, and as a result, the contents of these inventions and the patents claims were successfully and strategically applied.

Similarly, the author was able to verify PVAM at other leading corporations, and obtain the desired results (Amasaka, 2009a).

6.4.7 Conclusion

In this section, the qualitative appraisal method, PVAM, was proposed. By using this method, the difference in the patent evaluations made by three groups, development engineers, the intellectual property department staff, and chartered patent agents, who are all deeply involved in patent creation, was strategically visualized. Therefore, this method is considered to be a valid method of supporting the creation of strategic patents as a basis of advanced corporate strategy in Japan.

References

Amasaka, K. 1997. A construction of SQC intelligence system for quick registration and retrieval library, in *Lecture Notes in Economics and Mathematical Systems*, vol. 445, pp. 318–336, Springer-Verlag, Berlin.

Amasaka, K. 1998. Application of classification and related methods of to the SQC renaissance in Toyota Motor, in *Data Science, Classification, and Related Method*, vol. 445, pp. 684–695, Springer-Verlag, Tokyo.

Amasaka, K. 1999a. Management technology and partnering (keynote speech), in *JIMA Management Technology Symposium*, pp. 1–8, September 4, 1999, Osaka Institute of Technology, Osaka, Japan [in Japanese].

Amasaka, K. 1999b. A study on "Science SQC" by utilizing "Management SQC", *International Journal of Production Economics*, 60–61: 591–598.

Amasaka, K. 2000. Partnering chains as the platform for quality management in Toyota, in *Proceedings of the 1 World Conference on Production and Operations Management*, pp. 1–13, August 3, 2000, Seville, Spain (CD-ROM).

Amasaka, K. 2001a. Expectation and role of new TQM by next generation type smaller group activities, in *The 83rd Quality Symposium on Journal of the Japanese Society for Quality Control*, pp. 39–44, September 3, 2001, Tokyo, Japan [in Japanese].

Amasaka, K. 2001b. Proposal of "Marketing SQC" to revolutionize dealers' sales activities, in *Proceedings of the 16th International Conference on Production Research*, pp. 1–8, July 31, 2001, Prague, Czech Public, (CD-ROM).

Amasaka, K. 2002a. *New JIT*, A new management technology principle at Toyota, *International Journal of Production Economics*, 80: 135–144.

Amasaka, K. 2002b. Intelligence production and partnering for high quality assurance (keynote speech), in *The 94th of the Japanese Society of Mechanical Engineers Tokai*, pp. 35–42, July 12, 2002, Nagoya, Japan [in Japanese].

Amasaka, K. 2002c. Quality of engineer's work and significance of patent application. The Second Patent Seminar, June 1, 2002, Aoyama Gakuin University, Tokyo, Japan [in Japanese].

Amasaka, K. 2002d. Science TQM, a new principle for quality management, in *Proceedings of the 2nd Euro-Japanese Workshop on Stochastic Risk Modelling for France, Insurance, Production and Reliability*, pp. 6–14, September 20, 2002, Chamonix, France.

Amasaka, K. 2003a. Proposal and implementation of the "Science SQC" quality control principle, *International Journal of Mathematical and Computer Modeling*, 38(11–13): 1125–1136.

Amasaka, K. 2003b. The validity of "TJS-PPM" patent value appraisal method in the corporate strategy: Development of "Science TQM", a new principle for quality management (Part 3), in *The 17th Annual Conference of the Japan Society for Production Research*, pp. 189–192, March 16, 2003, Gakushuin University, Tokyo [in Japanese].

Amasaka, K. 2003c. Development of "New JIT", key to the excellence design "LEXUS"— The validity of "TDS-DTM", a strategic methodology of merchandise, in *Proceedings of the Production and Operations Management Society*, pp. 1–8, April 5, 2003, Savannah, GA, (CD-ROM).

Amasaka, K. 2003d. New application of Strategic Quality Management and SCM: A "dual total task management team" involving both Toyota and NOK in *Proceedings of the Group Technology/Cellular Manufacturing World Symposium-Year 2003*, pp. 265–270, July 29, 2003, Columbus, OH.

Amasaka, K. 2004a. Development of "Science TQM", a new principle of quality management: Effectiveness of strategic stratified task team at Toyota, *International Journal of Production Research*, 42(17): 3691–3706.

Amasaka, K. 2004b. *Science SQC, New Quality Control Principle: The Quality Strategy of Toyota*, Springer-Verlog, Tokyo.

Amasaka, K. 2004c. Global production and establishment of production system with high quality assurance, toward the next-generation quality management technology (Series 1), *Quality Management*, 55(1): 44–57 [in Japanese].

Amasaka, K. 2004d. New development of high quality manufacturing in global production, toward the next-generation quality management technology (Series 4), *Quality Management*, 55(4): 44–58 [in Japanese].

Amasaka, K. 2005. Constructing a customer science application system "CS-CIANS": Development of a global strategic vehicle "Lexus" utilizing New JIT, *WSEAS (World Scientific and Engineering and Society) Transactions on Business and Economics*, 3(2): 135–142.

Amasaka, K. 2007a. Proposal and validity of patent value appraisal model "TJS-PVAM": Development of "Science TQM" in the corporate strategy, in *Proceedings of the Risk, Quality and Reliability International Conference*, pp. 1–8, September 21, 2007, Technical University of Ostrava, Czech Republic, (CD-ROM).

Amasaka, K. (ed.). 2007b. *New Japan Model: Science TQM—Theory and Practice for Strategic Quality Management*, Study Group of the Ideal Situation on the Quality Management of the Manufacturing Industry, Maruzen Tokyo, Japan [in Japanese].

Amasaka, K. 2007c. Applying New JIT—Toyota's global production strategy: Epoch-making innovation in the work environment, *International Journal of Robotics and Computer-Integrated Manufacturing*, 23(3): 285–293.

Amasaka, K. 2007d. The validity of "*TDS-DTM*", a strategic methodology of merchandise—Development of *New JIT*, key to the excellence design "*LEXUS*", *The International Business and Economics Research Journal*, 6(11): 105–115.

Amasaka, K. 2008a. Science TQM, a new quality management principle: The quality management strategy of Toyota, *The Journal of Management and Engineering Integration*, June 2008, Editor-in-Cheif Naheel Yousef, Premiere Issue, 1(1): 7–22.

Amasaka, K. 2008b. Strategic QCD studies with affiliated and non-affiliated suppliers utilizing New JIT, in Putnik, G.D. and Cruz-Cunha, M.M. (eds), *Encyclopedia of Networked and Virtual Organizations*, Information Science Reference, Hershey, PA, pp. 1516–1527.

Amasaka, K. 2009a. Proposal and validity of patent value appraisal model "TJS-PVAM": Development of "Science TQM" in the corporate strategy, *The Academic Journal of China-USA Business Review*, 8(7): 45–56.

Amasaka, K. 2009b. The effectiveness of flyer advertising employing TMS: Key to scientific automobile sales innovation at Toyota, *The Academic Journal of China-USA Business Review*, 8(3): 1–12.

Amasaka, K. 2009c. Establishment of strategic quality management—Performance measurement model "SQM-PPM": Key to successful implementation of Science TQM, *China-USA Business Review*, 8(12): 1–11.

Amasaka, K. 2009d. An intellectual development production hyper-cycle model—*New JIT* fundamentals and applications in Toyota, *International Journal of Collaborative Enterprise*, 1(1): 103–127.

Amasaka, K. et al. 2000. AWD6P/J report of first term activity 1996–1999: Creation of 21st century production line in which people over 60's can work vigorously, pp. 1–93, Toyota Motor Corporation, Toyoto, Aichi-ken, Japan [in Japanese].

Amasaka, K. (ed.). 2012. Science TQM, New Quality Management Principle: The Quality Strategy of Toyota, Bentham Science, Sharjah, U.A.E.

Amasaka, K. and Osaki, S. 1999. The promotion of new statistical quality control internal education in Toyota motor, *European Journal of Engineering Education*, 24(3): 259–276.

Amasaka, K. and Sakai, H. 1996. Improving the reliability of body assembly line equipment, *International Journal of Reliability, Quality and Safety Engineering*, 3(1): 11–24.

Amasaka, K. and Sakai, H. 1998. Availability and Reliability Information Administration System "ARIM-BL" by methodology in "Inline-online SQC", *International Journal of Reliability, Quality and Safety Engineering*, 5(1): 55–63.

Amasaka, K. and Sakai, H. 2002. A study on "TPS-QAS" when utilizing inline-online SQC, in *Proceedings of the Production and Operations Management Society*, pp. 1–8, April 7, 2002, San Francisco, CA, (CD-ROM).

Amasaka, K. and Sakai, H. 2009. TPS-QAS, new production quality management model: Key to New JIT—Toyota's global production strategy, *International Journal of Manufacturing Technology and Management*, 18(4): 409–426.

Amasaka, K. and Sakai, H. 2010. Evolution of TPS fundamentals utilizing New JIT strategy—Proposal and validity of advanced TPS at Toyota, *Journal of Advanced Manufacturing Systems*, 9(2): 85–99.

Amasaka, K., Igarashi, M., Yamamura, N., Fukuda, S., and Nomasa, S. 1995c. The QA Network activity to prevent rusting of vehicle by using SQC, in *The 50th Technical Conference on Journal of the Japanese Society for Quality Control*, pp. 35–38, July 28, 1995, Nagaya, Japan [in Japanese].

Amasaka, K., Eri, Y., Mitoma, T., and Fukumoto, K. 1999a. Development of working environment for older worker in final assembly process, in *Proceedings of the 2nd Asia-Pacific Conference*, pp. 129–132, October 31, 1999, Kanazawa, Japan.

Anabuki, K., Kaneta, H., and Amasaka, K. 2011. Proposal and validity of patent evaluation method "A-PPM" for corporate strategy, *International Journal of Management and Information Systems*, 15(3): 129–137.

Anabuki, K., Kaneta, H., Yamaji, M., and Amasaka, K. 2007. A study of patent evaluation method "A-PPM" for corporate strategy, in *Proceedings of the 5th Asian Network Quality Congress*, pp. 670–678, October 18, 2007, Hyatt Regency, Incheon, Korea.

Amasaka, K., Kato, H., Yasusa, Y., and Kosugi, T. 1999b. Application of "Science SQC" for corporate management-"TQM-S" in Toyota group, in *The 61st Technical Conference of the Japanese Society for Quality Control*, pp. 57–60, May 29, 1999, Tokyo, Japan [in Japanese].

Amasaka, K., Kosugi, T., Maki, K., and Ishihara, S. 1998. A new principle "Science SQC" of TQM at Toyota—"TQM-S" in Toyota, in *The 28th Annual Technical Conference of the Japanese Society for Quality Control*, pp. 57–60, October 24, 1998, Hiroshima, Japan [in Japanese].

Amasaka, K., Maki, K., Koto, H., and Jittra, K. 1999c. SQC promotion activities in Thailand Toyota motors—The development of "Science SQC" in overseas company, in *The 62nd Technical Conference of the Japanese Society for Quality Control*, pp. 17–20, July 30, 1999, Nagoya, Japan [in Japanese].

Amasaka, K., Mitani, Y., and Tsukamoto, H. 1993. Studies on the rust preventive quality assurance of plated parts by using SQC, *Journal of the Japanese Society for Quality Control, Quality*, 23(2): 90–98 [in Japanese].

Amasaka, K., Tamoto, T., Oohashi, T., and Ihara, M. 1995a. A study of qualititative estimation of Q.C.C activities in *The 49th Technical Conference on the Journal of the Japanese Society for Quality Control*, pp. 49–52, May 27, 1995, Tokyo, Japan [in Japanese].

Amasaka, K., Nagaya, A., and Shibata, W. 1999d. Studies on design SQC with the application of Science SQC improving of business process method for automotive profile design, *Japanese Journal of Sensory Evaluations*, 3(1): 21–29.

Amasaka, K., Nakaya, H., Oda, K., Oohashi, T., and Osaki, S. 1996a. A study of estimating vehicle aerodynamics of lift-combining the usage of neural networks and multivariate analysis, *Transaction of the Institute of Systems, Control and Information Engineers*, 9(5): 227–235 [in Japanese].

Amasaka, K., Nitta, S., and Kondo, K. 1996b. An investigation of engineers' recognition and feelings about good patents by new SQC method, in *The 52nd Technical Conference on Journal of the Japanese Society for Quality Control*, pp. 17–24, June 1, 1996, Tokyo, Japan.

Asahi Shimbun. 2005. The manufacturing industry—Skill tradition feels uneasy, Tokyo, Japan, (May 3, 2005) [in Japanese].
Burke, W. W. and Trahant, W. 2000. *Business Climate Shifts*, Butterworth–Heinemann, Oxford.
Doz, Y. L. and Hamel, G. 1998. *Alliance Advantage*, Harvard Business School Press, Boston, MA.
Drucker, F. P. 1988. *The Coming of the New Organization*, Harvard Business Review, Boulder, CO.
Evans, R. J. and Lindsay, M. W. 2001. *The Management and Control of Quality*, South-Western, Mason, OH.
Eri, Y., Asaji, K., Furugori, N., and Amasaka, K. 1999. The development of working conditions for aging worker on assembly line (#2), in *The 10th Annual Technical Conference of the Japan Society for Production Management*, pp. 65–68, September 5, 1999, Fukuoka, Japan [in Japanese].
Gabor, A. 1990. *The Man Who Discovered Quality; How Deming W. E., Brought the Quality Revolution to America*, Random House, Inc., New York.
Gryna, F. M. 2001. *Quality Planning and Analysis*, McGraw-Hill Irwin, Columbus, OH.
Harada, T. 2000. *Knowledge Society Building and Organization Innovation*, Nikka-Giren, Tokyo, Japan [in Japanese].
Hayashi, T. and Kido, T. 1992. The proposal of comfortable workplace working environment-making, in *The 22nd Annual Technical Conference of the Japanese Society for Quality Control*, pp. 65–68, October 24, 1992, Tokyo, Japan [in Japanese].
Ishigaki, K. and Niihara, K. 2002. A study for objective evaluation of patent value—Proposal of A-PAT, Aoyama Gakuin University. Graduation Thesis. Amasaka-Laboratory Study Group.
Japan Patent Office. 2008. Available at: http://www.jpo.go.jp/.
J.D. Power and Associates. 2003. Available at: http://www.jdpower.com/.
Jeffrey, G. M. and Aleda, V. R. 1994. A taxonomy of manufacturing strategies, *Management Science*, 42(3): 285–304.
Joiner, B. L. 1994. *Forth Generation Management: The New Business Consciousness*, Joiner Associates, Inc., New York.
JUSE. 2006. Quality management survey: JUSE homepage. Available at: http://www.juse.or.jp/.
Kawakita, J., Kobayashi, S., and Noda, K. 1977. *Theory of Organization Worth Working*, Nippon Keiei Shuppan-kai, Tokyo, Japan [in Japanese].
Komuro, F. and Ito, S. 1992. A proposal of the appraisal method of the attachment, in *The 39th Technical Conference of the Japanese Society for Quality Control*, pp. 35–38, Tokyo, Japan [in Japanese].
Kevin, G.R. and David, K. 2001. *Rembrandts in the Attic: Unlocking the Hidden Value of Patents*, Harvand Business School, New York.
Kusune, K., Suzuki, Y., Nishimura, S., and Amasaka, K. 1992. Analysis of total curvature spring back in stamped parts with large curvature, *Journal of the Japanese Society for Quality Control, Quality*, 22(4): 24–30 [in Japanese].
Lagrosen, S. 2004. Quality management in global firms, *The TQM Magazine*, 16(6): 396–402.
Ljungström, M. 2005. A model for starting up and implementing continuous improvements and work development in practice, *The TQM Magazine*, 17(5): 385–405.
Manzoni, A. and Islam, S. 2007. Measuring collaboration effectiveness in globalised supply networks: A data envelopment analysis application, *International Journal of Logistics Economics and Globalization*, 1(1): 77–91.

Milo Media. 2006. WRI: MRO today homepage. Available at: http://www. Progressivedistributor. com/mro_today_body.htm.

Niemela, R., Rautio, S., Hannula, M., and Reijula, K. 2002. Work environment effects on labor productivity: An intervention study in a storage building, *American Journal of Industrial Medicine*, 42: 328–335.

Nihon Keizai Shimbun. 1999. Corporate survey (68QCS)-Symposium: Strict assessment of TQM, Tokyo, Japan, (July 15, 1999) [in Japanese].

Nihon Keizai Shimbun. 2000. Worst record: 40% Increase of vehicle recalls, Tokyo, Japan, (July 6, 2000) [in Japanese].

Nihon Keizai Shimbun. 2001. IT Innovation of manufacturing, Tokyo, Japan, (January 1, 2001) [in Japanese].

Nikkei. 2002. Corporate innovation through competitive patent (June 22, 2002).

Nonaka, Y. 1990. *The Management of the Knowledge Creation*, Nihon Keizai Shimbun-Sha, Tokyo, Japan [in Japanese].

Pires, S. and Cardoza, G. 2007. A study of new supply chain management practices in the Brazilian and Spanish auto industries, *International Journal of Automotive Technology and Management*, 7(1): 72–87.

Rivette, K. G. and Klein, D. 2000. *Discovering New Value in Intellectual Property*, Diamond Harvard Business Review, 78(1): 54–66.

Sakai, H. and Amasaka, K. 2002. V-MICS in production facilities administration by utilizing DB & CG for maintenance activities, in *Proceedings of the 2nd Euro-Japanese Workshop on Stochastic Risk Modelling*, pp. 1–10, September 20, 2002, Chamonix, France, (CD-ROM).

Sakai, H. and Amasaka, K. 2003. Construction of V-MICS-EM for equipment operation—Effectiveness of robot maintenance visual manual, in *Proceedings of the 17th International Conference on Production Research*, pp. 1–10, August 4, 2003, Blacksburg, VI, (CD-ROM).

Sakatoku, T. 2006. Partnering performance measurement for assembly makers and suppliers, master's thesis, Aoyama Gakuin University.

Stefan, S., Martin, M., and Maria, G. (eds). 2003. *Strategy and Organization in Supply Chains*, Physica-Verlag, Heidelberg, New York.

Suzumura, H., Sugimoto, Y., Furusawa, N., Amasaka, K., Eri, Y., Asaji, K., Furugori, N., and Fukumoto, K. 1998. The development of working conditions for aging worker on assembly line (#1), in *The 8th Annual Technical Conference of the Japan Society for Production Management*, pp. 136–143, September 5, 1998, Nagoya University, Aichiken, Japan [in Japanese].

Takaoka, T. and Amasaka, K. 1991. Analysis of factor for specific fuel consumption improvement, *Journal of the Japanese Society for Quality Control, Quality*, 21(1): 64–69 [in Japanese].

Taketomi, T., Hiraguchi, Y., and Hirabayashi, T. 1997. Evaluation of intellectual property value and recovery of invested resources, *R & D Management*, pp. 32–43.

Tanaka, T. 1990. Development of diesel engine emission odor evaluation method, in *The 20th Sensory Evaluation Symposium on Union of Japanese Scientists and Engineering*, pp. 183–188, June 3, 1990, Tokyo, Japan [in Japanese].

Taylor, D. and Brunt, D. 2001. *Manufacturing Operations and Supply Chain Management: Lean Approach*, 1st edn., Thomson Learning, Florence, KY.

The Japan Machinery Federation and the Japan Society of Industrial Machinery Manufacturers. 1995a. Research Report: Production system model considering aged workers, pp. 1–2 [in Japanese].

The Japan Machinery Federation and the Japan Society of Industrial Machinery Manufacturers. 1995b. Research report: Advanced technology introduction in machinery industry, pp. 82–114 [in Japanese].

Thompson, V.A., Simon, H.A., and Smithburg, D.W. (eds). 1977. *The Foundation Theory of the Organization and the Management*, Diamond-Sha, Tokyo, Japan [in Japanese].

Toyota Motor Corporation and Toyota Motor Kyushu Corporation. 1994. Development of a new automobile assembly line, (40th) Business report awarded with Ohkouchi Prize, 1993, pp. 377–381 [in Japanese].

Uchida, S. and Maki, K. 1991. The proposal of the desirable dining room environment at the employee restaurant, *Union of Japanese Scientists and Engineering, Total Quality Control*, 42(11): 420–425 [in Japanese].

Umezawa, K. 1999. Miscellaneous impressions of intellectual property rights. *Intellectual Property Management*, 49(3): 353–364.

Vecchio, D., Sasco, Jr. A., and Cann, I. C. 2003. Occupational risk in health care and research, *American Journal of Industrial Medicine*, 43: 369–397.

Yamaji, M. and Amasaka, K. 2007. Proposal and validity of global intelligence partnering model for corporate strategy, "GIPM-CS", in *Proceedings of the International IFIP TC 5.7 Conference on Advanced in Production Management Systems*, pp. 59–67, September 18, 2007, Linkoping, Sweden.

Yamaji, M., Sakatoku, T., and Amasaka, K. 2008. Partnering performance measurement "PPM-AS" to strengthen corporate management of Japanese automobile assembly makers and suppliers, *International Journal of Electronic Business Management*, 6(3): 139–145.

Chapter 7
Strategic Development of New JIT

7.1 Automobile Profile Design Concept Method Using Customer Science

7.1.1 Introduction

Customers normally react positively or negatively to existing products. However, in many cases, they do not have a clear idea of their future desire. Sales and service personnel are closest to the customers. When they convey customer needs to R&D and design division staff, whose languages are objective and numerically minded, scientific language rather than implicit language should be the common language.

Furthermore, product designers must accurately interpret the customers' expressions of their needs and develop an optimal drawing. As providing what customers desire before they notice their wants will become a more essential part of any successful manufacturing business, the author proposes *customer science* with the application of New Just in Time (JIT) for studying customer values (Amasaka, 2002, 2004a,b).

Customer science utilizing the Customer Information Analysis and Navigation System (CS-CIANS) (Amasaka, 2005) is proposed, employing a core-element of New JIT: Total Development System (TDS) (refer to Section 5.1). This is designed to analyze customer information incorporating the optimal scientific approach, Science SQC (Amasaka, 2003b), and to serve as a navigation system for the acquired analysis results (refer to Section 5.4).

For strategic product development, the author develops an Automobile Profile Design Concept Method (APD-CM) using customer science, which contributes to the excellence profile design of Lexus vehicles in Japan and overseas.

7.1.2 Need for a New Management Technology Principle

The production technology principle Japan contributed to the world in the latter half of the twentieth century was the Japanese-style production system typified by the Toyota Production System (Ohno, 1977), enhanced by the quality management technology principle generally referred to as JIT (Womack and Jones, 1994).

Today, however, improvements in the quality of Japanese-style management technology principles are strongly desired in the face of unexpected quality-related recall problems breaking out among industrial leaders, while at the same time delays in technical development cause enterprises to experience crises of existence (Hayes and Wheelwright, 1984; Goto, 1999; Nihon Keizai Shimbun, 2000, 2005).

To realize manufacturing that gives top priority to customers with good quality, cost, and delivery (QCD) in a rapidly changing technical environment, it is essential to create a core principle capable of changing the technical development work processes of development and design divisions. In addition, a new management technology principle linked with overall activities for higher work process quality in all divisions is necessary for an enterprise to survive (Hayes and Wheelwright, 1984; Amasaka, 2004c).

7.1.3 Customer Science: Studying Consumer Value Utilizing New JIT

7.1.3.1 Proposal of Customer Science

In this new century, in which the global marketing of products is the basis of management, it is necessary to manufacture products that bring increased value to customers in addition to matching the life stage and lifestyle of each customer (Amasaka, 2004a). However, customers generally evaluate existing products as good or poor, but they do not generally have concrete images of products they will desire in the future.

For new product development in the future, it is especially important to supply desirable products before customers desire them. For that purpose, it is important to precisely understand the vague desires of customers. The proposal of customer science shown in Figure 7.1 makes it possible to concretize customer desires.

Figure 7.1 shows how good new business processes are at creating "wants" indispensable to the development of attractive products. The image of customer's words (implicit knowledge) is translated first into common language (lingual knowledge) and then into engineering language (design drawings as explicit knowledge) by means of appropriate correlation. In other words, objectification of subjective information is important for future product development. It is also important to transform objective into subjective information through correlation to check that

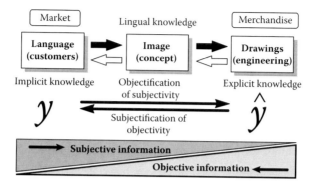

Figure 7.1 Schematic drawing of customer science.

engineering successfully reflects customer requirements. An approach based on customer science will make product planning and uncertain business processes more accurate, possibly increasing success rates and decreasing failure rates.

7.1.3.2 Strategic Implementation of New JIT

Today, customers select products that meet their lifestyle needs and fulfill a sense of value based on a value standard that justifies the cost. They are strict in demanding the reliability of enterprises through the utility values (quality, reliability) of products. For this reason, Amasaka (2002, 2004a, 2005) has proposed a new model of business management technology called New JIT by using customer science, as shown in Figure 5.1 (refer to Section 5.1), for use as a next-generation production methodology.

The primary aim of New JIT is to use hardware and software systems to strengthen management technologies into a next-generation management strategy. The hardware systems utilized are TMS, TDS, and TPS. These three core systems are indispensable for establishing new management technologies in the marketing, sales, development design, production engineering, and manufacturing divisions. Amasaka (2002, 2004b) has proposed total quality management promotion utilizing Science SQC (TQM-S) as the software system utilized in strategic quality management. The aim of this system is to improve the quality of business processes in all divisions.

7.1.4 Constructing a Customer Science Application System: CS-CIANS

7.1.4.1 New Model for Assisting the Conception of Strategic Product Development

Today, a growing number of companies in Japan and abroad are trying to assess the true desires of their customers from the viewpoint of customer-oriented business management and to reflect these desires in future product development. However, the actual behavioral patterns (conception methods) of designers (new product planners and designers) in trying to grasp latent customer desires depend heavily on the designers' empirical skills.

Accordingly, designers often worry that their current business approaches are likely to depend on job performing capabilities and on the sensitivity (intuition or knack) of individual persons, which will not improve the probability of success in the future, regardless of whether or not they have "lucky success" or "unlucky failure."

For the realization of strategic products, the collection as well as intellectual analysis of customer information for creating customer "wants" is the core essence for success in customer science. Figure 7.2 shows the levels of systematic utilization of customer information and the modes of intellectual information-sharing among the related divisions inside and outside the company that are necessary to achieve this objective.

In order to advance the level of execution of customer science activities, it is necessary to evolve customer information-sharing among the marketing/sales/service, product management/design, and engineering/production divisions from off-line to on-line (Amasaka, 2004c).

7.1.4.2 CS-CIANS: A Networking of the Customer Science Application System

For strategic product development, it is important to explore consumer values, which are the basis for creating "wants," through the collection/analysis of customer information and to reflect as well as exteriorize such values in product development. Against this background, CS-CIANS is proposed, as shown in Figure 7.3. This is designed to analyze customer information incorporating the optimal scientific approach, Science SQC, and to serve as a navigation system for the acquired analysis results.

Figure 7.3 shows the CS-CIANS networking system to which the customer science method is applied. As indicated therein, this system enables (1) the merchandise planning division, which explores customer values, and (2) the product

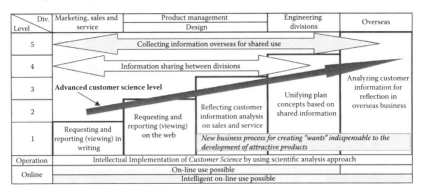

Figure 7.2 Systematic utilization levels of customer information for customer science activities.

Strategic Development of New JIT ■ 145

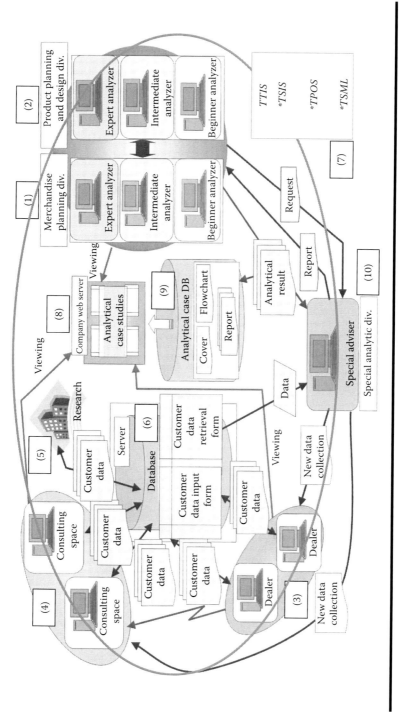

Figure 7.3 CS-CIANS, networking of customer science application system for strategic product development.

planning and design divisions to regularly receive customer data from (3) domestic and overseas dealers, which are exposed to the front line of customer desires through their marketing/sales/service activities.

Similarly, the collection of customer data is also possible through (4) consulting spaces, namely, the showrooms promoting the company's own products or public facilities for discussions and consultations with the customers. Moreover, (5) the domestic/overseas marketing research companies (research) conducting analyses of market trends and customer preferences are also a source of information for feeling out customer values.

All these sections are connected through on-line networking for building (6) a database via a server of the company's own information system division. Into this system, (7) the core system of SQC integration network system, Total SQC Technical Intelligence System (TTIS) (Amasaka, 2000), namely, Total SQC Intelligence System for Quick Registration and Retrieval Library (TSIS-QR) (Amasaka, 1995), which has been developed by the author and others and already utilized by Toyota, is incorporated so that registration or search of customer information can be carried out in real time using the data input/output form.

Another feature of CS-CIANS is that (1) the merchandise planning and (2) design divisions can conduct market investigations of the company's own products in a timely manner as well as benchmark competitors through cooperation with (3) dealers, (4) consulting spaces, and (5) market research companies via (8) an exclusive company web server. In particular, it can carry out an SQC analysis of the acquired data by incorporating (7) Total TQM Promotional Original SQC Soft (TPOS) (Amasaka, 2000), that is, the second core system developed by the author and others for creating "wants." In practice, an advanced analyzer skilled in SQC analysis (*expert*) gives instructions to an inexperienced analyzer *beginner*, while collaborating with a mid-level analyzer *intermediate analyzer* who has a certain level of experience (Amasaka, 2003b).

The core system of TTIS, such as (7) the Total SQC Manual Library (TSML) (Amasaka, 2000), the guidebook of SQC analytical results, and Total Technical Information System (TIRS) (Amasaka, 2000), the technical references and reports, serve as the guide for this scientific approach.

These are accessible for search and utilization from (9) the analytical case database. As seen in Figure 7.3, each database consists of (i) an A4-sized cover page with an overview summary (*cover*), (ii) stepflow charts of SQC analytic processes on sheets of the same size, called *flowcharts*, and (iii) the main contents of the report on customer information analysis, the *report*, designed to be very user friendly.

Particularly for highly complicated customer information analysis, cooperation requests for analysis can be submitted to (10) a *special adviser* in an exclusive SQC analysis division. The system is designed in such a way that the collection of analytical results created by (1) merchandise planning, (2) design, and (10) the special adviser is registered and kept in (9) an analytical case database for the successive development of analytical technology (Amasaka, 2003b).

7.1.5 Application: Development of Customer Science, Key to Excellence Design: Lexus

7.1.5.1 Lexus Design Profile Study

During the development of a model-change vehicle, close attention is paid to changes in the design in addition to selling points strengthened through functional improvements (Shinohara et al., 1996). The reason consumer goods such as vehicles are sometimes purchased on impulse is that customers are impressed with and attracted by their designs.

Psychographics of automobile design profile refers to obtaining an appropriate outline profile (proportion), which is the skeleton of an automobile, by scientifically studying the relationships between the profile and psychological factors in the development stage so as to match the customer's feel (sensibility). As a scientific approach, Science SQC is applied from the viewpoint of customer science (Amasaka et al., 1999; Okazaki et al., 2000).

7.1.5.2 Automobile Profile Design Concept Method for Developing Strategic Product

The author thinks that the analysis process itself that turns implicit knowledge, named designing, into explicit knowledge constitutes the secret to the conception. In this regard, the author has developed an Automobile Profile Design Concept Method (APD-CM) for developing strategic product. The conception support method for developing strategic product quotes the SQC Technical Method, which is a core method of the Science SQC in Figure 7.4 (Amasaka, 2003c, 2004a; Amasaka et al., 1999; Okazaki et al., 2000).

Figure 7.4 represents the implementation of Customer Science by employing

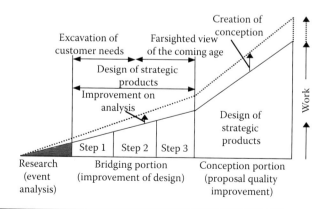

Figure 7.4 An Automobile Profile Design Concept Method (APD-CM) for developing strategic product.

Figure 7.4 represents the implementation of Customer Science by employing Figure 7.1 (refer to Sections 7.1.3 and 7.1.4).

The author will attempt to carry out the "bridging" to conception portion in the course of realization of strategic product design from the mainly research-oriented analysis as the event analysis according to the business process (Step 1 to Step 2 to Step 3 in the diagram). Figure 7.4 is intended to establish a new methodology for supporting conception, which will ascertain the field covered by conception adapted to the times and contribute to enhancing the proposal capability of designers. The application process of ICDM for the product development of a new model of Aristo/Celsior (U.S. name: Lexus GS400/LS430) is described in the sequence of Step 1, Step 2, and Step 3. In this regard, the author confirms the development of New JIT, the key to excellence design of Lexus.

7.1.5.3 Studies on Customers' Sense of Values Using Collages (Step 1)

A collage is something created by gluing various materials on a picture surface. It is a design process for creating images as represented by Picasso (Kawakami et al., 1972) with his work (*Still Life with Chair Caning*, 1912); such a work is called a collage panel. On designing a global strategic vehicle, Lexus, the matter of primary concern for the designer is how to catch the target customers' hearts.

Currently, the author uses collages created by the designer and searches for coincidence with the designer's images by investigating the customers' sense of values scientifically using the following methods:

7.1.5.3.1 Creation of Collage Panels for Researching Design Images

The author prepared for six of the following kinds, as an example shown in Figure 7.5: (i) types (an image by the designer): A, ethnic (soft); B: British (traditional); C, German (artificial); D: French (natural); E: Japanese (ceremonial); F, Italian (casual); and (ii) composition: decorative accessories, watches, rooms, interior decor, gadgets, clothing.

7.1.5.3.2 Customer Questionnaire for Image Survey by Generation

(i) Example: 372 Japanese subjects selected with an assumption of domestic sales. (ii) Generation: The name of the generation, the age, the number samples, and the origin of the name are as follows: (a) cinema generation (19 people aged 56–65), (b) baby boomers (30 people aged 50–55), (c) DC-baptized generation (53 people

Figure 7.5   Image example of the design of a collage panel.

aged 43–49), (d) new humans (*shinjinrui*) (52 people aged 37–42), (e) bananas (90 people aged 31–36), and (f) baby boomer juniors (128 people aged 25–30). (iii) Implementation of questionnaire (subjects selected at random): Out of the image-expressing design terminologies, the designer extracts 24 "image words" (such as 1, Elegant; 2, Cute; …; 23, Youthful; and 24, Individuality) popularly used by designers and easily understood by people in general. (iv) The subjects (a) select image words (multiple answer) they associate with the panel presented and (b) rank the panels in order of their preference (first place, 6 points; second place, 5 points; third place, 4 points; fourth place, 3 points; fifth place, 2 points; and sixth place, 1 point).

7.1.5.3.3 Analytical Example of Images of Preferences by Generation

As an analytical example of investigating customers' sense of values, Figure 7.6 shows the result of analysis of the main component of the questionnaire (order of image word preference by generation). From this figure, the primary component axis is interpreted as modern/formal and the secondary axis as soft/sharp. A (ethnic), D (Italian), and F (French) are positioned in the first quadrant; B (British) in the second quadrant; E (Japanese) in the third quadrant; and C (German) in the fourth quadrant.

In addition, according to the data given in Figure 7.6, the mean value of each group implies the general characteristics of each group as follows: the bananas (90 people aged 31–36; this generation tend to be sensible and prefer moderate personality. Sensuous characters in the work of the Japanese novelist Banana Yoshimoto are typical of this generation.) and new humans (52 people aged 37–42; when

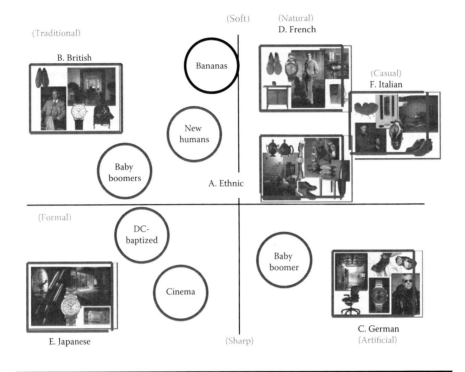

Figure 7.6 Collage panel image analysis by generation (principal component analysis/correlation method: scatter diagram of the principal ratings).

they were young, they were thought of as eccentric by adults) tend to like Panels A, D, and F; the baby boomers (30 people aged 50–55; their generation presents the maximum population in the postwar period) and DC-baptized generation (53 people aged 43–49; this generation is characterized by zeal for wearing fashionable designer and character (DC branded clothes) tend to like Panels B and E; the cinema generation (19 people aged 56–65; when they were in the bloom of youth, the movie industry was also in the golden period and at the center of youth culture) like Panels E; and baby boomer juniors (128 people aged 25–30; this generation is the children of baby boomers) like Panels C.

The author's new finding is that the panel images of the designer coincide with the preferences of the customers of the generation the designer belongs to, but do not necessarily coincide fully with those of other generations, as shown in Figure 7.7. From these analytical results, it is possible to surmise that strategic products can be developed in two ways: by the royal road, using a designer of the same generation as that of the target customers, or by using a designer of a different generation from the target customers to create an appealing design from a new point of view.

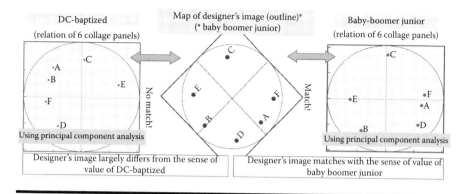

Figure 7.7 Relation of designer's images and customers' preferences using six collage panels.

7.1.5.4 Studies on Customers' Design Preferences for Vehicle Appearance (Step 2)

It is a well-known fact that the design of vehicle appearance has considerable weight in the customer's decision to make a purchase. Which part of vehicle appearance design do customers, domestic and overseas (regardless of age and sex), pay attention to? Professional automotive profile designers have a theory (rule of thumb) that, in general, Japanese customers tend to focus on the front design while North American customers look at the overall design of front, side, and rear. Our challenge here is to give an objective analysis of the theory. To the author's knowledge, there are no reports on the objective verification of this theory by research and analysis in the academic world.

Therefore, quantitative evaluation on the parts of vehicle appearance that customers are interested in will advance a customer design strategy. To assist in the design of the new Lexus GS400/LS430 for model change, 157 customers (young and old, and male and female panels) of various personalities were asked to evaluate the appearance of the four major competing models (BMW 850i/1990 model, Mercedes-Benz 300-24/1989 model, Legend coupe/1991 model, and Soarer 4.0GT/1991 model) and prioritize these three appearance factors: front, side, and rear views.

In Analysis 1, the author verified the correlation between the evaluation and priority by multiregression analysis. The three appearance factors were further divided into the design balance (profile) and detailed elements (4, 9, and 5 sections respectively) for a similar study on their causal relationships (Analysis 2). A preliminary cluster analysis showed that the customers could be stratified in terms of the overall liking of the vehicle appearance into "a group lower in age and annual income" and "a group higher in age and annual income" in their personalities for all four models.

Figure 7.8 shows an example of analytical results for a vehicle model specific to the group higher in age and annual income in Japan. In Analysis 1, the contributory factor adjusted for the degree of freedom (R^{*2}) representing the degree of influence on the overall evaluation of vehicle appearance (X_1) is 0.62, indicating a high causal relationship. The breakdown is as follows: The influence of the front view is even higher at $B_{fv}=0.59$, while the influence of the side and rear views (X_7 and X_{17}) are relatively low at $B_{sv}=0.18$ and $B_{rv}=0.17$ in analysis I. In Analysis 2, the analytical results are similar to those for the group with lower age and annual income, but the influence of bonnet (X_4) is high on the front view (X_2). It is verified that the vehicle appearance is evaluated over a wider range; for example, the influence of the line (X_{19}) from the rear to the trunk and the design balance (X_{22}) of the rear as a whole are high on the side view (X_7). This analytical trend also applies to the other three models.

A similar survey and analysis are conducted in the group lower in age and annual income in North American market. While the front view is generally a high

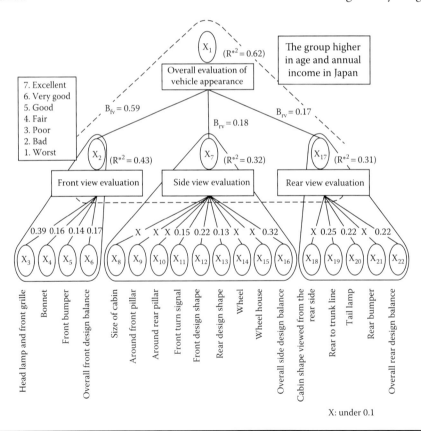

Figure 7.8 Causal relationship between customer satisfaction assessment and vehicle appearance assessment factors by multiple regression analysis.

priority in Japan, it is known that the front, side, and rear views are equally valued in North America.

Other noticeable input is that Japanese customers are likely to provide individual evaluation for each front, side, and rear view at a dealer, while North American customers evaluate the front view while looking at a moving car in the opposite lane, evaluate the side view while looking at a car driving past, and evaluate the rear view similarly on the street. The author confirmed that as the three appearance factors are evaluated, the focus is on the total balance of the design. Through this analytical study approach, designers have understood the need for the customer design strategy that gives consideration to the characteristics of each country.

These findings are the result of verification of the designer's theory (rule of thumb), which greatly contributed to the design of the global strategic vehicle Lexus GS400/LS430.

7.1.5.5 Studies on the Psychographics of the Lexus Profile Design (Step 3)

In the product development stage, a process of modifying the appearance profile (proportion: ratio) to a scientifically legitimate profile that matches the preferences of customers based on the relationship of the contour and psychological factors is called *profile design psychographics*. The significance of the present study does not lie in the fashionable research into the "newness" of a passing fad that places importance on form and surfaces (round, square).

Rather, its significance lies in the development of explicit knowledge of "evolutional newness," demanded as a matter of course. It is most typically expressed with proportion. A principal component analysis was conducted on 62 domestic and overseas vehicle models. Their model years (year of introduction), classes (selling price), and proportions (hood ratio, luggage compartment ratio, cabin ratio, roof ratio, front and rear overhang ratios, wheel base ratio, overall height ratio, overall breadth ratio, and the roof/cabin ratio in relation to overall length) were obtained autographically and their principal components were analyzed, respectively, by vehicle class (from class 5 for high class to class 1 for low class) and the year of introduction (three categories: before 1986, 1987–1991, and 1992 and onward).

When the scatter diagrams on these two principal components are overlapped, as shown in Figures 7.9 and 7.10, old, high-class vehicles with a coach-type cabin (with long hood and luggage compartment length), such as the famous Rolls Royce, Mercedes-Benz W123, BMW518, Jaguar X16, and so on, are in the second quadrant. Late-model vehicles with a long cabin and shorter hood and luggage compartment are positioned in the opposite, fourth quadrant. From the results of these analyses, it has been quantitatively clarified that seemingly "highest class" and "latest" are mutually contradictory elements.

The author then identified the general rule (this was not intentional but a natural consequence of designers' works) of the common proportional ratio (highest class)

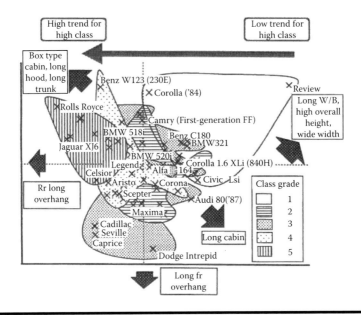

Figure 7.9 Classification of vehicle model by the class degree.

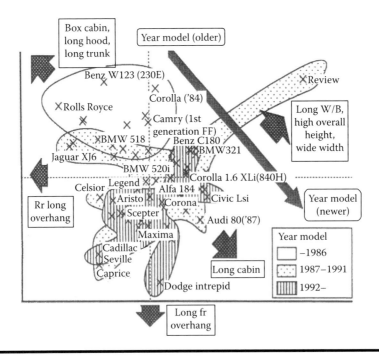

Figure 7.10 Classification of vehicle model by the year model.

that is inherited in the world's prestigious vehicles and insusceptible to change over 5 or 10 years. In addition to applying the general rule to the Lexus design profile, the author manifested advanced form and surface to realize the combination of "highest class" and "latest" for the design development of Lexus.

With this study, the author realized and embodied the combination of a profile of a reputable vehicle, advanced form, and surface through the development of the Lexus GS400/LS430. The market reputation of the Lexus in Japan and overseas has demonstrated the validity of Total Development System-Design Technical Methods (TDS-DTM) (Motor Fan, 1997, 2000; JD Power and Associates, 2004).

7.1.6 Conclusion

This study discussed the effectiveness of constructing a CS-CIANS for strategic implementation of New JIT. Furthermore, the author proposed the innovation of the business process in developing and designing attractive products by using an APD-CM as the method for strategic product development. ICDM was applied to the study of Lexus profile design to enable analysis of customers' values regardless of nationality from the customer science point of view.

7.2 Strategic Development Design Computer-Aided Engineering Model Employing TDS

7.2.1 Introduction

At present, advanced companies in Japan and overseas in the automobile and other industries are endeavoring to survive in today's competitive market by expanding their global production while also aiming to respond to worldwide quality competition (Amasaka, 2007c). Given this management situation, the author has recognized the need to make advancements in the product development system. A new area of interest has arisen in the study of a design management model that realizes high quality assurance in automobile development design.

This new area is the shift of business process management from experimental evaluation based on actual vehicles and tests to predictive evaluation based on highly reliable computer-aided engineering (CAE) analysis. Given this background, Amasaka (2002, 2007e) proposed the Strategic Development Design CAE Model employing TDS, which strategically deploys New JIT. This model consists of the Total QA High Cyclization Business Process System, the Stratified Intelligence CAE Management System, and the Intelligence CAE Development Design Approach System. The validity of this model has been verified by the author and will help realize the simultaneous achievement of QCD.

This model was created in an effort to reform the automobile development design process. The author used this model to investigate the transaxle oil seal leak mechanism that had been a bottleneck technological problem for the world's automobile manufacturers (Amasaka, 2007d,f, 2008). At the implementation stage, the author developed a visualization device that could capture the dynamic behavior of the oil leak with the cooperation of advanced car manufacturer A and its group (supplier B) (Amasaka, 2003a).

This knowledge was then combined with the CG Navigation function that employs computer graphics (CG) technology to create the Intelligence CAE Software: Oil Leakage Simulator and this made highly reliable CAE analysis possible. As a result of this outcome, precise improvement of designs and process management could be implemented. This then led to an even more dramatic effect, the achievement of high-reliability assurance of the automotive transaxle in the marketplace.

7.2.2 Expectations for Automotive Development Production and Simulation Technology

For manufacturers to be successful in the future global market, they need to develop products that make strong impressions on consumers and then supply those products in a timely fashion through effective corporate management. The mission of the automotive manufacturers in this environment of rapidly changing management technology is to be prepared for worldwide quality competition, so that they are not pushed out of the market, and to establish a new management technology model that enables them to offer highly reliable products of the latest design that are also capable of enhancing customer value (Amasaka, 2007d,e).

In the field of management technology for automobile development and production processes that are being considered here, prototyping, testing, and evaluation are excessively repeated to prevent the scale-up effect in the bridging stage between testing and mass production. This has resulted in increases in the development period and cost. Therefore, it is now necessary to reform the conventional development and production model (Amasaka, 2007e).

More specifically, it is increasingly vital to realize the simultaneous achievement of QCD, which satisfies the requirements of developing and producing high-quality products, while also reducing the cost and development period through incorporation of the latest simulation technology, CAE, and statistical science, called SQC (Amasaka, 2007f; Yamaji and Amasaka, 2008). In the vehicle development process employed in the past, after completing the designing process, problem detection and improvement were repeated mainly through the process of prototyping, testing, and evaluation.

In some current automotive development, a prototype of a vehicle body is not manufactured in the early stage of development due to the utilization of CAE and simultaneous engineering (SE) activities, and therefore the development period has been substantially shortened (first from 4 to 2 years, and then to 1 year at present) (Amasaka, 2007e,f). Given this background, it is clear that the conventional development process of repeated evaluation using prototypes is no longer capable of handling this task.

Collaboration between CAE and SE activities, which are now faster and more precise, will be indispensable for fully utilizing the accumulated knowledge database. As discussed so far, expectations are high for the realization of super-short-term development, which would be done through utilization of CAE. In other words, there will be a conversion from the development through real object confirmation and improvement to prediction evaluation-oriented development (Amasaka, 2007e,f; Tanabe et al., 2007).

7.2.3 Proposal of the Strategic Development Design CAE Model

The author will apply New JIT, a new principle of next-generation management technology, as shown in Figure 6.1 (refer to Section 6.1), in order to create and propose the Strategic Development Design CAE Model employing TDS in an effort to reform the business process of development design.

In an effort to reform the process of automobile development design, the author proposes the Total QA High Cyclization Business Process System, the Stratified Intelligence CAE Management System, and the Intelligence CAE Development Design Approach System as a part of the Integrated Intelligence Development Design CAE Model (Amasaka, 2008).

7.2.3.1 Total QA High Cyclization Business Process Model

As the first step, the author proposes the development design business process model. This model is devised from the standpoint of verification/validation (divergence of CAE from theory and divergence of CAE from testing) in order to make possible highly reliable CAE analysis that is consistent with the market–testing–theory profile. Amasaka (1999, 2007e) therefore recommends the introduction and utilization of the Total Quality Assurance (QA) High Cyclization Business Process System called Management SQC, which systematically and strategically realizes high quality assurance by incorporating the analysis using the core technologies of Science SQC, as shown in Figure 5.19 (refer to Section 5.4).

For example, in order to solve the pending issue of a technology problem in the market, it is necessary to create a universal solution (general solution) by clarifying the existing six gaps (1–6 in Figure 5.19) in the process consisting of theory (technological design model)–experiment (prototype to production)–calculation (simulation)–actual result (market), as shown in the lower left of Figure 5.19. To accomplish this, the clarification of the six gaps (1–6) in the business processes across the divisions, shown on the lower right of Figure 5.19, is of primary importance. By taking these steps, the intelligent technical information owned by the related divisions inside and outside the corporation will be totally linked, thus reforming the business process of development design.

7.2.3.2 Stratified Intelligence CAE Development Design System

Next, as the second step, Amasaka (2007e) proposes the Stratified Intelligence CAE Development Design System shown in Figure 7.11. This method contributes to high quality assurance and the simultaneous achievement of QCD. Among many of the automotive manufacturers, there is a gap between the actual vehicle testing results and the CAE analysis results in the development design stage.

Due to a lack of confidence in the CAE evaluation results, they tend to heavily rely on survey tests (Step I). Even among advanced manufacturers, the utilization of CAE stops at relative evaluation (Step II). The author recognized the dilemma that the utilization ratio of CAE compared to actual vehicle (prototype) and testing evaluation is about 25% for survey purposes and about 50% for relative evaluation. In other words, the effectiveness of CAE for the purpose of reducing the length of the development period has not been proven. This also revealed that the usual solution for technical problems, which are difficult to solve theoretically, is actual vehicle (prototype) and testing evaluation based on empirical or CAE evaluation conducted by trial and error using a makeshift modeling process.

To help improve this situation, Amasaka (1999) clarified the mechanism causing the problem by means of the research results accumulated in the intelligent technological

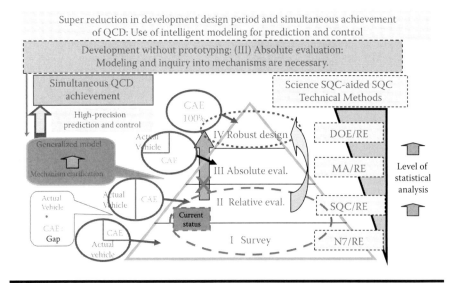

Figure 7.11 Stratified Intelligence CAE Development Design Model.

integration system.; in other words, the combination of visualization technology and Science SQC. Then, drawing on that knowledge, further studies focused on improving the precision of the CAE analysis. As a good example of CAE utilization, attention was focused on the effectiveness of SQC Technical Methods such as New Seven Tools (N7), reliability (RE), SQC, multivariate analysis (MA), and design of experiments (DE) (Amasaka, 2004d, 2007e,f, 2008; Yamaji and Amasaka, 2008).

These are capable of taking a functional approach to variable factor analysis of the real machine (actual vehicle) testing data and then feeding it back to the CAE analysis software through a deductive methodology in order to derive general solutions. Next, in Step III, the mechanism causing the pending technological problem was clarified by using visualization technology. Then, by creating a general model, the absolute evaluation (III) was made possible, an intelligent simulation could be realized, and the prediction and control of the mechanism could be made highly precise through the use of CAE analysis (Amasaka, 2007e). Based on this knowledge, in Step IV, a robust design method was employed to eliminate the reliance on actual vehicle and testing results.

This method allowed for a parameter study in which the influential factors and their effects, which are important to achieve optimal design, were reflected. Furthermore, this also led to prevention of the scale-up effect at the mass production stage and the realization of a rise in the CAE utilization rate (Amasaka, 2007e).

7.2.3.3 Intelligence CAE Management Approach System

In general, experienced development design staff and CAE engineers understand the mechanism that is causing the bottleneck technical problem as implicit knowledge (Magoshi et al., 2003; Hashimoto et al., 2005). The formulation of this implicit knowledge and know-how that is dependent on individual expertise is an essential step in refining CAE analysis as a problem-solving method. It is also a problem-solving approach that utilizes empirical rules and knowledge. The creation of highly reliable CAE software will allow this valuable implicit knowledge to be turned into explicit knowledge, which is why its creation is so important (Amasaka, 2007e; Tanabe et al., 2007).

Therefore, Amasaka (2007e) applied the previously mentioned Intelligence CAE Management System Approach Model and developed highly reliable CAE software in an effort to help solve the bottleneck technical problem, which had become a global technological issue. As an intelligent application method of this software, the author proposed the Intelligence CAE Management Approach System, as shown in Figure 7.12.

To accomplish this, first, it was important to visualize the dynamic behavior of the problem by using actual vehicles and carrying out testing (A). At this point, the expertise of specialists from both inside and outside the company was brought together through partnering activities. The most advanced SQC methods were used to analyze and investigate the complex cause-and-effect relationships.

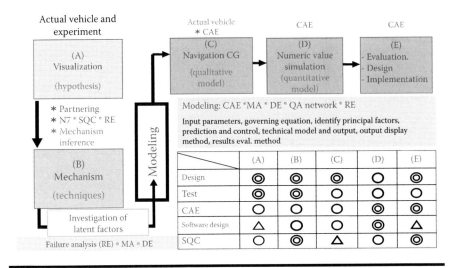

Figure 7.12 Intelligence CAE Management Approach System.

It was vital to deduce the fault mechanism. Next, in order to carry out (B), a precise fault analysis and factor analysis, N7, SQC, RE, MA, and DE were combined and utilized to search out and identify previously unknown or overlooked latent causes. In this way, a logical thinking process was used to carry out a logical investigation into the cause of the fault mechanism. Furthermore, all of this knowledge and information was then unified through (C), the creation of CAE navigation software that employs CG to reproduce the visualization of the actual vehicle and testing data so that it can be made consistent on a qualitative level.

At this stage, where CAE navigation software was being created, it was important to carry out actual vehicle and testing work so that a model (qualitative model) could be made for the cause-and-effect relationships of the unknown mechanism. It would then become extremely important to use this model to reduce the divergence (gap) between the results from the actual vehicle testing and the CAE absolute value evaluation.

In addition, at the stage of developing the highly reliable CAE software (D), exhaustive actual vehicle testing was carried out in order to convert the leak mechanism from implicit knowledge into precise explicit knowledge. The information gained from these work processes would then be unified and a highly credible numerical simulation (quantitative model) would be carried out to make absolute value prediction and control possible.

In the final stage (E), the CAE analysis results are then verified by comparing them to the actual vehicle testing results. In the case of a decentralized organization and business process (such as shown in Figure 7.12), it is essential that the specialists in the fields of design, testing, CAE analysis, CAE software development, and

SQC carry out cooperative team activities, partnering (◉ main, ○ sub, △ support), at each stage of the work process (A–E).

The author (Amasaka and Ohtaki, 1999; Amasaka and Osaki, 2002; Amasaka and Yamaji, 2006; Amasaka, 2007e) acted as the coordinator to promote integration of a cooperative team and as a result, a dramatic improvement in the number of claims from the market was achieved, as illustrated in Section 7.2.4.

7.2.4 Application Example: High Reliability Assurance of the Automotive Transaxle Oil Seal Leakage

In this section, advanced car manufacturer A and supplier B cooperative task team activity—high reliability assurance of the automotive transaxle oil seal leakage case study—will be presented. This case study applied the clarification of the mechanism of oil seal leakage and development of Intelligence CAE Software: Oil Leakage Simulator, and the validity of the Integrated Intelligence Development Design CAE Model proposed by the author was also verified (Amasaka, 2008).

7.2.4.1 Oil Seal Function

An oil seal on an automobile's transaxle prevents the oil lubricant within the drive system from leaking from the driveshaft. It is comprised of a rubber lip molded onto a round metal casing. The rubber lip grips the surface of the shaft around its entire circumference, thus creating a physical oil barrier. In this case the sealing ability of microscopic roughness on the rubber surface is of primary importance (Lopez et al., 1997).

The parameters for the sealing condition of the oil film involve not only the design of the seal itself, but also external factors such as shaft surface conditions, shaft eccentricity, and so on. Contamination of the oil by minute particles was found to be of particular importance to this problem since these are technical issues that involve not only the seal, but also the entire drivetrain of the vehicle (Fukuchi et al., 1998).

7.2.4.2 Understanding of the Mechanism through Visualization

In the case of an oil seal leakage, this was a pending problem where no progress was being made in the reduction of claims from the marketplace or the functional fault. At the time, no one knew where exactly the fault was occurring or what mechanism was causing it. It was important to search out the root cause in order to solve this technological problem (Sato et al., 1999; Kameike et al., 2000). According to supplier B, the oil leaks occurred due to wear.

The result of a wear test on the oil seals indicated that a running distance of 400,000 km (equivalent to 10 years or more of vehicle life) is regarded as sufficiently reliable for the oil seal design (Amasaka and Ohtaki, 1999). However, according to the fault repair records of advanced car manufacturer A for parts that had market claims, which make use of the Dynamic Assurance System (DAS) (Sasaki, 1972), there were sporadic cases of the oil leak problem occurring in vehicles that had not even reached half of the running distance set by supplier B.

Judging from the survey and analysis of parts returned from customers due to claims, the cause of the failure was identified as being due to the accumulation of foreign matter between the oil seal lip and the contact point with the transaxle shaft, resulting in insufficient sealing. Oil leaks were found not only during running, but also in new vehicles at rest. Thus, it was determined that the cause was poor foreign matter control during the manufacturing process, and that it was vital to improve the production quality in this process.

The established theory used to be that fine metal particles (on the order of microns in size) would not adversely affect the lip sealing effect of supplier B (NOK Corporation, 2000). However, when these particles combine to produce larger particles, do they then affect the sealing effect? Also, what about the effect of alignment between the driveshaft and the oil seal (fixing eccentricity) during assembly? In addition, if oil leakage occurs due to foreign matter accumulation on the oil seal lip during transaxle assembly, what is the minimum particle size that causes the problem? The answers to these questions were all unknown since the dynamic behavior of the oil leakage had not yet been visualized. This meant that the true cause also had yet to be clarified.

Consequently, a device was developed to visualize the dynamic behavior of the oil seal lip, as shown in Figure 7.13, in order to turn this unknown mechanism into explicit knowledge (Amasaka and Ohtaki, 1999; Amasaka and Osaki, 2002). As shown in the figure, the oil seal was immersed in the lubrication oil in the same

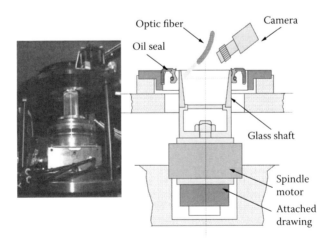

Figure 7.13 Oil seal visualization equipment.

manner as the transaxle, and the driveshaft was changed to a glass shaft that rotated eccentrically via a spindle motor so as to reproduce the operation that would occur in an actual vehicle. The sealing effect of the oil seal lip was then visualized using an optical fiber.

It was conjectured that in an eccentric seal with one-sided wear, the foreign matter becomes entangled at the place where the contact width changes from small to large. Three trial tests were carried out to ascertain if this was true or not. Based on the examination of faulty parts returned from the market and the results of the visualization experiment, it was observed that very fine foreign matter (which was previously thought to not impact the oil leakage problem) grew at the contact section, as shown in Figure 7.14 (Test 1).

It was also confirmed from the results of the component analysis that the fine foreign matter was a powder produced during gear engagement inside the transaxle gearbox. This fine foreign matter on top of microscopic irregularities on the lip sliding surface resulted in microscopic pressure distribution that eventually led to degradation of the sealing performance (Figure 7.15, Test 2).

Also, the presence of this mechanism was confirmed from a separate observation that foreign matter had cut into the lip sliding surface, thereby causing aeration (cavitations) to be generated in the oil flow on the lip sliding surface. This caused deterioration of the sealing performance, as shown in Figure 7.16 (Test 3). The figure indicates that cavitations occur in the vicinity of the foreign matter as the speed of the spindle increases, even when the amount of foreign matter that has accumulated on the oil seal lip is relatively small.

As the size of the foreign matter increases, the oil sealing balance position of the oil seal lip moves further toward the atmospheric side and causes oil leaks at

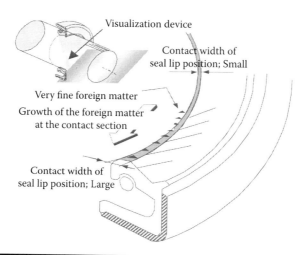

Figure 7.14 Oil leakage mechanism (Test 1).

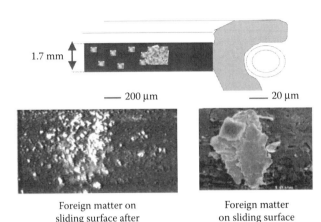

Figure 7.15 Oil leakage mechanism (Test 2).

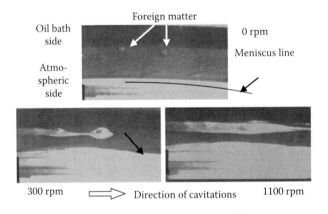

Figure 7.16 Oil leakage mechanism (Test 3).

low speeds or even when the vehicle is at rest. This fact was unknown prior to this study and therefore was not incorporated into the original design of the oil seals (Amasaka and Ohtaki, 1999; Whaley et al., 2000).

7.2.4.3 Fault and Factor Analyses

Before studying the mechanism of the oil seal leaks described in Section 5.3, both NOK and Toyota believed that the wear on leaking oil seal lips would follow a typical pattern. The empirical knowledge based on the results of individual oil seal reliability tests was that the unit axle is highly reliable and would ensure

400,000 km or more in B10 life (the period of time in which <10% of the items fail). It was thought that the oil seal lip should wear gradually because of smooth contact between the oil seal lip and the rotating driveshaft and also because of an oil film in between the two rough surfaces (Fukuchi et al., 1998).

As a result of the study and investigation discussed, however, it was found that metal particles generated from the gears in the differential case accelerated the eccentric wear of the oil seal lip, making the expected design life unobtainable. Since the wear pattern was not simple, it had to be confirmed that the oil leak problem could be reproduced with the faulty oil seals returned due to customer claims.

At this point, Amasaka (1999, 2003a, 2004d, 2007e) performed a search on the research that advanced car manufacturer A had performed up to now using TTIS. The SQC technical method was also applied and the information obtained up to now was further classified and summarized using N7 (affinity diagrams and association charts among others) to promote the fault analysis and factor analysis.

First, in addition to defective oil seals, nondefective ones were collected on a regular basis to check if the oil leak could be reproduced and for comparison through visual observations. Next, transaxle units from vehicles, both with and without oil leak problems, were also collected on a regular basis to check if the leak could be reproduced in the same way. Integrating the results from transaxles both with and without defective oil seals confirmed that the defect could be reproduced and in all of these tests, the oil leaks were reproduced as expected. Based on these test results, a Weibull analysis was then conducted as described below.

The plot of the results (based on defective items that resulted in claims) is shown in Figure 7.17. It clearly shows a bathtub-shaped failure rate for the oil seal failures. The three shape parameter (m) values correspond to the three different failure modes. This analysis resulted in the following new knowledge:

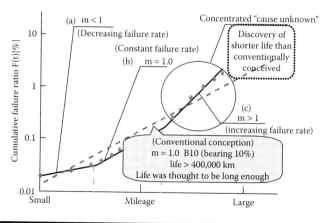

Figure 7.17   Result of Weibull analysis.

1. In the initial period, the failure rate decreases (slope (m) < 1), in the middle period it is constant (slope = 1), and in the latter period it increases (slope > 1), indicating a bathtub-shaped failure rate. The failure rate in each of the three sections can be modeled by a different Weibull distribution, so that the failures can be modeled by a sectional Weibull model.
2. The initial failures (where the failure rate decreases) occur up to a running distance of 50,000 km. Failures in the intermediate range (where the failure rate is constant) occur up to 120,000 km. Finally, failures occurring above this value (where the failure rate increases) are due to wear.
3. The B10 mode life is approximately 220,000 km, about half the value stated as the design requirement.

To confirm the reliability of these results, subsequent claims were analyzed using the Toyota DAS system. Within the warranty period (number of years covered by warranty), the total number of claims classified by each month of production (total number of claims from the month of sale to the current month for vehicles manufactured in the same month) divided by the number of vehicles manufactured in the respective month of production is about twice the design requirement.

This agrees with the result of the above reliability analysis. The influence of five dominating wear-causing factors (period of use, mileage, margin of tightening, hardness of rubber, and average width of lip wear) was studied by two-group linear discriminate analysis using both leaking and nonleaking parts collected in the past. The result showed high positive discriminate ratios of 92.0% and 91.7% for both Group 1 (leaking parts) and Group 2 (nonleaking parts) (Amasaka, 2007d; Amasaka and Osaki, 2002).

From the partial regression coefficients of the explanatory variables in the linear discriminate function obtained, the most significant influence was found to be the hardness of the rubber of the oil seal lip. The influence ratios for the five factors were obtained by means of an orthogonal experimental design (L27), with three level values, which were thought to be technically reasonable in consideration of the nonlinear effects assigned to each of them (Steinberg, 1996).

Figure 7.18 shows the influence ratios of each factor contributing to the discrimination. It shows that the hardness factor of the rubber is highly influential. This analytical result was also convincing in terms of inherent technologies. To test the validity of this result, the lip rubber hardness and the degree of wear on the other collected oil seals were examined further.

As a result, it has been confirmed that eccentric wear is more likely to shorten the seal life because the rubber hardness at the lip portion decreases. This result is consistent with the established theory and empirical knowledge (empirical rules) obtained up to now. This survey and analysis could not have been carried out successfully by the conventional and separate investigation activities of advanced car manufacturer A or supplier B (Amasaka and Ohtaki, 1999; Fukuchi et al., 1998).

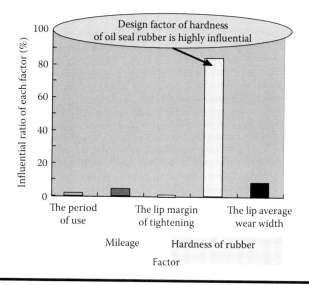

Figure 7.18 Influential effect of each factor.

7.2.4.4 CG Navigation and Intelligence CAE Software: Oil Leakage Simulator

The author combined the CG Navigation function that explains the dynamic behavior of the oil leak with the technological knowledge examined and acquired above to create the Intelligence CAE Software: Oil Leakage Simulator (Amasaka, 2007e, 2008; NOK Corporation, 2000). Figure 7.19 shows a typical example of the modeling of the sliding surface condition that has been created for the purpose of reducing the weight of the sliding surface of the oil seal contact part. Judging from what has been observed up to this point, it is necessary to have the sliding surface minutely irregular and the parts that are actually in contact biased toward the oil side. This is done in order to maintain a good sealing condition that will prevent oil leaks from occurring at the contact part of the oil seals.

As shown in Figure 7.19, the upper section of the sliding surface is the oil side and the lower section is the air side. The darkest black part indicates the areas that are actually in contact. Among the conditions of characteristic values necessary for sealing, the minute roughness of the sliding surface or the small black area representing the actual contact area can be described in this way. Next, another condition is that this black area is biased toward the oil side, which can be incorporated in the sliding surface model like this.

Here, the two black areas are not completely parallel, but rather the upper ends are found to be pointing inward. This takes into account the condition of a real oil seal. The actual sliding surface of the oil seal consists of countless tiny projections,

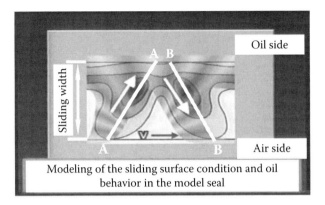

Figure 7.19 CG navigation and intelligence CAE software: oil leakage simulator. (From Amasaka, K., *The 2nd Annual Technical Conference Transdisciplinary Science and Technology Initiative*, pp. 321–326, 2007e).

which are represented by the black area, pointing in random directions. However, statistically speaking, the directional orientation of these projections shows counterbalancing characteristics.

In this model, such factors have been taken into consideration. In other words, the two model projections representing the random projections are arranged to face each other at the same angle, so that a directionless model is presented. The author (Amasaka, 2007e; Kameike et al., 2000) actually photographed an oil seal reproducing this model sliding surface and observed the behavior of the oil. The upper section in the figure is the oil side and the lower section is the air side, while the rotating axis of the driveshaft (called the shaft hereinafter) rotates in the direction of the arrows (→). As the shaft rotates, a flow of oil in the same direction as the rotation is generated and it flows along the two tiny projections.

With this situation in mind, let us consider the cross sections of the tiny projections, A–A and B–B. First of all, let us look at the cross section A–A. At the oil inlet, the angle between the shaft and the microscopic projection is small. Because of this, a strong, hydrodynamic wedge effect is produced, causing the oil film to become thicker and increasing the amount of flow to the oil side. On the other hand, at the cross section B–B, the angle at the oil inlet is larger. Consequently, the wedge effect is small and the oil film does not get thick, resulting in a smaller amount of flow to the air side. Comparing the inlet flow and outlet flow here, the flow rate into the oil side is larger and achieves sealing. This leak prevention phenomenon has been reproduced and confirmed by an actual oil seal having the same characteristic values as the above model, and therefore the validity of this sliding surface model has been verified.

It is this phenomenon that creates a circulation of the oil flowing in and out of the sliding surface against the direction of the shaft rotation (V) when sealed, as shown in Figure 7.19. This circulation, which is promoted by the tiny projections, is the very factor that separates the lip sliding surface and the shaft and maintains a favorable fluid lubrication condition. This is in line with the phenomenon explained at the beginning and explains why the wear on the oil seals is limited. The series of discussions to this point have sufficiently explained why the newly designed oil seal suffers little wear and maintains its sealing effect for a long period of time. The author has confirmed the leak-proof phenomenon utilizing actual model seals. The validity of the Sliding Surface Model–Sealing Mechanism Analysis was verified against the results of actual vehicles and tests with a difference rate of 2% (Amasaka and Yamaji, 2006).

This clarified concept of Numerical Simulation by CAE—the Sliding Surface Model has been applied to the development design engineering of high-precision oil seals. That is to say, as a result of incorporating the Intelligence CAE Software, the minute roughness on the sliding surface has been controlled by regulating the composition of the materials. The next factor concerning the biased distribution of roughness toward the oil side can be interpreted as the bias of contact pressure distribution toward the oil side. Therefore, this factor has been controlled by shape designing technology used for designing the seal lip.

The result obtained from incorporating CG Navigation and Intelligence CAE Software for Oil Leak Analysis has helped to identify and refine the high-precision sealing mechanism of oil seals. Furthermore, the study conducted by the author has established this as a predictive engineering method for the functional design of oil seal parts.

7.2.4.5 Design Changes and Process Control for Improving Reliability

From the comprehensive knowledge gained in Sections 7.2.4.3 and 7.2.4.4, the following facts were learned: (1) The results of the Weibull analysis and visualization tests showed that some gears in the transaxle units had low surface hardness, and that there was a lot of wear during meshing (causing the generation of minute metal particles) leading to an unusually short operational life. It was recognized that it was necessary to prolong the life of these gears. (2) The study confirmed that there was considerable variation in the oil seal lip rubber hardness and this also had to be controlled. Consequently, Amasaka (2003a, 2007e) carried out the following improvements in order to ensure high quality assurance for the transaxle.

1. At advanced car manufacturer A, (i) improvement in wear resistance was achieved by increasing the gear surface hardness through changes to the gear material and heat treatment. Furthermore, (ii) for transaxles, improvements

in the roundness and surface smoothness of the driveshaft (resulting in the reduction of metal particles caused by the wear of gears in the differential case) were achieved.
2. At supplier B (iii) the mean value of the oil seal lip rubber hardness was increased and the specification allowance range narrowed. This, in combination with improvements in oil seal lip production technology (including in the rubber compound mixing process to suppress deviation between production lots), led to improved process capability. In addition, (iv) the higher coaxial centers of metal oil seal housings, the alignment of coil springs and seal lips, the contact width of the oil seal lips, and the thread profile identified during the design modifications were properly monitored and controlled during the production process to ensure high quality of the oil seals.
3. Furthermore, in order to control the generation of foreign matter, the new information that the oil starts leaking when fine metal particles of approximately 75 μm in size are present (caused by the yarn dust from gloves during work, rubbish, powder dust, etc.) was publicized. Based on this new knowledge about when the oil leaks start, both advanced car manufacturer A and supplier B promoted work improvements at their production sites and reduced the cases of early faults in the market.

Due to these comprehensive reliability improvements, the B10 life was increased to >400,000 km. As a result, the cumulative number of market claims per produc-

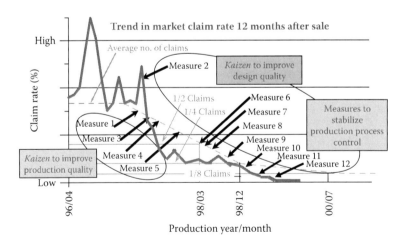

Figure 7.20 Effectiveness of market claim rate reduction.

tion month was reduced to <1/20th of the previous level and the desired effect was achieved, as shown in Figure 7.20 (Amasaka, 2008).

7.2.5 Conclusion

With a view to helping corporations survive worldwide quality competition, the author brought about reform of the development design business process. This reform entailed the change from conventional development methods that use experimental evaluation based on actual vehicles and tests to predictive evaluation based on development methods that use highly reliable CAE analysis. Given this background, the author has proposed the Strategic Development Design CAE Model employing TDS.

In order to demonstrate its effectiveness, the author devised the Intelligence CAE Software–Oil Leakage Simulator, which incorporated CG Navigation for the purpose of preventing oil seal leakage. This has contributed to a remarkable reduction in market claims regarding this problem, and a substantial result has been achieved in the field of ensuring high reliability assurance at advanced car manufacturer A and supplier B.

7.3 Process Layout CAE System for Intelligence TPS

7.3.1 Introduction

Recent Japanese enterprises have been promoting global production to realize uniform quality worldwide and production at optimal locations for survival through severe competition. Based on the trend and the production environment surrounding those manufacturing enterprises, the author proposes New JIT as a new management technology principle for manufacturing in the twenty-first century). The important mission of New JIT with TDS, TPS, and TMS is to achieve successful global production and high quality assurance and productivity, as shown in Figure 5.1 (refer to Section 5.1).

Nowadays, the operational procedure at production workshops has been transformed by digital engineering and computer simulation. The author has established strategic manufacturing technology as the key to Intelligence TPS for the Lean system. So far, the author has established strategic manufacturing technology for the Process Layout CAE System using Process Layout Analysis Simulation (TPS-LAS) for the establishment of Intelligence TPS (Sakai and Amasaka, 2006b, 2008). TPS-LAS contains the three-core system with logistics investigation simulation (LIS), digital factory simulation (DFS), and workability investigation simulation (WIS). The author has implemented the established TPS-LAS at advanced car manufacturer A and achieved the expected results. They have also made preparations for global production strategy.

7.3.2 Proposal of the Process Layout CAE System Using TPS-LAS

In the future, reform of production workshops by transforming production processes and facilities, and evolution of the TPS for improved quality assurance utilizing digital engineering will be required.

7.3.2.1 Necessity of Digital Engineering and Computer Simulation in the Manufacture of Motor Vehicles

Digital engineering is generally interpreted as a process that uses information technology to handle three-dimensional data. This technology is used as a tool for concurrently innovating processes by visualizing essential problems latent in conventional systems and processes and then simplifying these and turning them into efficient systems and processes (Sackett and Williams, 2003). Here, the author describes how digital engineering is applied in the case of advanced car manufacturer A.

Firstly, in 1996, digital engineering was applied to the investigation of vehicles, products, and production facilities, contributing to epoch-making reductions in lead time, from product development to the production preparation processes. Next, in 1998, it was applied to vehicle development to realize a computer-supported process that enabled problems to be detected and solved on the drawing board, without using a prototype. In the case of production facilities, digital engineering has contributed to cost reductions in manufacturing stamping dies and realization of a lean production system for new body welding lines, and so on.

Finally, since 2000, in order to succeed in global production, advanced car manufacturer A has put those technologies into practice more by combining and integrating them.

When implementing the application, the following technical tasks require solutions:

1. Making information easily understandable visually
2. Sharing information in real time
3. Turning implicit knowledge into explicit knowledge (properly feeding high-level skills in manufacturing back to the production design) and so on.

On the other hand, computer simulation is not applied practically. For instance, concerning *kanban* especially, it consists of two types (ordering quantity model and ordering point of time model) and has several kinds of systems (pull system, push system, SLAM II, CONWIP) and references (Kimura and Terada, 1981; Spearman et al., 1990; Spearman, 1992; Hiraki et al., 1992; Tayur, 1993; Takahashi et al., 1994).

However, these are not applied because the actual line processes do not have the ideal availability and reliability. Therefore, it is essential to make a practical model for global production.

7.3.2.2 Concept of Process Layout CAE System Using TPS-LAS

The author proposes the Process Layout CAE System using TPS-LAS as a new principle that will contribute to the establishment of Intelligence TPS (based on reformation of production processes for high quality assurance and shortening of production preparation) by implementing and integrating digital engineering and computer simulation that consist of the three elements (i–iii) shown in Figure 7.21.

This would realize the addition of fundamental requirements, such as renewing production management systems appropriate for digitized production; establishing highly reliable production systems; creating attractive workshop environments tailored to increasing the number of older and female workers; reforming the work environment for enhancement of intelligent productivity; developing intelligent operators (skill level improvement); and establishing an intelligent production operating system.

Figure 7.21 shows the TPS-LAS using a Process Layout CAE System for Intelligence TPS (Sakai and Amasaka, 2006b, 2008). The authors have successfully proved the three-core system consisting of (i) logistics investigation simulation (LIS) (Leo et al., 2004), (ii) digital factory simulation (DFS) (Byoungkyu et al., 2004; Kesavadas and Ernzer, 2003), and (iii) workability investigation simulation (WIS) (Strassburger et al., 2003). The three-core system is explained in the following subsections.

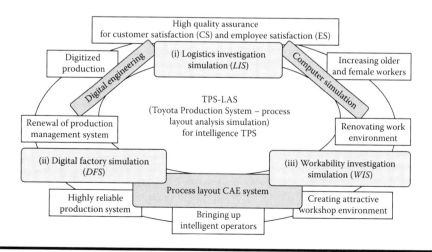

Figure 7.21 TPS-LAS using Process Layout CAE System.

7.3.2.2.1 Logistics Investigation Simulation (LIS)

A failure such as an empty buffer number may have been occurred, since buffer or transfer number between production processes in an automobile production line is determined by individual knowledge and know-how of production engineering personnel. For example, line stops resulting from shortage of transfer career may lessen the line productivity. Therefore, an optimum design for buffer number is necessary in designing a production system. Although buffer number is calculated on a theoretical basis, equipment defects will cause a line stop easily in theoretical design. Therefore, it has less meaning than expected in an actual case such as this.

The author has developed logistics investigation simulation (LIS). In practical terms, an optimum buffer number is simulated by the computer simulation under the conditions of an actual production situation (according to the equipment defect occurrence ratio or interval).

7.3.2.2.2 Digital Factory Simulation

When production facilities are placed in production processes, various repetitions occur. For instance, the equipment may not be set due to the narrow gap between the equipment and the ceiling or due to the interference toward piping and so on. In addition, not only these problems of static interference but also more dynamic ones may occur. When all equipment has operated through the designated cycle for the first time, any problems of dynamic interference between the equipment will be found. For minimum repetition, it is necessary to confirm not only static interference but also dynamic interference at the earliest stage possible.

Digital factory simulation (DFS) is a system for simulating the dynamic interference; three-dimensional data of production facilities are set up temporarily in the virtual process line using digital engineering. When the designated operation is simulated in a computer simulation, various interferences can be found and the production processes can be redesigned in advance.

7.3.2.2.3 Workability Investigation Simulation

In some manufacturing plants, there are some operators who are not available to work in actual production processes due to a mismatch between their working image and their work environment. It is necessary to inform them about the actual work environment (operational procedure, working posture, etc.) and the workload with quantitative analysis before they are hired.

In the case of taking parts of factory products away from an operational table, this system—workability investigation simulation (WIS)—will describe the posture of the operator with three-dimensional data on the equipment and operators in a virtual scene by using digital engineering. It will simultaneously calculate the Toyota Verification Assembly Load (TVAL) for quantitative evaluation of ergonomic workload (Amasaka and Sakai, 1998; Amasaka, 2007a).

This paper focuses especially on LIS and outlines the results of applying the Process Layout CAE System to several applications in order to improve productivity. To be specific, the process layout was designed through a virtual scene on a computer, contributing to epoch-making reductions in lead time, from product development to production preparation processes. The author has implemented the established TPS-LAS model at Toyota and achieved the expected results while also making preparations for implementing the model in a global production strategy of advanced car manufacturer A. The process layout CAE system is roughly explained in Section 7.3.2.3.

7.3.2.3 Development of TPS-LAS

With recent developments in automation and technology, production systems have become extremely large and complicated. This has led to increased items to address when carrying out process planning and line drawing, for example, transfer routes, buffering numbers, the number of transfer equipment, and so on. These are the parameters that compose the line specification and layout. Although buffer number is calculated by the simulation on a theoretical basis and an automobile production line is designed, equipment defects resulting from equipment trouble may cause line stops, and theoretical design is not significant for explaining an actual situation such as this. The conventional theoretical method has a standardized character that is not suitable for the actual conditions.

For this reason, the author developed TPS-LAS. Production processes based on practical dynamic conditions are confirmed with virtual scenes by using digital engineering and computer simulation. Solutions with high accuracy and the most suitable conditions for the actual circumstances should be selected, no matter how many trial numbers are required to support between theory and practice. Furthermore, there is also a need to plan production processes using concurrent engineering in order to reduce preparation lead time.

The functions and requirements of TPS-LAS, especially for the assembly line, are shown in Table 7.1.

Table 7.1 Functions and Requirements for TPS-LAS

Functions	Requirements
1. Quantative analysis of availability/ determination of optimum buffering number	a. Grasp the real situation of equipment defects and measurement according to the practical line
2. Measuring bottleneck process	b. Visualize the logistics situation in real time
3. Implementing concurrent engineering for process planning job	c. Combine conventional CAD/CAM systems and integrate them

The actual line process, which consists of many robots, a circulatory system of vehicle hangers, and the logistics route, is much too long. The necessary functions in the model are (1) quantitative analysis of availability according to practical production conditions and determination of the optimum buffering number between each subline process; (2) measurement of bottleneck processes, specifically, a surplus or shortage of vehicle hangers at the joint process; and (3) implementation of concurrent engineering for process planning to reduce lead time, from product development to production preparation processes.

According to these functions, this system is incorporated by (a) grasping the real situation of equipment defects, depending on the defect model, by using reliability technology; (b) visualizing the logistics situation in real time; and (c) combining conventional CAD/CAM systems and integrating them. The authors, therefore, have perceived the necessity of these functions. They have devised the knowledge and key to reliability technology (Amasaka and Sakai, 1996, 1998) concerning (1). The remainder of this chapter focuses particularly on (2) and explains the visualization of the bottleneck process (Sakai and Amasaka, 2007a, 2008).

As a system of defect prevention and of defect recurrence prevention by quick pursuit of the causes, the availability and reliability of information administration systems in domestic and overseas plants have systematically been networked as shown in Figure 7.22a to ensure high reliability and maintainability using digital engineering. Furthermore, as shown in Figure 7.22b, equipment reliability and maintainability are improved by collecting the production line operation information through Andon control devices into the defect control monitor for real-time Weibull analysis according to the Inline-Online SQC methodology (Amasaka and Sakai, 1998).

The remainder of this study focuses particularly on (2) and explains the visualization of the bottleneck process. In order to summarize this idea, the authors have revealed that the key element is the relation model between availability and buffer number (Amasaka and Sakai, 2009).

First of all, the availability and reliability of the information administration system makes it possible for it to collect the number of equipment defects occurring and to output the operational availability according to each production process. While the problem occurrence rate of the operation (i.e., failure ratio) is set up randomly, the production number and simulate the buffer number will be calculated timely. Finally, the authors determine the optimum buffer number due to the calculation of the maximum buffer number.

As the same layout database is shared between the conventional layout CAD system and the new simulation system, process planning and line drawing in concurrent engineering are accomplished with certainty and efficiently. By parallel processing of all planning and drawing in the optional time period, productivity can be improved and efficient planning of production preparation processes can be achieved.

As well as (3), the system is integrated with all functions ((1) through (3)) by using information technology (IT).

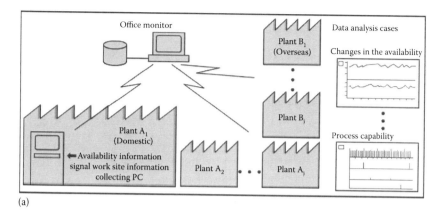

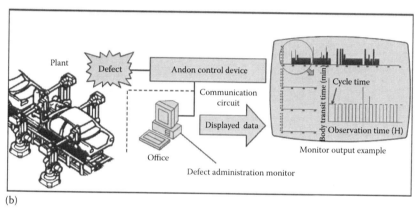

Figure 7.22 Networking of Availability and Reliability Information Administration System: (a) networking using digital engineering; (b) defect data collection method.

7.3.3 Relation Model between Availability and Buffer Number

The authors have revealed that the key element is the relation model between availability and buffer number. First, the authors consider the model of this system (Amasaka and Sakai, 1995, 1998): it has a forward process (failure ratio: $\lambda 1$), a backward process (failure ratio: $\lambda 2$), and a buffer B between the processes, as shown in Figure 7.23.

Figure 7.24 clarifies the change of the necessary buffer number according to the operational availability; in other words, the equipment failure. The horizontal line shows the time. As for the vertical line, from the top, the first four layers explain the equipment failure occurrences and the rest explain the necessary buffer number.

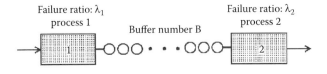

Figure 7.23 Basic system that consists of line processes.

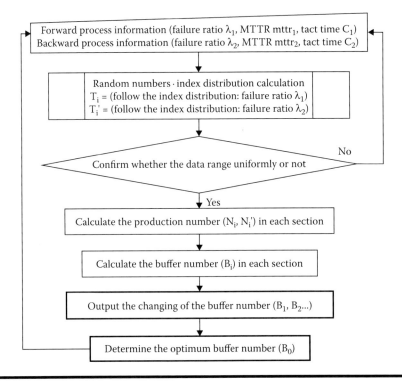

Figure 7.24 Changing of buffer number.

This means the calculation of the maximum buffer number B_0, while the failure stoppage time of each process is $F_1, F_2, \ldots, F_1', F_2', \ldots$ and the initial time between the failures of each one is $T_1, T_2, \ldots, T_1', T_2', \ldots$.

Figure 7.25 shows the flowchart for calculating the optimized buffer number. While the problem occurrence rate of the operation (i.e., failure ratio) is set up at a random time, the production number and simulated buffer number will be calculated exactly.

Finally, the authors have determined a concrete algorithm for calculating the optimum buffer number.

The buffer number B_i up to N_i ultimately calculates B_0 (initial buffer number), which forms the following formula (Equation 7.1).

$$B_i = N_i - N_i' + B_{(i-1)} > 0 \quad (i = 1, 2, 3 \ldots) \tag{7.1}$$

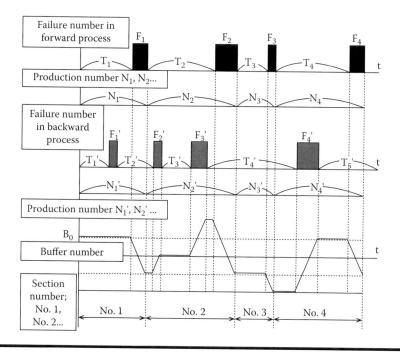

Figure 7.25 Flowchart for calculation of optimized buffer number.

Table 7.2 Production Number in Each Section

	No. 1	No. 2	No. 3	No. 4
N_i	T_1/C_1	T_2/C_1	T_3/C_1	T_4/C_1
N_i'	$\dfrac{T_1+F_1-F_1'}{C_2}$	$\dfrac{T_2+F_1-F_2'-F_3'}{C_2}$	$\dfrac{T_3+F_3}{C_2}$	$\dfrac{T_4+F_4-F_4'}{C_2}$

N_i and N_i' are the production numbers of the forward and backward processes; they are leaded from Table 7.2. C_1, C_2 are the tact times of each individual process.

7.3.4 TPS-LAS Application

7.3.4.1 Result of TPS-LAS Application at Advanced Car Manufacturer A

Based on the actual results of TPS-LAS applications, an examination proved the efficiency of a main body transfer route at a domestic plant. Figure 7.26 shows an example of this route.

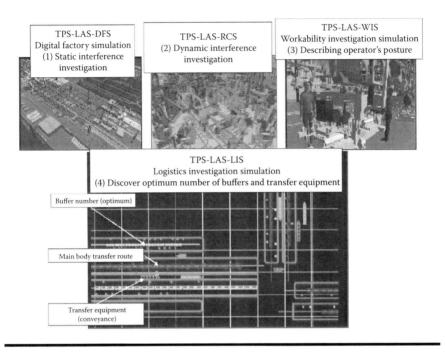

Figure 7.26 Result of simulation for the main body transfer route.

The necessary production machinery is modeled, and a hypothetical production line is setup within a "digital factory" on a computer.

1. TPS-LAS-DFS is then used to reproduce the flow of people and parts within the production site. Transfer equipment or store racks are set up temporarily in the virtual process line using three-dimensional data provided by the TPS-LAS digital factory simulation (DFS). The optimum location for store racks for each process and the transfer route can be determined.
2. TPS-LAS-RCS for the optimum placement of welding robots for the mainbody to ensure that no interference occurs. Simulating the dynamic interference using TPS-LAS-RCS, three-dimensional data of all robots and their attachments are installed into the virtual production process. While the designated operating work is simulated, various interferences between the welding robots or attachments and so on can be discovered, their arrangement can be verified, and the production process can be designed at an earlier stage (Taj and Berro, 2006).
3. Advance verification is performed using TPS-LAS workability investigation simulation (WIS) to ensure the predetermined cycle time with no waste (muda) or overburburdening (muvi). The posture of the operator is described by the TPS-LAS-WIS and verified TVAL for quantitative evaluation of ergonomic work load as well as the workability simultaneously. If the workload is higher than the designated level, the equipment will be redesigned or the

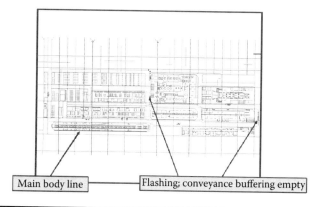

Figure 7.27 TPS-LAS-LIS: optimum buffer number (conveyance) simulation.

operator's standing position will be changed. Furthermore, each standardized work is proved within the limited production cycle time. Temporary workers are informed of the actual work environment (operational procedure, working posture, etc.) and the workload with quantitative analysis in advance of hiring them; this is part of their contracts.

4. The key topic of this section: logistics is investigated by conducting a TPS-LAS-LIS logistics investigation simulation. Similar production lines are searched for and an available and reliable information administration system is applied. The number of equipment defects was collected and the accidental failure mode was confirmed. Operational availability was output according to each production process. While the operational availability is set up randomly, the production number and buffer number are calculated exactly. When the buffer number is empty, relating the shortage of buffers or collisions of transfer equipment, they flash in real time.

This enables us to predict causes of overtime work and declines in the availability of the assembly line. Therefore, by applying this latest method, the author was able to discover the optimum number of buffers and transfer equipment and develop an optimum transfer route, as shown in Figure 7.27.

These process plans and line drawings in concurrent engineering are accomplished certainly and efficiently by these simulations in advance of planning of production preparation processes.

7.3.4.2 Effectiveness of TPS-LAS Application

The first core element, TPS-LAS-DFS, enabled the author to accomplish an improvement in availability on the assembly line of 8% on average. By finding various interferences between the facilities, verifying their arrangement, and redesigning the production process at an early stage, repeated works were reduced by half.

The second TPS-LAS-WIS enabled the author to verify the workload and the workability simultaneously in advance. The facilities were redesigned. The operator's standing position was changed, and a 10% improvement in line availability was achieved. Furthermore, the ratio of temporary workers voluntarily resigning could be reduced by providing them with a virtual scene of their work environment and workload in advance, before they are hired.

Regarding overtime work issues, the third TPS-LAS-LIS enabled prediction of causes of overtime work and declines in the availability of the assembly line in process planning and line drawing stages relating the concurrent engineering. Therefore, the author was able to find an optimum number of buffers and transfer equipment and develop an optimum transfer route. As a result, they have reduced the amount of instances of one hour of overtime work.

New JIT enabled the author to achieve the same high productivity and quality at both domestic and overseas plants and reduce lead time and overseas labor force support in global production processes throughout the world.

The results of this section have been developed in the global production strategy of advanced car manufacturer A, with its effectiveness verified by the excellent reputation of recent Toyota vehicles for their reliability and common workability in Europe and the United States (JD Power and Associates, 2004).

7.3.5 Conclusion

In the environment of worldwide quality competition, establishing a new management technology that contributes to business processes is an urgent goal being pursued in the global manufacturing industry. In this report, the author perceived the need to develop Intelligent TPS for the Lean system as the key to the development of the global strategy of advanced car manufacturer A and presented the strategic production technology. In the course of its implementation, the author also created the Process Layout CAE System using TPS-LAS as a strategic production quality management model, which consists of (LIS), (DFS), and (WIS), and verified that it will contribute to the renovation of management technologies when made organic.

7.4 Robot Reliability Design and Improvement Method Utilizing TPS

7.4.1 Introduction

To survive amid fierce competition, Japanese enterprises have recently been promoting global production, to achieve uniform quality worldwide, and increased production at optimal locations. This trend and a particular production environment have led Amasaka (2002) to propose TPS as a core principle of New JIT for manufacturing in the twenty-first century.

This section concentrates on the strategic application of TPS and the creation of a highly reliable production system based on it (Amasaka and Sakai, 2009). Nowadays, the automobile industry uses many robots, and a practical scientific approach of reliability-enhancing activities is the key to the realization of a highly reliable production system with faster-operating production robots.

The reliability of robots is improved not only by hardware designed to be more reliable on the supplier side (Craig, 1988), but also by collecting and analyzing users' observations on the production line over a long period of time (Kitano, 1996; Atkeson and McIntyre, 1985; Atkeson et al., 1986).

Sakai and Amasaka (2007a) established a strategic manufacturing technology for robots called Robot Reliability Design and Improvement Method (RRD-IM), a method for robot reliability-enhancing activities. RRD-IM contains three core methods of evaluating robot reliability: initial failure (IF), random failure (RF), and wear failure (WF). The authors implemented RRD-IM at advanced car manufacturer A and achieved the anticipated results. Then, based on these three failure modes, they implemented the reliability design to ensure better line availability.

The authors also suggested a methodology to increase reliability by defining the target value of mean time between failure (MTBF) in preparation for a global production strategy. They simultaneously achieved a QCD program enabling manufacturers to offer high product assurance.

7.4.2 Evolution of the TPS

In recent years, the technical manufacturing capability in the Japanese manufacturing industry has declined as a result of production transfers to overseas locations. Yet high quality assurance at parts per million (PPM) level is needed in domestic plants, and the following requirements have emerged:

1. Worldwide quality assurance
2. Uniform productivity all over the world

7.4.2.1 TPS as the Key to Success in Global Production

Amasaka (2002) addressed these two requirements and organically integrated four core technologies to develop TPS as an effective manufacturing technology for global production (Figure 5.3, refer to Section 5.1). The mission of TPS is to achieve epoch-making improvements in productivity, cost, quality, and workability. The four core elements are (a) production by information, (b) production by management, (c) production by technology, and (d) production by partnership.

In the future, the reform of production workshops by the production facilities' administration, ideal maintenance, and the advancement of TPS for high quality assurance and a highly reliable production system will be required.

7.4.2.2 Need for a Scientific Approach

The requirements for a highly reliable production system are (1) fast repair of facilities in the event of a breakdown (maintainability) and (2) speed of production (productivity) (Kelly, 1997). Regarding maintainability, the primary objective of maintenance is to provide quality products promptly (Wang and Marvin, 2000). For this, it is essential to improve the administration of the production facilities to enhance availability (Smith, 1993; Modarres et al., 1999).

Amasaka and Sakai (1996, 1998) therefore structured and developed Availability Reliability Information Monitor-Body Line (ARIM-BL) at advanced car manufacturer A, with excellent results. ARIM-BL consists of two systems: (a) a body measurement system and (b) a line information administration system, which allows data to be accessed in real time from the facility.

Regarding the first issue, maintenance instructions are one way of conveying maintenance skills, but their written nature gives them disadvantages such as complexity and the difficulty in finding specific information quickly. It is crucial to establish a new universal means of transmitting skills easily. Using database technology and CG, the authors have developed new and more general approaches.

The second issue, productivity, benefits from reliability-enhancing activities that improve the operation rate of production robots. Therefore, the rest of this chapter contains a proposal for a model for robot reliability-enhancing activities.

7.4.3 RRD-IM as the Key to TPS

In order to develop a highly reliable production system for global purposes, the author proposes the RRD-IM system shown in Figure 7.28.

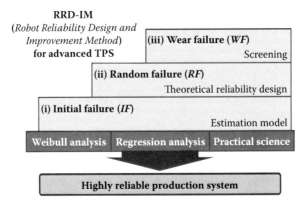

Figure 7.28 Robot Reliability Design and Improvement Method (RRD-IM).

7.4.3.1 Concept and Structural Elements

This incorporates Weibull analysis, multiple linear regression analysis, and practical science, and is composed of three main elements corresponding to the three failure modes, as shown in Figure 7.28. The model proceeds through the following stages: (i) initial failure (IF), (ii) random failure (RF), and (iii) wear failure (WF). There are three basic countermeasures for the failure forms, as follows:

1. Eliminate the initial failure factors by screening before placing into line operation.
2. Evaluate the MTBF for random failures and theoretically predict the relation between the number of units to be introduced to the line and the availability of the production line. When the production line availability requirement is not satisfied, revisit the hardware design of the robot to improve the MTBF.
3. To address wear-out failure, accurately specify the worn-out parts and the expected lifetime of the parts. Carry out part replacement and robot overhauls as preventative measures.

To detect defective elements that could lead to initial failure, the author generally recommends screening via heat-accelerated testing.

Using a theoretical examination of the MTBFs in the random failure mode, the number of units to be introduced, and availability and an investigation of the validity of MTBF of 3000 h, we implement a design review on the use of the elements if necessary. The author proposes a more accurate degradation model that focuses on gear wear in the wear-out failure mode and construct the part-life estimation system based on this model.

7.4.3.1.1 Initial Failure

In screening for the defective elements, we must first find appropriate conditions for the acceleration test on the printed circuit board in the robot. Screening is difficult under gentle conditions, but, on the other hand, even normal elements are destroyed under excessively severe conditions. We therefore carried out an accelerated life test to find the conditions under which failure forms do not change.

Taking $T = 20°C$ as the environmental temperature for initial failure, a Weibull form modulus $m = 0.58$ is obtained as a result of Weibull analysis of the holding time and accumulated failure rate $F(t)$. Heating to $T = 60°C$ shows that the form modulus stays at $m = 0.6$ up to the accumulated failure rate of 70%, demonstrating that since the m value is almost constant, acceleration is possible. However, the gradient of the line gradually increases after $F(t) = 70\%$, which indicates the possibility of a change in the failure form. We therefore conclude that the time showing the

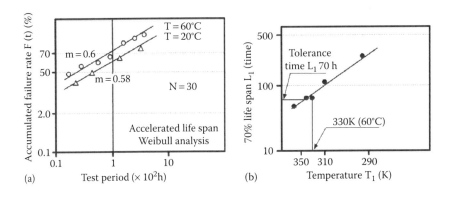

Figure 7.29 (a) Accelerated life via high temperature; (b) Arrhenius plot between temperature and holding time.

accumulated failure rate of 70% is the life span of the defective parts in the initial failure form. The results mentioned above are shown in Figure 7.29a.

Next, to find the proper heating and holding time, the author made an Arrhenius plot (Shiomi, 1970) of life span against temperature change. The results, shown in Figure 7.29b, reveal the life span. This allows us to determine the tolerance time (L_1) in the acceleration test for temperature T_1, which is determined by the heating equipment.

7.4.3.1.2 Random Failure

7.4.3.1.2.1 Reliability Design for the Robot Line
— In the random failure mode, robots have a constant failure rate regardless of their operating time. Hence, it is possible to theoretically predict the availability of a production line using multiple robots. First, we implemented the theoretical reliability design for the line, using robots to define the MTBF. Then, the author tried to find the countermeasures through the comparison body assembly line adopting the series model (Matsubara, 1981) where multiple units of robots are arranged, as shown in Figure 7.30. The line availability is therefore decided by the number of robots introduced.

Where the number of robots is N, the MTBF of a robot is t, the failure recovery time of a robot is R, the total number of robots in failure is n, and the line operating time is T, the robot availability A will be calculated as follows:

$$A = \frac{(T - n \times R)}{T} \tag{7.2}$$

where the average number of failures in the line is n.

$$m = \frac{N \times T}{t} \tag{7.3}$$

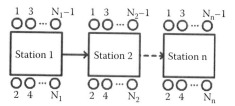

Figure 7.30 Body assembly model line.

The number of failed robots n in random failure mode follows the Poisson distribution (Nakamura et al., 1997). P is the probability of n failed robots:

$$P = \frac{m^n \times e^{-m}}{n!} \tag{7.4}$$

The line availability A is secured if the failure number is less than n, so the condition to secure A with a significance level α is calculated as follows:

$$\sum_{0}^{n} P(n) < 1 - \alpha \tag{7.5}$$

The expression (7.5) can be changed by using a chi-squared distribution as follows:

$$m = \frac{\chi_\varphi^2 (1-\alpha)}{2} \tag{7.6}$$

$$N \times T = \frac{1}{2t} \times \chi_{2(T(1-A)/r+1)}^2 (1-\alpha) \tag{7.7}$$

where χ is the chi-squared distribution and φ is the degree of freedom, $\varphi = 2(n+1)$. Furthermore, the expression (7.7) can be changed as follows:

$$t = \frac{2N \times T}{\chi_{2(T(1-A)/r+1)}^2} (1-\alpha) \tag{7.8}$$

7.4.3.1.2.2 Reliability Improvement for Robots — A design review is necessary for some defective elements that appear in the random failure mode. It is well known that the failure rate of a system of elements in random failure mode is the sum of the failure rate of each element (Taki, 1994). Hence, a design review is possible by finding the failure rate and the number of elements used for the robot controller (Valter, 1993; Balestrino et al., 1983; Koivo and Guo, 1983; Slotine and Li, 1987).

7.4.3.1.3 Wear Failure

As mentioned above, wear failure in robots is caused by gear wear. The author recommends periodic overhaul as the appropriate preventative measure. Nevertheless, failures still occur during line operation, and it is necessary to make a life estimation based on deterioration due to wear, to implement an effective overhaul regime. The author has therefore decided to investigate the wear phenomenon and propose a life estimation based on the deterioration model.

The author implemented an acceleration test under oil-less conditions to investigate the wear phenomenon on the robot gear, as shown in Figure 7.31. This shows that the wear developed rapidly after 200 h of operation, and we also found the carbonized layer did not remain. A ferrograph revealed a flake structure in the wear powder, confirming that wear was caused by adhesive wear.

It is well known that adhesive wear can be found by the expression (7.9), where the load on the gear tooth is P, the slippage of the tooth face is L, and the amount of wear is Δ:

$$\Delta = k \times P \times L \tag{7.9}$$

If the wear depth of the gear is δ and the engaging time is S, expression (7.9) can be changed as follows:

$$\delta = a \times \ln(P \times S) + b \tag{7.10}$$

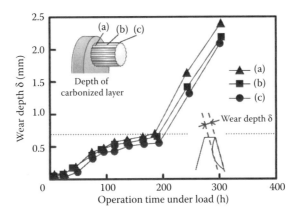

Figure 7.31 Checking wear under no lubrication conditions.

"a" and "b" in the above expression are coefficients that vary depending on the lubrication condition (Nakamura et al., 1997).

7.4.4 RRD-IM Application Example

In the robot system used on the production line shown in Figure 7.32, the processing tool is mounted on the main body of the control unit. The processing tool in the body assembly line is surveyed to evaluate the reliability of the facility. A line that has a high maintenance coefficient has difficulty achieving a high line availability (Shiomi, 1970).

The result of the evaluation shows that the robot body has an overwhelmingly high maintenance coefficient. Thus, the reliability of the production line depends on the reliability of the robot body.

7.4.4.1 Practical Robot Failures

The author investigated the failure rate of the robot body in the 1980s over an 8-year period without overhaul since introduction and found the results showed a typical bathtub curve. The result after Weibull analysis of this curve is shown in Figure 7.33a and indicates that the initial failure period ($m = 0.6$) is 500 h, and the random failure period ($m = 0.97$) comes after that time, although this is not always

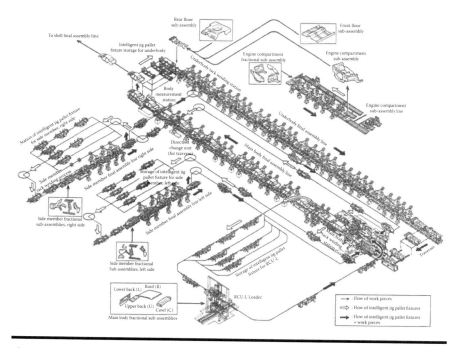

Figure 7.32 Overview of body assembly line utilizing many robots.

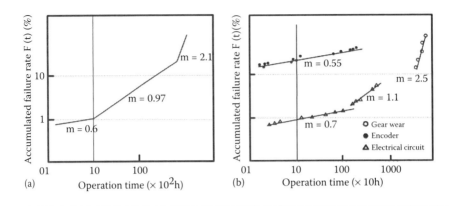

Figure 7.33 (a) Weibull analysis results for robots in the 1980s; (b) Weibull analysis results of robot defect factors.

clear because it is affected by the working conditions of the robots. Also, the MTBF at the random failure period is found to be 8000 h.

The author carried out a Weibull analysis for each robot defect factor. The results showed that the initial failure (m = 0.55, 0.7) factor was mostly defective elements such as an IC chip mounted in the print circuit board of the robot's controller or position detector. Defective elements are found in the random failure (m = 1.1) factors too. For this, the design factor will be estimated. The wear failure (m = 2.5) factor is gear wear, resulting from incomplete preventative maintenance. These results are shown in Figure 7.33b.

7.4.4.2 Result of RRD-IM Application in Body Assembly Line

7.4.4.2.1 Initial Failure

To assess the effectiveness of the countermeasures for initial failure and random failure, we compared the failure rates of conventional robots with new ones via the bathtub curve that we have used since the robot introduction in 1987. This proves that the initial failure in conventional robots that occurred at approximately 500 h has almost been solved, and the MTBF for the random failure period has satisfied the required 30,000 h.

7.4.4.2.2 Random Failure

7.4.4.2.2.1 Reliability Design for Line Robots — Where the operating time is 1 month T = 400 h, the significance level α = 0.05, the failure recovery time r = 1.2 h, and the required line availability A = 98%, the relationship between the number of robots N and the robot MTBF t is found from expression (7.8) and is plotted in Figure 7.34a.

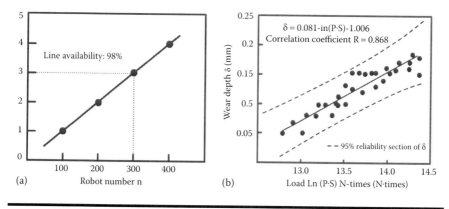

Figure 7.34 (a) Robot number and necessary MTBF; (b) linear line showing the relation between load and wear depth.

If the number of robots to be introduced to the body assembly line is 300 units, the graph shows that the required MTBF is 30,000 h, so it is necessary to improve the MTBF of 3000 h per robot by one order of magnitude.

7.4.4.2.2.2 Reliability Improvement for Robots — The author investigated the total failure rate of the elements used for the robot controller and decided to stop the use of elements with a high failure rate. At the same time, it was attempted to reduce the element number from 1100 to 770 by integration of the control circuit.

Furthermore, by studying the printed circuit boards with failed elements, it was found that at high frequency, metal parts evaporated and the failed element was destroyed and blackened by overvoltage.

The mechanism generating the overvoltage in the line was partially determined through fault tree analysis (FTA) and included wraparound of the surge voltage, defect in the self-discharge circuit, and mistaken operation (Takada, 1999), but the underlying cause was difficult to ascertain. So, after investigating the voltage margin for all elements, the author installed an electronic circuit using a Zener diode to protect against overvoltage.

7.4.4.2.3 Wear Failure

The depth of the gear wear δ can be estimated by using expression (7.10) as the deterioration model. When this value is over the depth of the carbonized layer, the wear failure can be predicted, making it possible to apply preventative maintenance measures.

To find the a and b coefficients in expression (7.10), the relationship of δ, P, and S is found by using a particular robot developed to improve lubrication by the

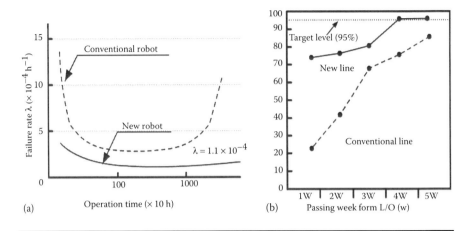

Figure 7.35 (a) Reliability achieved situation; (b) line availability tendency.

attachment of an automatic oil feeder. The relationship is plotted in Figure 7.35b and given by the empirical formula below:

$$\delta = 0.081 \times \ln(P \times S) - 1.006 \qquad (7.11)$$

This shows that the load exerted on the gear P and the engaging time S can be found by measuring the phase between the motor current and the position transducer of the robot. However, this estimation is made at the start of the operation. Recently, robot programming has often been carried out off-line. The off-line programming system simulates the motion of the robot using mechanical modeling of its kinematics and dynamics. Therefore the P and S loads on each tooth of the robot's gear are easy to predict.

The author decided to construct a system to estimate the life span using expression (7.11) during programming, to ascertain the overhaul time, and then use this for preventative maintenance after robot introduction (Walker and Orin, 1982).

7.4.4.3 Evaluation on the Actual Line

The author evaluated the robot on the line used to practice reliability design and improvement methodology discussed in Section 7.4.3.1.2.

To evaluate the countermeasures for initial failure and random failure, the failure rates for conventional robots and new ones were compared using the bathtub curve. This proves that the initial failure in conventional robots that occurred at approximately 500 h has almost been solved, and the MTBF for random failure has achieved the required 30,000 h.

In addition, we applied the overhaul time regime calculated from loads P and S for each robot, through the life-estimation system presented in Section 7.4.3.1.3. As a result, the wear failure has also been solved, as shown in Figure 7.35a.

Furthermore, we found that we could reduce the overhaul cost by half by changing from the conventional method of setting wear depth at 0.2 mm as the gear replacement standard to the new method of replacing only the gears predicted to reach their limit of use (carbonized layer thickness, generally 0.7 mm), when shifting the vehicle model year.

The change in the availability of the line that introduced 300 units of robots is shown in Figure 7.35b. This indicates that the target level for line availability, of 95%, was achieved successfully soon after starting the line operation.

7.4.5 Conclusion

Through long-term observation and analysis of robot failure in the vehicle body assembly line, the author achieved a large improvement in reliability. The author considers it significant that these activities have been driven by robot users who are involved in the management of the production line, rather than just by the robot vendor. The conclusions drawn from this study include:

1. A strategic manufacturing technology has been established for robots called Robot Reliability Design and Improvement Method (RRD-IM), a model for robot reliability-enhancing activities. RRD-IM contains the three core methods of initial failure (IF), random failure (RF), and wear failure (WF).
2. As a countermeasure for initial failure (RRD-IM-IF), how to find an effective screening procedure for the elements was demonstrated, and its effectiveness was confirmed.
3. As a countermeasure for random failure (RRD-IM-RF), it was shown theoretically that it is necessary to increase MTBF by one order of magnitude (to 30,000 h). This was done by installing a circuit to protect against overvoltage, by integration, and by adopting elements with low failure rates.
4. As a countermeasure for wear failure (RRD-IM-WF), the author found that gear wear is caused by adhesive wear and proposed a wear deterioration model and a life estimation system. Concerning the life estimation, the author determined the overhaul time through the off-line programming system when the robot was introduced. This reduced the overhaul cost by half, as well as improving preventative maintenance against wear failure.

According to the schedule, the author has fulfilled the simultaneous realization of QCD, which enables manufacturers to offer high product assurance. In the future, the author will systematize this technology into a professional development for manufacturing (Sakai and Amasaka, 2005, 2006a, 2007b).

7.5 Scientific Mixed Media Model Employing TMS

7.5.1 Introduction

Recently, Amasaka (2002) has touched on the development of New JIT, a new management technology principle, and its validity as a new management technology for twenty-first-century manufacturing (refer to Section 5.1). New JIT innovates business processes in each division, encompassing sales, development, and production, by utilizing a customer science that employs the Science SQC approach (Amasaka, 2003b, 2004b, 2005).

In light of recent changes in the marketing environment, the author believes it is now necessary to develop innovative business and sales activities that adequately take into account the changing characteristics of customers who are seeking to break free from convention. If they are to be successful in the future, those involved in global marketing must develop a marketing system of the highest quality.

A marketing management system needs to be established particularly so that business, sales, and service divisions, which are developing and designing appealing products and are also closest to customers, can organically learn customer tastes and desires by means of the continued application of objective data and scientific methodology. However, a system for applying scientific analytical methods to customer data has not yet been satisfactorily established. In some cases, its importance has not even been recognized.

In this section, therefore, the author discusses the validity of a TMS as a core principle of New JIT, which contributes to the construction of a Scientific Mixed Media Model (SMMM) (Amasaka et al., 2013) for boosting automobile dealer visits. A model that enables the sales, marketing, and service divisions nearest the customers to systematically identify their tastes and desires is critical. The aim is an evolution of market creation through innovative advertisements promoting dealer sales activities by utilizing the scientific approach of TMS (Amasaka, 2009, 2011).

To achieve this goal, the author presents SMMM, which takes the form of strategic marketing and has four core elements: Videos that Unite Customer Behavior and Manufacturer Design Intentions (VUCMIN), Customer Motion Picture–Flyer Design Method (CMP-FDM), Attention-Grabbing Train Car Advertisements (AGTCA), and Practical Method Using Optimization and Statistics for Direct Mail (PMOS-DM).

The effectiveness of SMMM using four core elements has been applied to a dealership representing advanced car manufacturer A, where its effectiveness was verified.

7.5.2 Need for a Marketing Strategy That Considers Market Trends

Today's marketing activities require more than just short-term strategies by the business and sales divisions. In a mass-consumption society, when the market was growing in an unchanging way, sales increases were achieved by means of simple mass marketing through huge corporate investments in advertising (Nikkei Business, 1999; Amasaka, 2005). However, after the collapse of the bubble economy, the competitive market environment changed drastically. Since then, companies that have implemented strategic marketing quickly and aggressively have been the only ones enjoying continued growth (Okada et al., 2001).

On close examination, it was determined that strategic marketing activities must be conducted as company-wide, core corporate management activities that involve interactions between each division inside and outside the company (Rayport and Jaworski, 2005). Therefore, a marketing management model needs to be established so that business, sales, and service divisions, which develop and design appealing products and are also closest to customers, can organizationally learn customer tastes and desires (Shimakawa et al., 2006). Specifically, pursuing improvements in product quality by means of the continued application of objective data and scientific methodology is increasingly important (Fitzsimmons and Fitzsimmons, 2004; Amasaka, 2005).

At present, the organizational system and rational methodology that allow them to analyze data on each customer using a scientific analysis approach have not yet been fully established in these divisions; in some cases, the importance of this system and methodology has not even been widely recognized (Niiya and Matsuoka, 2001; Lilien and Rangaswamy, 2003; Ikeo, 2006; Amasaka, 2007b).

7.5.3 Evolution of Market Creation Employing TMS

7.5.3.1 Significance of TMS: The Key to Application of New JIT

To create attractive, customer-oriented products that satisfy customers, the various divisions of a manufacturer must share a common language, ensuring unity and proper direction. This is necessary for all divisions, including business, sales, service, planning, development, design, production engineering, manufacturing, logistics, administration, and management.

Thus, Amasaka (2002, 2009) proposed New JIT with three core principles (TMS, TDS, and TPS) as a new management technology principle for manufacturer activities in the next generation, as shown in Figure 5.1. The aim of TMS is to promote market creation as shown in Figure 5.4 and to realize quality management through scientific marketing and sales, not by sticking to conventional concepts (refer to Section 5.1).

As shown in Figure 5.4, in order to realize market creation with an emphasis on the customer, TMS is composed of these technological elements: (a) market-creation activities through collection and utilization of customer information; (b) strengthening of merchandise power based on the understanding that products are supposed to retain their value; (c) establishment of marketing systems from the viewpoint of building bonds with customers; and (d) realization of the customer information network for customer satisfaction (CS), customer delight (CD), and customer retention (CR) elements needed for the corporate attitude (behavior norm) to enhance customer values.

7.5.3.2 Developing Customer Science Using Science SQC

Supplying products that satisfy consumers (customers) is the ultimate goal of companies that desire continuous growth. Customers generally evaluate existing products as good or poor, but they do not generally have concrete images of products they will desire in the future. For new product development, it is important to precisely understand the vague desires of customers. To achieve this goal, Amasaka (2002, 2005) proposed customer science to help systematize TMS, as shown in Figure 5.5 (refer to Sections 5.1 and 7.1).

To plan and provide customers with attractive products is the mission of companies and the basis of their existence. It is particularly important to convert customer opinion (implicit knowledge) to images (linguistic knowledge) through market-creation activities and to accurately reflect this knowledge in creating products (drawings, for example) using engineering language (explicit knowledge). This refers to the conceptual diagram that rationally objectifies subjective information (y) and subjectifies objective information ($\hat{y}$) through the application of correlation technology.

Amasaka (2003b) applied the statistical science methodology to Science SQC, which has four core principles (Scientific SQC, SQC Technical Methods, Integrated SQC Network (TTIS), and Management SQC) as shown in Figure 5.6 and which is designed to develop customer science in business and sales divisions to make changes to marketing process management (refer to Sections 5.1 and 7.1).

7.5.4 Constructing a Scientific Mixed Media Model for Boosting Automobile Dealer Visits

7.5.4.1 Publicity and Advertising as Automobile Sales Promotion Activities

For many years, automobile dealers have been employing various publicity and advertising strategies in cooperation with automobile manufacturers in order to encourage customers to visit their shops.

Figure 7.36 is a graphical representation of the relationship between publicity and advertising media—a relationship that helps draw customer traffic to dealers (Amasaka, 2009). Area A represents multimedia advertising (the Internet, CD-ROMs, etc.), Area B represents direct advertising (catalogs, direct mail [DM], handbills [directly handed to customers], telephone calls, etc.), and Area C represents mass media advertising (TV and radio broadcasting, flyers, public transportation [train cars], newspapers, magazines, etc.).

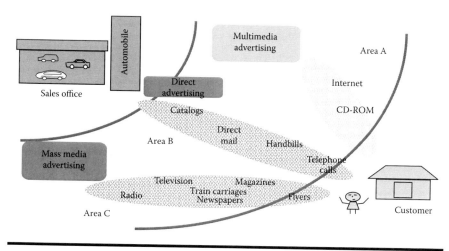

Figure 7.36 Graphical representation of customer motives for visiting automobile dealers.

There appear to be few cases where scientific research methods have been applied to the effect of mixed media (Areas A–C) and used to study the ways in which such sales activities actually draw customers to automobile dealers (Kubomura and Murata, 1969). However, the rational effects of the media mix are insufficient as advertisement methods, and the author therefore considers the need to scientifically promote a new advertisement mixed media model (Melewar and Smith, 2003; Amasaka, 2007b; Smith, 2009; Ogura et al., 2012).

7.5.4.2 Proposal of Scientific Mixed Media Model for Boosting Automobile Dealer Visits

As part of an organization's market-creation activities, it is important to gain a quantitative understanding of the effect of publicity and advertising, which are the principal methods involved in sales promotion and order taking, in order to aid the development of future business and sales strategies (Kobayashi and Shimamura, 1997; Kishi et al., 2000; Shimizu, 2004; Ferrell and Hartline, 2005; Amasaka, 2007b, 2009, 2011).

After recent changes in the marketing environment, what is needed now is to develop innovative business and sales activities that are unconventional and correctly identify the characteristics of and changes in customer tastes. There has never been a greater need for careful attention and practice in customer contact, and in order to continuously offer an appealing and customer-oriented marketing strategy, it is important to evolve current market-creation activities to strengthen commercial viability and reform office/shop appearance and operations using the customer science approach. Therefore, the author wants to construct a Scientific

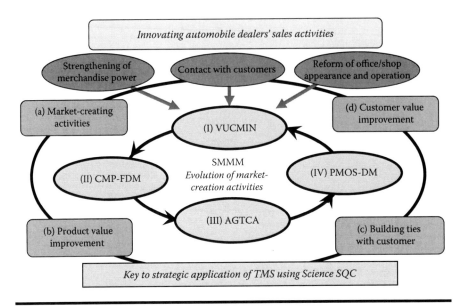

Figure 7.37 A Scientific Mixed Media Model for boosting automobile dealer visits.

Mixed Media Model (SMMM) for boosting automobile dealer visits employing TMS, as shown in Figure 7.37.

SMMM aims to achieve a high cycle rate for market-creation activities and is composed of four core elements: (I)–(IV). Core elements (I) and (III), Video that Unites Customer behavior and Manufacturer Design Intentions (VUCMIN) and Attention-Grabbing Train Car Advertisements (AGTCA), are for improving mass media and multimedia advertising (TV, train cars, radio, Internet, CD-ROM, etc.) through (b) product value improvement and (d) customer value improvement— and these are particularly important. These elements constitute the basis for the innovation of (a) market-creation activities and (c) building ties with customers.

At a certain stage in (a) market-creation activities and (c) building ties with customers, it becomes particularly important to develop (II) Customer Motion Picture–Flyer Design Method (CMP-FDM) and (IV) Practical Method using Optimization and Statistics for Direct Mail (PMOS-DM) to achieve a high cycle rate for improving mass media and direct advertising (flyers, magazines, catalogs, DM, handbills, etc.). These four core elements aim to provide an up-to-date inquiry as to deep-seated customer wants in terms of customer behavior analysis (Blackwell et al., 2006). Disparate behaviors by gender and age segment will be clarified in the analysis of standard behaviors.

SMMM for the evolution of market-creation activities improves innovative automobile dealer sales activity know-how regarding repeat users of various manufacturers' vehicles. Its characteristics are described below.

7.5.4.3 VUCMIN

The proposed VUCMIN uses video advertisements and is developed based on scientific approaches and analyses that focus on the standard behavioral movements of customers who visit dealers when choosing an automobile (Hifumi, 2006; Murat et al., 2008; Yamaji et al., 2010). This method, which is based on the different approaches identified in target customer profiles, aims to make customers more eager to visit automobile dealers. After creating this video advertisement, customers are verified as having a positive opinion toward visiting dealers with a plan to purchase the vehicle featured in the video.

More concretely, based on the research approach outlined in the previous chapter, the framework of VUCMIN was established as in Figure 7.38. In this figure, standard behaviors and disparate behaviors by gender are identified and classified. After classifying the subjects by age, the details of disparate behaviors are identified mainly in terms of the front seat of the vehicle (driver's seat features, passenger rearview mirror, etc.) and the rear seat (not shown in figure). This knowledge of customer behaviors and knowledge of the parts that product planning and designers wish to show to customers are taken into consideration as the basis for the VUCMIN model framework.

7.5.4.4 CMP-FDM

In this study, the author establishes a method of creating attractive flyer designs while using customer behavior analysis with videos that help dealers attract customers and aim to reform conventional marketing activities. Firstly, CMP-FDM analyzes how customers see flyers, and the author creates attractive designs that

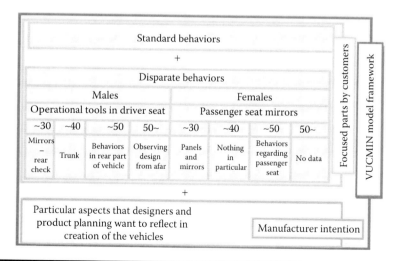

Figure 7.38 Frame of VUCMIN.

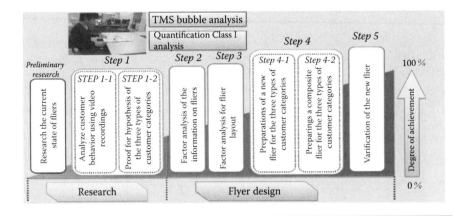

Figure 7.39 Steps of establishing CMP-FDM.

guarantee each customer's satisfaction. Next, the author integrates the design elements into one that will satisfy all types of customers (universal type) by organizing the design features (design elements) and then validating the method (Koyama et al., 2010).

More concretely, the author shows Steps 1–5 for establishing CMP-FDM, as shown in Figure 7.39. The proposed CMP-FDM utilizing SQC Technical Methods (Amasaka, 2003b) consists roughly of two processes: research and flyer design. In the first process, the author checks the current state of flyers as preliminary research.

In Step 1, the author analyzes customer behavior toward the flyers using video recordings to understand how the materials are actually viewed. Then, the author proves that customers can be classified into three types: active customers, collection-first customers, and indifferent customers.

In Step 2, to address problems with the current flyer design (information appearing on the flyer, such as exterior photos, price, car name, loan information, and interior photos), the author clarifies what each customer type wants to know from the content. One problem is the provision of a lot of unnecessary information and a lack of necessary information.

In Step 3, to solve problems in flyer layout, the author clarifies what kind of layout each customer group wants. One problem is typeface that is too small to see or information that is too varied to understand. In Step 4, based on the results of Steps 2 and 3, the author incorporates the design elements into one flyer that it is attractive to all customer types. In Step 5, the author conducts a survey to compare the composite flyer developed in Step 4.

7.5.4.5 AGTCA

This study deals with train car advertisements (hanging posters, above-window posters, and sticker ads), which have become increasingly popular in recent years.

Focusing on transit advertising, which has a good contact rate and provides long-term contact, the author decided to examine customer relationships and how they relate to train car advertising with the aim of defining the ideal format for this type of media. The goal was to first quantify the way passengers pay attention to train car advertisements and then propose the ideal form that in-car train advertising should take based on a visual representation of passenger information.

More concretely, the purpose of AGTCA is to examine the correlations between passenger information and riding conditions in train car advertising in order to discover the ideal way to advertise inside passenger trains, using the same research steps as CMP-FDM employing SQC Technical Methods (Ogura et al., 2013). These steps are as follows.

Step 1: Look at overall trends in passenger information using a cross-tabulation method that focuses on whether passenger attention turned to hanging posters, above-window posters, sticker advertisements, or others.

Step 2: Perform a cluster analysis on riding conditions and group the results using Quantification Theory Type III. Then, look at the relationship between (1) the riding conditions grouped in the cross-tabulation and (2) whether passenger attention turned to hanging posters, above-window posters, or sticker advertisements.

Step 3: Research the grouped riding condition data and basic passenger information to determine how they relate to attention rates established for the three types of advertising using a categorical automatic interaction detector (CAID) analysis (Murayama et al., 1982; Amasaka et al., 1998; Amasaka, 2011).

7.5.4.6 PMOS-DM

No clear processes are used at car dealers when deciding target customers for DM campaigns, and individual sales representatives tend to rely on their personal experience when making such decisions (Bult and Wansbeek, 2005; Jonker et al., 2006; Bell et al., 2006; Beco and Jagric, 2011). This means that dealer strategies lose their effectiveness and dealers fail to achieve the desired increase in customer visits.

Thus, for this study, the author established a practical method using PMOS-DM as a method of deciding the most suitable target customers for DM campaigns (Ishiguro and Amasaka, 2012). Specifically, in order to both clarify the dealer's target customer types and increase the number of customer visits, the author applied mathematical programming (combinatorial optimization) using statistics to establish a model for determining the most suitable target customers for DM campaigns.

More concretely, this model was created using the same research steps as CMP-FDM employing SQC Technical Methods: a three-pronged approach to resolving dealers' current problems with DM activities.

Step 1 is to increase the response rate, or the percentage of customers who visit the dealer as a result of receiving DM. To achieve this, the PMOS-DM uses statistical analysis to determine which customers are most likely to respond.

Step 2 is to reflect dealer aims in the recipient selection process. This is achieved by using a simulation driven by mathematical programming to optimize the selection of target customers (refer to Appendix 7.1). Finally, Step 3 is to clarify the recipient selection process by providing dealers with a model that outlines a specific approach.

Following an explicit model informed by statistics and mathematical programming keeps inconsistency among salespeople to a minimum. The three-pronged approach proposed in this study therefore provides a DM method that allows dealers to target their desired customer segment and boost response rates at the same time.

In short, the PMOS-DM uses statistics and mathematical programming to create an objective decision-making process that does not rely on the current selection methods used by salespeople, which are based on personal knowledge and experience and are therefore vague and implicit. At the same time, the model aims to boost the DM response rate in line with dealer targets.

7.5.5 Example Applications

This section validates the effectiveness of SMMM for effective advertising designed to bring customers into auto dealerships by the use of four core elements (VUCMIN, CMP-FDM, AGTCA, PMOS-DN) and a customer science approach to quantitatively assess the effectiveness of various advertising (mass media, direct advertising, and multimedia).

7.5.5.1 Visualizing Causal Relationships in Customer Purchase Behavior

SMMM was developed in order to draw more attention to the vehicle, spark interest in the vehicle, and make customers want to visit the dealer. In order to achieve the purpose of this research, a field survey on vehicle advertising was conducted to identify the core elements of each media type and to visualize the relationship between those elements and the media as well as the causal relationships between each media type and (a) vehicle awareness, (b) vehicle interest, and (c) desire to visit dealers (sales shops).

A survey was conducted in order to better understand the causal relationships among different types of media, media elements, and customer (consumer) purchase behavior. The advertising and marketing division at advanced car manufacturer A, Japanese Dealer Y of advanced car manufacturer A, and the Z market survey company helped to conduct an in-person survey on advertising and marketing by visiting male and female licensed drivers age 18 and older living in Tokyo, Fukuoka, and Sapporo. A total of 318 valid responses (197 male and 121 female, generally uniform age balance) were collected. The investigation period was the

5 months leading up to the release of the new Q model by advanced car manufacturer A.

Based on the author's existing research and knowledge (Amasaka, 2007b, 2009, 2011), they were able to identify media mix effects in each form of media using a purchasing action model, TV ads, radio ads, and newspaper ads (early June 2005), as well as Internet ads (early July) and train car (transportation) ads (mid-August) before the new car sale, and flyer and magazine ads (late August), DM ads (early September), and DH ads (mid-September) issued by Japanese Toyota Dealer Y. Participants were shown TV commercials and newspaper ads promoting Japanese Toyota's new Q car and then were asked questions inquiring about their purchase behavior and about the media and media elements.

The collected data were analyzed and the causal relationships between media, media elements, and consumer purchase behavior were outlined (Ogura et al., 2012). The questions that the author used in the survey are listed in Table 7.3. The questionnaire was multiple choice and asked respondents to describe their opinion (item (1) was yes–no, and the five-point scale in items (3) and (4) was converted into binary data).

Elements in each form of media were identified using multivariate analysis (cluster analysis, Quantification Theory Type III) and other statistical methods shown in Figure 7.40. As the example in the figure shows, the critical elements in terms of generating the distinct promotional outcomes that consumers expect are impact, contact frequency, newsworthiness, informativeness, and memorability (Ogura et al., 2013).

Figure 7.40 also positions TV ads, transportation ads, Internet ads, newspaper ads, and DM in order to identify the expected advertising effectiveness of each. Finally, the insights gained through scientific analysis are used to describe specific characteristics of the four core elements of the proposed SMMM (VUCMIN, AGTCA, CMP-FDM, and PMOS-D) so that more customers would be drawn to visit auto dealers.

Table 7.3 Survey Questions

(1)	Are you aware of the Japanese Toyota Q model?
(2)	What media did you see advertising this vehicle?
(3)	Are you interested in this vehicle?
(4)	Do you want to or did you actually visit a dealer to inquire about this car?
(5)	What kind of influence does each type of media have on you in terms of your attention, interest, and desire?
(6)	What was your impression of the advertisement?
(7)	Which advertising elements do you consider most important?

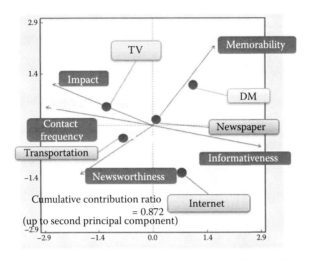

Figure 7.40 A scatter diagram with the principal component scores using Quantification Theory Type III.

7.5.5.2 Application of VUCMIN

In previous research, the author identified behavioral patterns of customers as they focus on the exterior of a vehicle. Insights gained during this research were used to explain the influence of product planning and designer intentions in VUCMIN creation (Yamaji et al., 2010). The Mark X was used as a target vehicle in design inquiries. According to common opinions from designers and product planning at advanced car manufacturer A, the parts that are given the most attention when the vehicle is shown to customers are the (i) front proportions, (ii) streamlined side proportions, (iii) tri-beam headlamps (lenses), (iv) widened console box, and (v) sharpened rear.

More concretely, the video created for the target profile of males in their 50s is explained using the timetable in Figure 7.41. The timetable shows video time as set to 90 s. Video shooting order is composed specifically of scenes from 1 to 11 starting from the front, driver seat, side, and rear of the vehicle. Scenes are (1) direct front scene, (2) diagonal front view scene, (3) driver side door opening scene, (4) entire driver seat view scene, (5) console box and shift lever scene, (6) steering wheel scene, (7) driver seat operational scene, (8) side view scene from driver's seat, (9) rear side view scene, (10) entire view of vehicle from rear, and, lastly, (11) moving from back to front, the entire view of the vehicle scene.

The scenes that form the VUCMIN video were composed on the basis of the standard and disparate behaviors of customers. Example photos representing these scenes (1–11) are shown in Figure 7.42. Using the same approach, a VUCMIN was created for each age and gender. In this section, customer surveys were executed in

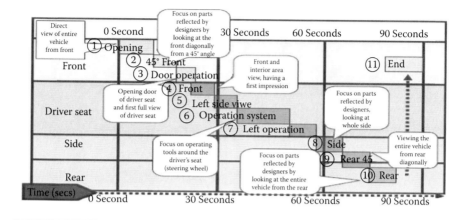

Figure 7.41 VUCMIN creation timetable (males in their 50s, Mark X).

order to test the validity of VUCMIN. This was done by asking customers, "After seeing the Mark X video, when do you think you might visit advanced car manufacturer A dealer to consider purchasing one?" to verify their desire to visit dealers (high, low).

According to the survey results, the desire to visit dealers (early consideration of Mark X purchase) not only increased for current Toyota vehicle owners but also for customers who owned vehicles from other manufacturers. The author is currently promoting the results of this research as part of the strategic advertising method VUCMIN, which utilizes an Internet interface in collaboration with universities and industrial players.

7.5.5.3 Application of CMP-FDM

Flyers are a form of advertising media important for raising the customer-attraction effect. However, the results of interviews by the author at six dealers (national and foreign-affiliated) and two advertising agencies specializing in flyers showed that the dealers did not think that designing flyers was important—their only priority was distribution, and they outsourced the design. Moreover, they did not understand actual customer behavior (how customers looked at flyers and what they paid attention to). Therefore, customer behavior is not reflected in the creation of flyers, and the resulting design is not attractive to customers who want to visit dealers.

The author studied how customers view flyers by analyzing browsing behavior (Koyama et al., 2010). In order to resolve the problems with the information contained in current flyer designs, the author identified which information each consumer type focuses on. In Steps 1 and 2, the purchase group is taken up as an example of the factor analysis results, with the Text Mining Studio corresponding bubble analysis results focusing on the purchase group shown in Figure 7.43.

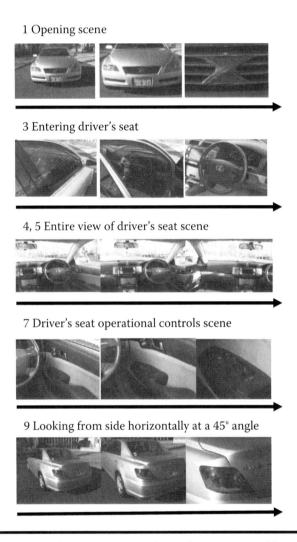

Figure 7.42 Example of representative photos for VUCMIN video.

From this figure, it can be seen that the purchase group strongly correlates with viewing behaviors in the following order: (1) looking at a picture of the car, (2) looking at the price, (3) looking at the car name, (4) looking at the loan information, (5) looking at a picture of the interior, (6) looking at dealer gifts, and (7) looking at the logo and company name. Through this analysis, we were able to clearly reveal the information on flyers that consumers actually focus their attention on, which dealers had previously been unable to grasp.

In Step 3, in order to resolve problems with the flyer layout, the author clarified the position and size of the information on flyers that each consumer type focuses

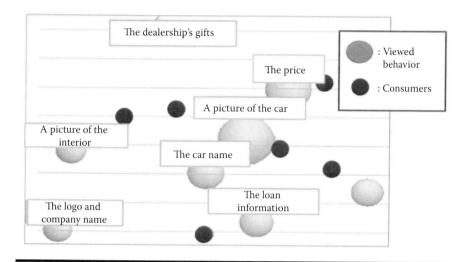

Figure 7.43 Results for the information on flyers for the purchase group using the text mining studio corresponding to bubble analysis.

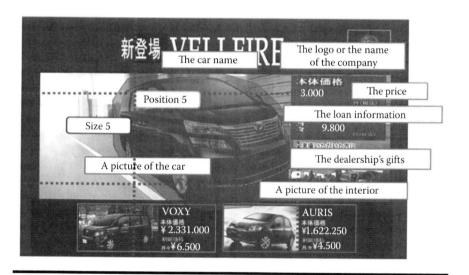

Figure 7.44 Example of new attractive flyer design.

their attention on. In Step 4, an attractive flyer design is created based on the knowledge gained in Steps 1–3.

An example of a new attractive flyer design for Toyota's new vehicle is shown in Figure 7.44. This design was intended to be appealing for a universal type of consumer. The effectiveness of the CMP-FDM method for creating the appealing flyer design shown in Figure 7.44 was confirmed by the survey procedures and the analysis of Steps 1–5, as well as the acquired results.

7.5.5.4 Application of AGTCA

Firstly, the author researched the causal relationship between basic passenger information and ad awareness (Ogura et al., 2012). In Step 1, the author performed a cross-tabulation on the survey data and researched the correlations between whether passengers notice each form of train car advertising and passenger information (age and gender) in order to see how passenger information relates to attention rates.

Secondly, the author researched the causal relationship between riding conditions and ad awareness of passengers. In Step 2, the author used the survey data gathered to represent current in-train advertising conditions and subjected it to a cluster analysis of group riding conditions, as shown in Figure 7.45. Since the first group consisted of standing passengers who ride the train for 0–15 min, they were labeled *short-distance passengers*. The second group rode the train for a longer period of time and tended to sit, so this group was called the *long-distance passengers*.

Thirdly, the author researched the causal relationships among basic passenger information, riding conditions, and ad awareness of passengers. In Step 3, Figure 7.46 shows the results of the awareness rate for hanging posters among short-term riders, which was used as a criterion variable in the CAID analysis.

The results indicate that the highest awareness rate in this group is among men in the youngest age category (15–25 years). Barring a few exceptions, the results indicate an overall trend where awareness rates are higher among younger people. Comparing the two groups, the author found that the short-term passengers (who tended to stand when riding) had higher awareness rates in general.

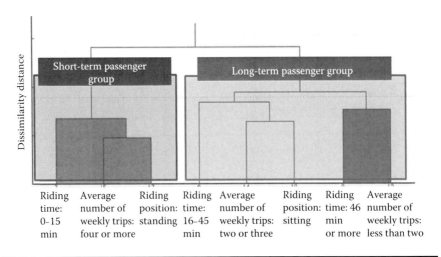

Figure 7.45 Cluster analysis of the riding conditions.

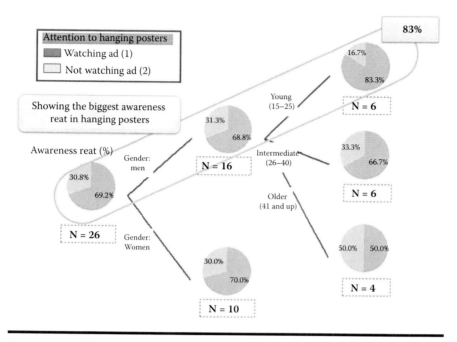

Figure 7.46 The results of the CAID analysis on the short-term passenger group.

Passengers who sat, on the other hand, had more opportunities to engage in different activities during their ride, such as reading or doing work, which probably contributed to their paying less attention to advertisements than the passengers who were standing. The analysis revealed that women passengers 26 years and older in particular did not look at in-car advertisements. It also indicated that older passengers frequently paid attention to advertisements located above windows.

Based on these conclusions, the author's first recommendation is for existing train car advertising. Because hanging posters, stickers, and other in-car advertisements are likely to attract younger riders, this space should be used to advertise weekly manga magazines, fashion magazines, sales, or other products likely to appeal to this generation. Another important consideration is using popular celebrities to catch the eye of these passengers. Above-window advertising space, on the other hand, may be better used to appeal to those of the older generation. These passengers are more likely to be married and have children, so it may be beneficial to feature family-friendly topics.

Specifically, posters advertising events for families or travel may be ideal in this location. Also, because it was found that standing passengers tend to look at advertisements frequently, riders may pay attention not only to ads that help them pass the time while standing, but also those that stimulate their interest or desire. Instead of showing just a picture, a magazine ad, for example, could feature headlines or other clever designs aimed at stimulating purchase behavior. It is important that other advertisements do not simply catch the eye, but encourage viewers to linger.

Secondly, the author suggests that trains adopt new forms of advertising media. The analysis results indicated that passengers who stand tend to have high awareness rates when it comes to in-car advertising, but advertisements on the floor may be easier for sitting passengers to see. Riders who sit naturally allow their eyes to fall downward, making a floor advertisement an eye-catching option. Those who sit and read are also looking downward, increasing the chances that they may see these advertisements.

Focusing on train car ads, which have a good contact rate and long-term contact, the author decided to examine those relationships and how they relate to train car advertising with the aim of defining the ideal format for this type of media.

7.5.5.5 Application of PMOS-DM

7.5.5.5.1 Putting POMS-DM to Work

The researchers teamed up with Company M to guide direct mailing efforts in conjunction with an event showcasing multiple new vehicle models (Kojima et al., 2010; Ishiguro and Amasaka, 2012). The following steps show how optimal selection using a model formula was applied with the formulas shown in (7.12–7.16) (refer to Appendix 7.1).

7.5.5.5.1.1 (Step 1) Organizing Customer Information — First, participating dealers had information on a total of 391 customers, which included data on sex (male/female), age (20s, 30s, 40s, 50s, 60+), and age of current vehicle (3–5 years, 6–8 years, 9+ years). There were a total of 391 values assigned to j (customer number in the formula), and a total of 10 different values assigned to m (customer attributes: e.g., sex, age, age of current vehicle). A binary code (0 or 1) was then assigned to the collected customer information to analyze it. This resulted in values for the f_j^m (indicates whether or not customer j has attribute m (0 or 1)) variable. Recipients of direct mailing could now be determined based on the dealers' customer information.

7.5.5.5.1.2 (Step 2) Determining Response Likelihood — The next step was to conduct a survey and analyze the data to determine which customer attributes were most likely to lead customers to visit a dealer as a result of receiving DM. The survey method used in this study was to ask customers of varying attributes (sex, age, vehicle age, etc.) whether receiving DM had ever caused them to visit the dealer. Once the results were collected, they were quantified and subjected to the multivariate analysis (Quantification Theory Type II) analysis to determine which customers had the highest likelihood of responding to DM.

This made it possible to analyze customer attributes in terms of whether or not they were likely to lead to a dealer visit in terms of an external standard. These response likelihood values were then assigned the variable E^m (effect of customer attribute (m) on the likelihood that the customer will visit the dealer). Formula (7.12) shows the results of this analysis. The discriminant ratio for the analysis results was 77.36%, indicating that they were fairly reliable. The linear discriminant formula produced from the analysis results is as follows:

$$y = 0x_{11} - 1.4x_{12} + 0x_{21} + 0.9x_{22} + 2.1x_{23} + 3.7x_{24} + 1.7x_{25}$$
$$+ 0x_{31} - 1.3x_{32} - 0.9x_{33} - 1.4 \qquad (7.12)$$

If the linear discriminant is >0, the customer is likely to visit the dealer as a result of receiving DM. If it is <0, it indicates that they are not likely to visit. Therefore, the coefficient produced by this formula indicates the response likelihood for the customer attribute as expressed by E^m. The results of this analysis, which allowed us to identify which customers were likely to visit the dealer, are summarized in Table 7.4.

7.5.5.5.1.3 (Step 3) Selecting DM Recipients — First, the customer information collected in Step 1 is plugged into f_j^m, and the information on response likelihood for each customer attribute is plugged into E^m. The number of direct mailings to be sent is plugged into C (total number of direct mailing sent). The upper and lower limits for the percentage of direct mailings to go to customers with each

Table 7.4 Response Likelihood by Attribute

Customer Attribute	Response Likelihood
Men	0
Women	−1.4
22–29 years old	0
30–39 years old	0.89
40–49 years old	2.1
50–59 years old	3.7
60+ years old	1.7
Currently driving a vehicle 3–5 years old	0
Currently driving a vehicle 6–8 years old	−1.3
Currently driving a vehicle 9+ years old	−0.9

attribute is set at the dealer's discretion using the variables H^m (upper limit for the percentage of direct mailings sent to customers with attribute m) and L^m (lower limit for the percentage of direct mailings sent to customers with attribute m). Once all the parameters are set, the simulation is carried out.

During this process, formulas (7.16) through (7.18) (refer to Appendix 7.1) are solved as a weighted constraint satisfaction problem. In the weighted constraint satisfaction problem, the weighted constraints are moved to the target function as in (7.13), where they are added as a way of minimizing the level of deviation outside the given limits. Even if a feasible solution that satisfies the constraints does not exist, the formula allows dealers to come as close as possible to meeting the constraints.

Here, in constraining the number of mailings sent to customers with the attributes defined in formula (7.18), it is difficult to set customer attributes L^m and H^m, ensuring that a feasible solution is more likely to exist. Therefore, when approaching the issue as a weighted constraint satisfaction problem, it is best to find the solution that best satisfies formula (7.18). In other words, this allows dealers to send DM to those customers most likely to come into the shop based on dealer strategy.

$$\text{Min} \quad -\sum_m \left(E^m \sum_j \left(f_j^m x_j \right) \right) + \sum_m W^m \left(L^m C - f_j^m x_j \right)$$
$$+ \sum_m W^m \left(H^m C - f_j^m x_j \right) \quad (7.13)$$

7.5.5.5.1.4 (Step 4) Evaluating the Results — Once the recipients of direct mailings are selected based on the simulation, whoever is sending out the DM checks the simulation results to make sure that they accurately reflect the dealer's marketing strategy. If the desired results are not achieved, the causes for the discrepancy are identified, the parameters are adjusted, and the simulation is run again.

7.5.5.5.2 Effectiveness of PMOS-DM

The effectiveness of the PMOS-DM was assessed by comparing the response rates (percentage of DM recipients who visited the dealer as a result) when salespeople selected DM recipients based on personal knowledge and experience and when recipients were selected using the model. Five new models were showcased at the event held. Four of the design concepts targeted female buyers, and one targeted male buyers. As a result, the dealer's marketing strategy was to target women in particular throughout a wide range of age groups. This strategy was thus taken into account when verifying the effectiveness of the model. These verification results are summarized in Table 7.5.

The response rate when DM recipients were selected on the basis of personal knowledge and experience of the sales staff was 19%. When selection was made

Table 7.5 Verification Results (All)

	Dealer	PMOS-DM
Number of direct mailings sent	269	269
Number of resulting dealer visitors	51	59
Response rate	19.0%	20.4%

Table 7.6 Verification Results (Women)

	Dealer	PMOS-DM
Number of direct mailings sent	48	61
Number of resulting dealer visitors	2	12
Response rate	4.2%	19.8%

using the PMOS-DM model, the rate was 20.4%. Table 7.6 shows the same information for female customers only (those targeted in the dealer's marketing strategy). Salespeople generated a 4.2% response rate using their personal knowledge and experience, while the model generated a 19.8% response rate, signaling a significant improvement. The effectiveness of the model was thereby verified in this study.

7.5.5.5.3 Verification Results

Using the analysis results obtained in the previous section (Visualizing Causal Relationships in Customer Purchase Behavior), a follow-up survey using questionnaire data from Table 7.3 was then conducted to verify whether the research achieved its aim of bringing more customers into the dealers by means of raising the percentage of people affected.

Figure 7.47 shows the verification results from application of new mixed media by a SMMM for raising the percentage of people affected. SMMM is effective due to its composition of four core elements (VUCMIN, CMP-FDM, AGTCA, and PMOS-DM) and use as a new strategic advertisement in nine media elements (TV, radio, newspapers, internet, train cars, flyers, magazines, DM, and handbills) designed by the author.

The figure shows the result of a follow-up survey using SMMM, where 16 people (percentage of people affected: 11.8%) actually visited the dealer, while 8 people signed a sales contract. Comparative verification was done by looking at the results of the usual experience of mixed media when the dealer in the figure announced the old model Q 4 years ago in a survey of similar size. In this case, the percentage of people affected was just 1.1%, thus validating the effectiveness of SMMM.

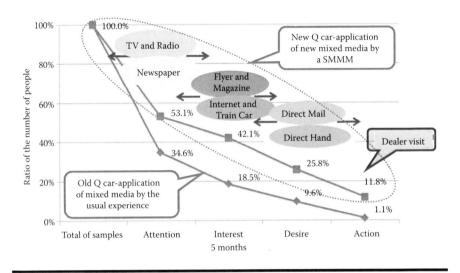

Figure 7.47 Verification results.

7.5.6 Conclusion

The aim of this research study was to bring more customers into auto dealers. In order to achieve this, a SMMM was developed as a way to improve the quality of the consumer purchase behavior model in terms of vehicle awareness, vehicle interest, and desire to visit dealers. The collected research results are now being widely distributed as part of Toyota's current sales strategy.

Appendix 7.1 Optimal Selection Using a Model Formula

Using numerical simulation, the PMOS-DM model uses a mathematical formula to select target customers for DM. In coming up with a formula to determine who should be targeted by DM, the author referred to the formulas shown in (7.14) and (7.15), which were developed by Kojima et al. (2010) and Ishiguro and Amasaka (2012).

$$\min L^m C \le \sum_j f_j^m x_j \le H^m C \qquad (7.14)$$

$$\text{subject to } \sum_m \left(E^m \sum_j \left(f_j^m x_j \right) \right) \qquad (7.15)$$

where

    m = customer attributes (e.g., sex, age, age of current vehicle)
    j = customer number
    W^m = weighting for customers with customer attributes m in DM target group
    f_j^m indicates whether or not customer j has attribute m (0 or 1)
    x_j marks customer j for direct mailing (0 or 1)
    R^m = ideal percentage with customer attributes m in the DM target group
    C = total number of direct mailings sent
    L^m = lower limit for the percentage of direct mailings sent to customers with attribute m
    H^m = upper limit for the percentage of direct mailings sent to customers with attribute m

The target function of formula (7.14) is to minimize the gap between the ideal number of direct mailings sent to customers with attribute m (CR^m) and the number actually sent to customers with that attribute ($\Sigma f_j^m x_j$). In other words, the formula expresses the concept of setting a target value when sending out DM. Accordingly, the formula can be adapted to cases where a clear, rational target value can be set. However, the formula cannot be used when it is difficult to set a logical target value for the number of direct mailings to be sent—and a dozen or so of the dealers that the author studied did not set one.

For those dealers, the author sets up a formula that would clarify the process that senior sales staff used to determine who should be targeted by a given DM campaign. In the process of conducting interviews, the author learned that senior sales staff use an abstract method of targeting those customers who seem like they would have an easy time coming into the dealer. The author then constructed a makeshift definition of this group of customers as follows.

Each group of customers defined by a given attribute (male, female, 20s, 30s, etc.) has different preferences that would motivate them to come into the dealer. Each customer's willingness to come in can be assigned a cumulative value based on that person's attributes. Those with a high cumulative value can be considered the ones who are likely to come into the shop. With this line of thinking, the author developed a formula for calculating the total willingness for customers targeted by DM. They then constructed a model for optimizing those values. Finally, the author came up with a set of constraints in order to put limits on the number of mailings dealers would send, with the aim of maximizing the effectiveness of those that were sent.

A.7.1.1 Model Formula

This is the model formula used in the numerical simulation.

$$\max \sum_{m \in M} W^m \left(\sum_{j \in J} f_j^m x_j - CR^m \right)^2 \qquad (7.16)$$

$$\text{subject to } \sum_j x_j = C \qquad (7.17)$$

$$L^m C \leq \sum_j f_j^m x_j \leq H^m C \qquad (7.18)$$

where:
    m = customer attributes (e.g., sex, age, age of current vehicle)
    j = customer number
    E^m = effect of customer attribute (m) on the likelihood that the customer will visit the dealer
    f_j^m indicates whether or not customer j has attribute m (0 or 1)
    x_j marks customer j for direct mailing (0 or 1)
    C = total number of direct mailings sent
    L^m = lower limit for the percentage of direct mailings sent to customers with attribute m
    H^m = upper limit for the percentage of direct mailings sent to customers with attribute m

This mathematical formula is designed to determine a value for the variable x_j. If the value is 1, mailings should be sent to the customer number indicated by j. If it is 0, a direct mailing should not be sent. The other variables are parameters that must be given values before solving the formula. C, L^m, and H^m are set at the discretion of whoever is sending out the DM. The value f_j^m is determined based on the customer information that the dealer has. E^m is determined later by statistical analysis. The roles of the individual formulas are as follows. The objective function in formula (7.16) is used to maximize the customer response rate (the percentage of customers that come to the dealer as a result of the DM).

The constraint in formula (7.17) determines the number of direct mailings that are to be sent out. The constraint in formula (7.18) determines how many direct mailings are to be sent to each customer segment, which is how dealer aims are incorporated into the model. The mathematical formula is designed so that the number of customer attributes it handles (m) can be increased at will. Depending on what customer information dealers have, they can limit these attributes to basic life stages or expand them to include hobbies, preferences, and other lifestyle characteristics.

A.7.1.2 Recipient Selection Process

The author describes the procedure for using the mathematical formula provided to select DM recipients. First, a response likelihood value must be set for each customer using the variable E^m. The list of customers is then reordered with those with the highest likelihood of responding at the top. The purpose of the objective function in formula (7.16) is to order customers according to their likelihood of responding (visiting the dealer as a result of DM). Next, this list is used to select the number of customers equal to the number of direct mailings (the constraint) to be sent out, starting with those most likely to respond. For example, if 50 direct mailings are to be sent, they would be sent to the top 50 customers most likely to respond to them. This is the basic principle behind the development of the formulas.

In addition, when the dealer has a specific aim in mind (e.g., sending a large number of direct mailings to women), the constraint function in formula (7.18) can be used to incorporate that aim in the calculations. For example, if the dealer wanted at least 60% of the 50 mailings to go to women, the women customers would be listed in order of response likelihood and the top 30 customers would be selected to receive DM. The remaining 20 recipients would be selected from the entire pool of target customers in order of their response likelihood as well. The purpose of this function is to allow dealers to use their marketing strategies to boost the response rate.

References

Amasaka, K. 1995. A construction of SQC intelligence system for quick registration and retrieval library, in *Lecture Notes in Economics and Mathematical Systems*, vol. 445, pp. 318–336. Springer-Verlag, Berlin.

Amasaka, K. 1999. A study on "Science SQC" by utilizing "Management SQC": A demonstrative study on a new SQC concept and procedure in the manufacturing industry", *International Journal of Production Economics*, 60–61: 591–598.

Amasaka, K. 2000. A demonstrative study of a new SQC concept and procedure in the manufacturing industry, *An International Journal of Mathematical and Computer Modeling*, 31(10–12): 1–10.

Amasaka, K. 2002. New JIT, a new management technology principle at Toyota, *International Journal of Production Economics*, 80: 135–144.

Amasaka, K. 2003a. New application of strategic quality management and SCM—A "Dual Total Task Management Team" involving both Toyota and NOK, in *Proceedings of the Group Technology/Cellular Manufacturing World Symposium*, pp. 265–270. July 29, Columbus, OH.

Amasaka, K. 2003b. Proposal and implementation of the "Science SQC" quality control principle, *International Journal of Mathematical and Computer Modelling*, 38(11–13): 1125–1136.

Amasaka, K. 2003c. Development of New JIT, key to the excellence design "Lexus": The validity of TDS-DTM, a stategic methodology of merchandise, in *Proceedings of the Production and Operations Management Society*, pp. 1–8. April 5, Georgia, USA.

Amasaka, K. 2004a. *Customer Science*: Studying consumer values, in *The 32nd Annual Conference, Japan Journal of Behaviormetrics Society*, pp. 196–199, Kanagawa-ke.

Amasaka, K. 2004b. Development of Science TQM, a new principle of quality management: Effectiveness of strategic stratified task team at Toyota, *International Journal of Production Research*, 42(17): 3691–3706.

Amasaka, K. 2004c. Applying New JIT, a management technology strategy model at Toyota, in *Proceedings of the 2nd World Conference on Production and Operation Management*, pp. 1–12. May 1, Mexico, (CD-ROM).

Amasaka, K. 2004d. *Science SQC, New Quality Control Principle: The Quality Strategy of Toyota*, Springer-Verlag, Tokyo.

Amasaka, K. 2005. Constructing a Customer Science application system "*CS-CIANS*"—Development of a global strategic vehicle "Lexus" utilizing New JIT, *WSEAS (World Scientific and Engineering and Society) Transactions on Business and Economics*, 3(2): 135–142.

Amasaka, K. 2007a. Applying New JIT—Toyota's global production strategy: Epoch-making innovation in the work environment, *Robotics and Computer-Integrated Manufacturing*, 23(3): 285–293.

Amasaka, K. 2007b. The validity of Advanced TMS, a strategic development marketing system utilizing New JIT, *The International Business and Economics Research Journal*, 6(8): 35–42.

Amasaka, K. 2007c. Strategic QCD studies with affiliated and non-affiliated suppliers utilizing New JIT, *Encyclopedia of Networked and Virtual Organizations*, III(PU-Z): 1516–1527.

Amasaka, K. 2007d. High linkage model "Advanced TDS, TPS & TMS"—Strategic development of "New JIT" at Toyota, *International Journal of Operations and Quantitative Management*, 13(3): 101–121.

Amasaka, K. 2007e. Final report of WG4's studies in JSQC research activity of simulation and SQC (Part-1)—Proposal and validity of the high quality assurance CAE model for automobile development design, in *The 2nd Annual Technical Conference, Transdisciplinary Science and Technology Initiative*, pp. 321–326. November 29, Kyoto University, Japan [in Japanese].

Amasaka, K. 2007f. Highly reliable CAE model, the key to strategic development of New JIT, *Journal of Advanced Manufacturing Systems*, 6(2): 159–176.

Amasaka, K. 2008. An integrated intelligence development design CAE model utilizing New JIT, application to automotive high reliability assurance, *Journal of Advanced Manufacturing Systems*, 7(2): 221–241.

Amasaka, K. 2009. The effectiveness of flyer advertising employing TMS: Key to scientific automobile sales innovation at Toyota, *The Academic Journal of China-USA Business Review*, 8(3): 1–12.

Amasaka, K. 2011. Changes in marketing process management employing TMS: Establishment of Toyota sales marketing system, *The Academic Journal of China and USA Business Review*, 10(7): 539–550.

Amasaka, K., Kishimoto, M., Murayama, T., and Ando, Y. 1998. The development of marketing SQC for dealers' sales operation system, in *The 58th Technical Conference, Journal of the Japanese Society for Quality Control*, pp. 76–79. May 31, Tokyo, Japan [in Japanese].

Amasaka, K., Nagaya, A., and Shibata, W. 1999. Studies on design SQC with the application of Science SQC, *Japanese Journal of Sensory Evaluations*, 3(1): 21.

Amasaka, K., Ogura, M., and Ishiguro, H. 2013. Constructing a Scientific Mixed Media Model for boosting automobile dealer visits: Evolution of market creation employing TMS, *International Journal of Business Research and Development*, 3(4): 1377–1391.

Amasaka, K. and Ohtaki, M. 1999. Development of new TQM by partnering—Effectiveness of TQM-S-P by collaborating total task management team activities, in *The 10th Annual Technical Conference, the Japan Society for Production Management*, pp. 69–74. September 5, Kyushu Sangyou University, Fukuoka, Japan.

Amasaka, K. and Osaki, S. 2002. Reliability of oil seal for transaxle—A Science SQC approach in Toyota, in Wallace, R. B. and Murthy, D. N. P. (eds), *Case Studies in Reliability and Maintenance*, pp. 571–588. Wiley, Hoboken, NJ.

Amasaka, K. and Sakai, H. 1996. Improving the reliability of body assembly line equipment, *International Journal of Reliability, Quality and Safety Engineering*, 3(1): 11–24.

Amasaka, K. and Sakai, H. 1998. Availability and Reliability Information Administration System ARIM-BL by methodology in inline-online SQC, *International Journal of Reliability, Quality and Safety Engineering*, 5(1): 55–63.

Amasaka, K. and Sakai, H. 2009. TPS-QAS, new production quality management model: Key to New JIT: Toyota's global production strategy, *International Journal of Manufacturing Technology and Management*, 18(4): 409–426.

Amasaka, K. and Yamaji, M. 2006. *Advanced TDS*, total development design management model: Application to automotive intelligence CAE methods utilizing New JIT, in *Proceedings of the International Applied Business Research Conference*, pp. 1–15. March 23, Cancun, Mexico (CD-ROM).

Atkeson, C. G., An, C. H., and Hollerbach, J. M. 1986. Estimation of inertial parameters of manipulator links and loads, *International Journal of Robotics Research*, 5(3): 101–119.

Atkeson, C. G. and McIntyre, J. 1985. Robot trajectory learning through practice, in *Proceedings of IEEE International Conference on Robotics and Automation*, pp. 1737–1742. April 8, San Francisco.

Balestrino, A., Maria, G. D., and Sciavicco, L. 1983. An adaptive model following control for robot manipulators, *ASME Journal of Dynamic Systems, Measurement, and Control*, 105: 143–151.

Beco, G. H. and Jagric, T. 2011. Demand models for direct mail and periodicals delivery services: Results for a transition economy, *Applied Economics*, 43(9): 1125–1138.

Bell, G. H., Ledoher, J., and Swersey, A. J. 2006. Experimental design on the front lines of marketing: Testing new ideas to increase direct mail sales, *International Journal of Research in Marketing*, 23(3): 309–319.

Blackwell, R. D., Miniard, P. W., and Engel, J. F. 2006. *Consumer Behavior*, 10th edn. Thomson Business and Economics, Mason, OH.

Bult, J. R. and Wansbeek, T. 2005. Optimal selection for direct mail, *Marketing Science*, 14(4): 378–394.

Byoungkyu, C., Bumchul, P., and Ho, Y. R. 2004. Virtual factory simulator framework for line prototyping, *Journal of Advanced Manufacturing Systems*, 3(1): 5–20.

Craig, J. J. 1988. *Adaptive Control of Mechanical Manipulators*, Addison-Wesley, Massachusetts.

Ferrell, O. C. and Hartline, M. 2005. *Marketing Strategy*, Thomson South-Western, Mason, OH.

Fitzsimmons, J. A. and Fitzsimmons, M. J. 2004. *Service Management: Operations, Strategy, and Information Technology*, McGraw-Hill, New York.

Fukuchi, H., Arai, Y., Ono, M., Suzuki, S. T., and Amasaka, K. 1998. A proposal TDS-D by utilizing Science SQC: An improving design quality of drive-train components, in *The 60th Technical Conference, the Japanese Society for Quality Control*, pp. 29–32. July 2, Nagoya, Japan [in Japanese].

Goto, T. 1999. *Forgotten Management Origin*, Seisansei-shuppan [in Japanese].

Hashimoto, H. et al. 2005. *Automotive Technological Handbook, Design and Body*, Chapter 6, CAE, pp. 313–319, Society of Automotive Engineers of Japan, Tosho Shuppan-sha, Tokyo [in Japanese].

Hayes, R. H. and Wheelwright, S. C. 1984. *Restoring Our Competitive Edge: Competing through Manufacturing*, Wiley, New York.

Hifumi, S. 2006. A study on a strategic propaganda advertisement (animation): The proposal of "A-MUCMIN" which raises the volition of coming to the auto shop, Aoyama Gakuin University, Master of Science and Engineering, Master's thesis of Management Technology Course.

Hiraki, S., Ishii, K., Takahashi, K., and Muramatsu, R. 1992. Designing a pull-type parts procurement system for international co-operative knockdown production systems, *International Journal of Production Research*, 30(2): 337–351.

Ikeo, K. (editing chairperson). 2006. Feature—Marketing innovation, *Japan Marketing Journal* [in Japanese].

Ishiguro, H. and Amasaka, K. 2012. Proposal and effectiveness of a highly compelling direct mail method—Establishment and deployment of PMOS-DM, *International Journal of Management and Information System*, 16(1): 1–10.

Jonker, J. J., Pielsma, N., and Potharst, R. 2006. A decision support system for direct mailing decision, *Decision Support System*, 42: 915–925.

Kameike, M., Ono, S., and Nakamura, K. 2000. The helical seal: Sealing concept and rib design, *Sealing Technology International*, 77: 7–11.

Kawakami, J. et al. 1972. *Collage Method—Esprit of Today*, Shibundou, Tokyo [in Japanese].

Kelly, A. 1997. *Maintenance Strategy*, Butterworth-Heinemann, Oxford.

Kesavadas, T. and Ernzer, M. 2003. Design of an interactive virtual factory using cell formation methodologies, *Journal of Advanced Manufacturing Systems*, 2(2): 229–246.

Kimura, O. and Terada, H. 1981. Design and analysis of pull system, a method of multistage production control, *International Journal of Production Research*, 9(3): 241–253.

Kishi, S., Tanaka, H., and Shimamura, K. 2000. *Theory of Modern Advertising*, Yuhikaku Publishers, Tokyo [in Japanese].

Kitano, M. 1996. New flexible automation: A process planner's view, in *Proceedings of the Japan/USA Symposium on Flexible Automation*, vol. 1, pp. 1–8. ASME, New York.

Kobayashi, T. and Shimamura, K. 1997. *New Edition New Ad.*, Dentsu, Tokyo [in Japanese].

Koivo, A. J. and Guo, T. H. 1983. Adaptive linear controller for robotic manipulators, *IEEE Transactions on Automatic Control*, AC-28(2): 162–171.

Kojima, T., Kimura, T., Yamaji, M., and Amasaka, K. 2010. Proposal and development of the direct mail method "PMCI-DM" for effectively attracting customers, *International Journal of Management and Information Systems*, 14(5): 15–22.

Koyama, H., Okajima, R., Todokoro, T., Yamaji, M., and Amasaka, K. 2010. Customer behavior analysis using motion pictures: Research on attractive flyer design method, *The Academic Journal of China–USA Business Review*, 9(10): 58–66.

Kubomura, R. and Murata, S. 1969. *Theory of PR*, Yuuhikaku, Tokyo [in Japanese].

Leo, J. V., Amos, H. C., and Jan, O. 2004. Simulation-based decision support for manufacturing system life cycle management, *Journal of Advanced Manufacturing Systems*, 3(2): 115–128.

Lilien, G. L. and Rangaswamy, A. 2003. *Marketing Engineering: Computer-Assisted Marketing Analysis and Planning*, Pearson Education, Pearson PLC, New Jersey.

Lopez, A. M., Nakamura, K., and Seki, K. 1997. A study on the sealing characteristics of lip seals with helical ribs, in *Proceedings of the 15th International Conference of British Hydromechanics Research Group Ltd., Fluid Sealing*, pp. 239–249.

Magoshi, M. et al. 2003. Simulation technology applied to vehicle development, *Journal of Society of Automotive Engineers of Japan*, 57(3): 95–100 [in Japanese].

Matsubara, K. 1981. *Tripology*, Sangyo Library.
Melewar, T. C. and Smith, N. 2003. The internet revolution: Some global marketing implications, *Marketing Intelligence and Planning*, 21(6): 363–369.
Modarres, M., Kaminskiy, M., and Krivtsov, V. 1999. *Reliability Engineering and Risk Analysis*, Marcel Dekker, New York.
Motor Fan. 1997. All of new-model "Aristo", Vol. 213, pp. 24–30 [in Japanese].
Motor Fan. 2000. All of new-model "Celsior", Vol. 268, pp. 23–24 [in Japanese].
Murat, S. M., Hifumi, S., Yamaji, M., and Amasaka, K. 2008. Developing a strategic advertisement method VUCMIN to enhance the desire of customers for visiting dealers, in *Proceedings of the International Symposium on Management Engineering*, pp. 248–257. October 28, Waseda University, Kitakyushu, Japan.
Murayama, Y. et al. 1982. *Analyzing VAID Marketing Review, Japan Research Center*, pp. 74–86. Tokyo, Japan [in Japanese].
Nakamura, H., Sakai, H., Miura, H., and Sakamoto, Y. 1997. Construction of life estimation based on deterioration model, *Precision Technology Bulletin*, 63(11): 1620–1624.
Nihon Keizai shimbun. 2000. Worst record: 40% increase of vehicle recalls (July 6, 2000) [in Japanese].
Nihon Keizai shimbun. 2005. The manufacturing industry—Skill tradition feels uneasy (May 3, 2005) [in Japanese].
Niiya, Y. and Matsuoka, F. (eds). 2001. Foundation lecture on the new advertising business, Senden-kaigi, Tokyo.
Nikkei Business. 1999. Renovation of shop, product and selling method: Targeting young customer by nets, pp. 46–50 [in Japanese].
NOK Corporation. 2000. Promotion video: The history of NOK's oil seal—Oil seal mechanism, Tokyo, Japan [in Japanese].
Ogura, M., Hachiya, T., Masubuchi, K., and Amasaka, K. 2012. Attention-grabbing train car advertisements, in *Proceedings of the 2nd International Symposium on Operations Management and Strategy 2012*, pp. 89–102. Aoyama Gakuin University, Tokyo, Japan.
Ohno, T. 1977. *Toyota Production System*, Diamond-sha, Tokyo [in Japanese].
Okada, A., Kijima, M., and Moriguchi, T. (eds.). 2001. *The Mathematical Model of Marketing*, Asakura-shoten, Tokyo [in Japanese].
Okazaki, R., Suzuki, M., and Amasaka, K. 2000. Study on the sense of values by age using design SQC, in *The 11th Annual Technical Conference, the Japan Society for Production Management*, pp. 139–142. March 26, Okayama, Japan.
Sackett, P. J. and Williams, D. K. 2003. Data visualization in manufacturing decision making, *Journal of Advanced Manufacturing Systems*, 2(2): 163–185.
Rayport, J. F. and Jaworski, B. J. 2005. *Best Face Forward*, pp. 62–77. Harvard Business Review Press, Boston, MA.
Sakai, H. and Amasaka, K. 2005. V-MICS, advanced TPS for strategic production administration: Innovative maintenance combining DB and CG, *Journal of Advanced Manufacturing Systems*, 4(6): 5–20.
Sakai, H. and Amasaka, K. 2006a. Strategic HI-POS, intelligence production operating system, *WSEAS Transactions on Advances in Engineering Education*, 3(3): 223–230.
Sakai, H. and Amasaka, K. 2006b. *TPS-LAS* model using process layout CAE system at Toyota, Advanced TPS, key to global production strategy New JIT, *Journal of Advanced Manufacturing Systems*, 5(2): 1–14.

Sakai, H. and Amasaka, K. 2007a. The Robot Reliability Design and Improvement Method and the advanced production system, *International Journal of Industrial and Service Robotics,* 6(8): 310–316.

Sakai, H. and Amasaka, K. 2007b. Development of a robot control method for curved seal extrusion for high productivity in an advanced Toyota Production System, *International Journal of Computer Integrated Manufacturing,* 20(5): 486–496.

Sakai, H. and Amasaka, K. 2008. Verification of process layout CAE system TPS-LAS at Toyota, pp. 71–82, *The Grammar of Technology Development,* H. Tsuhaki, K. Nishina, and S. Yamada, (eds), Springer, Printed in Japan.

Sasaki, S. 1972. Collection and analysis of reliability information on automotive industries, in *The 2nd Reliability and Maintainability Symposium, Union of Japanese Scientists and Engineers,* pp. 385–405, Tokyo, Japan.

Sato, Y., Toda, A., Ono, S., and Nakamura, K. 1999. A study of the sealing mechanism of radial lip seal with helical ribs—Measurement of the lubricant fluid behavior under sealing contact, *SAE Technical Paper Series,* 1999-01-0878.

Shimakawa, K., Katayama, K., Oshima, K., and Amasaka, K. 2006. Proposal of strategic marketing model for customer value maximization, in *The 23th Annual Technical Conference, the Japan Society for Production Management,* pp. 161–164. March 19, Osaka City University, Osaka, Japan [in Japanese].

Shimizu, K. 2004. *Theory and Strategy of Advertisement,* Sousei Publishers, Tokyo [in Japanese].

Shinohara, A. et al. 1996. *Invitation to Kaansei Engineering,* Morikita-shuppan, Tokyo [in Japanese].

Shiomi, H. 1970. *Introduction to Destructible Physics,* Union of Japanese Scientists and Engineers, Tokyo.

Slotine, J. J. and Li, W. 1987. On the adaptive control of robot manipulators, *International Journal of Robotics Research,* 6(3): 49–59.

Smith, A. M. 1993. *Reliability-Centered Maintenance,* McGraw-Hill.

Smith, D. A. 2009. Online accessibility concerns in shaping consumer relationships in the automotive industry, *Online Information Review,* 33(1): 77–95.

Spearman, M. L. 1992. Customer service in pull production systems, *Operation Research,* 40(2): 948–958.

Spearman, M. L., Woodruff, D. L., and Hopp, W. J. 1990. CONWIP: A pull alternative of Kanban, *International Journal of Production Research,* 28(5): 879–894.

Strassburger, S., Schmidgall, G., and Haasis, S. 2003. Distributed manufacturing simulation as an enabling technology for the digital factory, *Journal of Advanced Manufacturing Systems,* 2(1): 111–126.

Steinberg, D. M. 1996. Robust designs: Experiments for improving quality, Chapter 7 in Ghosh, S. and Rao, C. R. (eds), *Handbook of Statistics,* p. 13. Elsevier, Amsterdam.

Taj, S. and Berro, L. 2006. Application of constrained management and lean manufacturing in developing best practices for productivity improvement in an auto-assembly plant, *International Journal of Productivity and Performance Management,* 55(3/4): 332–345.

Takada, S. 1999. Life cycle maintenance, *Precision Technology Bulletin,* 65(33): 349–353.

Takahashi, K., Hiraki, S., and Soshiroda, M. 1994. Pull-push integration in production ordering systems, *International Journal of Production Economics,* 33(1): 155–161.

Taki, T. 1994. Influence of differences in material on the endurance limit of gearing, in *Proceedings of the JSLE, Trib Conference,* May 10, Tokyo, Japan.

Tanabe, T., Mitsuhashi, T., and Amasaka, K. 2007. On intellectualization and accuracy improvement for the development of high reliable CAE software, *Quality, Journal of the Japanese Society for Quality Control*, 38(1): 52–56 [in Japanese].

Tayur, S. R. 1993. Structural properties and heuristics for Kanban-controlled Serial Lines, *Management Science*, 39(11): 1347–1368.

Valter, L. 1993. *Handbook in Design of Reliable Equipment*, DELTA Danish Electronics, Light & Acoustics.

Walker, M. W. and Orin, D. E. 1982. Efficient dynamic computer simulation of robotic mechanisms, *ASME Journal of Dynamic Systems, Measurement, and Control*, 104(3): 205–211.

Wang, J. X. and Roush, M. L. 2000. *What Every Engineer Should Know about Risk Engineering and Management*, Marcel Dekker, New York.

Whaley, R. C., Petitet, A., and Dongarra, J. J. 2000. Automated empirical optimization of software and the ATLAS project, Technical report, Department of Computer Science, University of Tennessee, Knoxville, TN.

Womack, J. P. and Jones, D. 1994. From lean production to the lean enterprise, *Harvard Business Review*, pp. 93–103, March–April.

Yamaji, M. and Amasaka, K. 2008. CAE analysis technology for development design utilizing statistical sciences, *The Open Industrial and Manufacturing Engineering Journal*, 1: 1–8.

Yamaji, M., Hifumi, S., Sakalsis, M. M., and Amasaka, K. 2010. Developing a strategic advertisement method "VUCMIN" to enhance the desire of customers for visiting dealers, *Journal of Business Case Studies*, 6(3): 1–11.

Chapter 8

Advanced Total Development System, Total Production System, and Total Marketing System
Key to Global Manufacturing Strategy

8.1 High-Linkage Model: Advanced Total Development System, Total Production System, and Total Marketing System

8.1.1 Introduction

At present, leading companies are promoting global production strategies to realize international quality standards and simultaneous production startup worldwide (Amasaka, 2004a). Advanced companies in particular are eagerly looking for a new quality management method to supply new attractive product models ahead of their competitors to ensure their survival in the worldwide market (Doos et al., 1991; Womack and Jones 1994; Taylor and Brunt, 2001). A future successful global marketer must develop an advanced management system that impresses users and continuously provides excellent products of high quality in a timely manner through corporate management (Nezu, 1995; Amasaka, 2007a). To realize manufacturing that places top priority on customers with good quality, cost, and delivery (QCD) in a rapidly changing technical environment, it is essential to create a new core principle that is capable of changing the work process quality of all divisions in order to reform super-short-term development production (Amasaka, 2004a).

Therefore, a high-linkage model employing an advanced Total Development System (TDS), Total Production System (TPS), and Total Management System (TMS) for the strategic development of management technology in advanced companies that utilize the author's New Just in Time (JIT) (Amasaka, 2002, 2007a) is hereby proposed. This is the traditional JIT system for not only manufacturing, but also for customer relations, business and sales, merchandise planning and engineering design, production engineering, administration, and management for enhancing business process innovation and the introduction of new concepts and procedures. The effectiveness of the proposed structured integrated triple management technologies system—Advanced TDS, TPS, and TMS—was verified at advanced car manufacturer A.

8.1.2 Need for a New Global Management Technology Model for New JIT Strategy

At present, advanced companies in the world, including in Japan, are shifting to global production to realize uniform quality worldwide and production at optimum locations for survival in an environment of fierce competition (Amasaka, 2004a). To attain successful global production, technical administration, production control, purchasing control, sales administration, information systems, and other administrative departments should maintain close cooperation with clerical and indirect departments. At the same time, they should establish strategic cooperative and creative business linkages with individual development, production, and sales departments, as well as with outside manufacturers (suppliers). For a manufacturer to accurately grasp customer intentions in order to proceed with production that satisfies the demands of the times, it is important that all the departments play a leading role in company-wide advanced management.

Because realizing production at optimum locations with the same quality worldwide ahead of competitors is the key to successful global production, strategic partnering between engineering, production, and sales operations as well as suppliers is essential (Amasaka, 2007a,b,h). In order to manufacture attractive products, it is necessary first for each of the marketing, sales, development design, and manufacturing divisions to manage themselves with successful internal linkages and linkages to other divisions to organically connect with administrative (technical administration, production control, purchasing administration, information systems, and quality assurance) and indirect clerical divisions through systematic, organizational activation of human resources in individual divisions. The author, therefore, recommends the urgent establishment of a new global management technology model for the next generation.

8.1.3 High-Linkage Model: Advanced TDS, TPS, and TMS for Global Manufacturing Strategy

A future successful global marketer must develop an advanced management system that impresses users and continuously provides excellent products of high quality in a timely manner through corporate management. Since providing what customers want before they are aware of those desires will become a more essential part of any successful manufacturing business, the author has constructed a customer science principle (Amasaka, 2002, 2005a) that utilizes New JIT. In this chapter, the author (Amasaka, 2007d) proposes a high-linkage model that is composed of a *triple management technology system*. This system combines Advanced TDS, Advanced TPS, and Advanced TMS, as shown in Figure 8.1.

8.1.3.1 Advanced TDS: Total Development Design Model

Currently, to continuously offer attractive, customer-oriented products, it is important to establish a new development design model that predicts customer needs. In order to do so, it is crucial to reform the business process for development design (Amasaka, 2007a). Manufacturing is a battle against irregularities, and it is imperative to renovate the business process in the development design system and to create a technology so that serious market quality problems can be prevented in advance by means of accurate prediction/control. For example, as a solution to technical problems, approaches taken by design engineers, who tend to unreasonably rely on their own past experience, must be clearly corrected. In the business process from development design to production, the development cost is high and time period is prolonged due to the "scale-up effect" between the stages of experiments (tests and prototypes) and mass production.

In order to tackle this problem, it is urgently necessary to reform the conventional development design process. Focusing on the successful case mentioned above, the author (Amasaka and Yamaji, 2006) deems it a requisite for leading manufacturing corporations to balance high-quality development design with lower cost and shorter development time by incorporating the latest simulation computer-aided engineering (CAE) and Science Statistical Quality Control (SQC) (Amasaka, 2003a, 2007d,f). Against this background, it is vital not to stick to the conventional product development method, but to expedite the next-generation development design business process in response to a movement toward digitizing design methods. Having said this, the author (Amasaka, 2007b) proposes the Advanced TDS, Total Development Design Model, as described in Figure 8.1a,

228 ■ *New JIT, New Management Technology Principle*

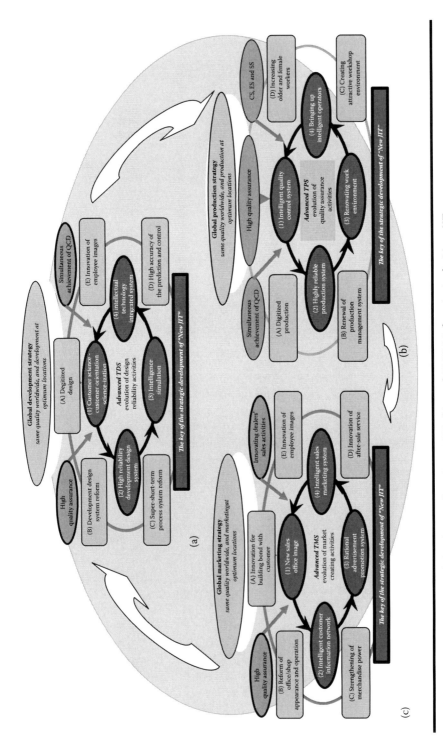

Figure 8.1 High-linkage model: advanced TDS (a), TPS (b), and TMS (c) for strategic New JIT.

and further updates TDS, a core technology of New JIT. New JIT is aimed at the simultaneous achievement of QCD by high-quality manufacturing, which is essential to realizing customer satisfaction (CS), employee satisfaction (ES), and social satisfaction (SS).

The following are necessary to realize this: (1) Customers' orientation (subjective implicit information) must be scientifically interpreted by means of customer science (Amasaka, 2002, 2005a), namely, converting the implicit information to explicit information by objectifying the subjective information using Science SQC, so as to (2) create a highly reliable development design system, thereby (3) eliminating prototypes with accurate prediction and control by means of *intelligence simulation*. To this end, it is important to (4) introduce the Intellectual Technology Integrated System, which enables knowledge sharing and all related divisions to possess the latest technical information.

8.1.3.2 Advanced TPS: Total Production Management Model

As digital engineering transforms manufacturing in workshops, a reduction in the engineering capability of members is often a result. This weakens the scientific production control that ensures that quality is incorporated in processes. Therefore, despite conventional success from the viewpoint of global production, it is an urgent task to strategically advance TPS (Advanced TPS) (Amasaka, 2007a). The author, considering the necessity of including and organically integrating these four elements in the strategic application of Advanced TPS toward global production, has clarified the Total Production Management Model, as described in Figure 8.1b. This model is an advanced production management principle designed to be applied as a global production technology and management model. The mission of Advanced TPS in the global deployment of New JIT is to realize CS, ES, and SS through production with high quality assurance.

In implementing New JIT for uniform quality worldwide and production at optimal locations (concurrent production), the fundamental requirements are (i) the renewal of production management systems to accommodate digitized production (see (A) and (B) in Figure 8.1b) and (C) the creation of attractive workshop environments tailored to (D) the increasing number of older and female workers (see (C) and (D) in Figure 8.1b). In more definite terms, what is needed is to (1) strengthen process capability maintenance and improvement by establishing an intelligent quality control system, (2) establish a highly reliable production system for high quality assurance, (3) reform the work environment in order to enhancement intelligent productivity, and (4) develop intelligent operators (skill level improvement) and establish an intelligent production operating system. Accomplishing these objectives will achieve higher-cycled next-generation business processes, enabling earlier implementation of uniform quality worldwide and production at optimum locations.

8.1.3.3 Advanced TMS: Strategic Development Marketing Model

With regard to recent changes in the marketing environment, what is needed now is to develop innovative business and sales activities that are unconventional and correctly grasp the characteristics and changes of customer tastes. Contact with customers has never called for more careful attention and practice. To offer an appealing, customer-oriented marketing strategy, it is important to evolve current market creation activities (Amasaka, 2005a). Therefore, the author proposes an Advanced TMS, Strategic Development Marketing System, as described in Figure 8.1c, which further updates the TMS. Advanced TMS (Amasaka, 2007c,h) is aimed at the implementation of a successful global marketing strategy by developing the same quality worldwide and marketing at optimum locations.

As shown in the figure, Advanced TMS aims to achieve a high cycle rate for market creation activities and is composed of four core elements: (1) A new vehicle sales office image to achieve a high cycle rate for market creation activities by (A) innovative bond building with the customer and (B) shop appearance and operation is particularly important. These constitute the basis for the innovation of (C) business talk, (D) after sales service, and (E) employee image. At a certain stage of execution, for example, it is more important to construct and develop (2) an intelligent customer information network, (3) a rational advertisement promotion system, and (4) an intelligent sales marketing system that systematically improve customer information software application know-how about users who patronize vehicles of various makes.

This information network turns customer management and service into a science by utilizing TMS according to customers' involvement with their vehicles in daily life. The strategic new marketing model that applies the proposed Advanced TMS is presented in the next section.

8.1.3.4 Advanced TDS, TPS, and TMS Driven by Science SQC

Supplying products that satisfy consumers (customers) is the ultimate goal of companies that desire continuous growth. Customers generally evaluate existing products as good or poor, but they do not generally have concrete images of products they will desire in the future. For new product development in the future, it is important to precisely understand the vague desires of customers. Proposal of customer science makes it possible to concretize customer desires, shown as "wants" in Figure 7.1 (refer to Section 7.1). To realize this, further expansion of Advanced TDS, TPS, and TMS, and Science SQC by utilizing the four core principles of New JIT strategy as shown in Figure 5.1.5 (refer to Section 5.4) is a new principle for a next-generation quality management technique for the manufacturing business, aiming at providing a universal "general solution" and thus creating a technology for problem solving (Amasaka, 2000, 2003a, 2005a, 2007a).

The first of the four core principles, *scientific SQC*, is a scientific quality control approach, and the second core principle, *SQC technical methods*, is a methodology for problem solving. The third core principle, Total SQC Technical Intelligence System (TTIS), is an integrated SQC network by using technical wealth. The fourth core principle, Management SQC, interprets the gap between the theory and reality of technical problems, such as the problems existing between departments and organizations, and verbalizes the implicit understanding inherent in the business process, thus further presenting it as explicit knowledge and as a general solution for the technical problem.

8.1.4 Application: Putting into Practice and Verifying the Validity of Advanced TDS, TPS, and TMS

In this section, the author introduces some typical research examples of how Advanced TDS, TPS, and TMS improved the management technology at advanced car manufacturer A and others.

8.1.4.1 Effectiveness of Advanced TDS

Some characteristic research cases that contributed to the establishment of Advanced TDS, the core technology of development design, are as follows: (i) the business process method for the Automobile Exterior Color Design Development Model and Automobile Form and Color Design Approach Model to support the development designer's conceptual process (Ando et al., 2008; Muto et al., 2011; Takebuchi et al., 2012a) and (ii) the creation of a New Software Development Model to assess the success of information sharing (Nakamura et al., 2011).

Moreover, through the cooperation of the development design, production, sales, service, and purchasing procurement departments with the suppliers, (iii) the application of advanced and accurate CAE for the Highly Reliable Development Model (Amasaka, 2007b), which shortened the period needed for highly reliable design and development, (iv) the failure mechanism of such worldwide technical issues as the Optimal CAE Design Approach Model for preventing oil leakage in the drive train oil seal (Amasaka, 2012), and (v) the automotive product design and CAE for bottleneck solution of loosening bolts (Amasaka et al., 2013b) were also clarified. As for this research, by utilizing the acquired technical results for improving the prediction accuracy of the CAE numerical simulation (Amasaka, 2007d), a substantial quality improvement was successfully achieved.

8.1.4.2 Effectiveness of Advanced TPS

Some characteristic research cases that contributed to the establishment of Advanced TPS, the core technology for production engineering and manufacturing, are

(i) the implementation of the global production compatible New Japan Global Production Model (NJ-GPM), a system designed to achieve worldwide uniform quality and production at optimal locations (Amasaka and Sakai, 2011), and (ii) the New Global Partnering Production Model to increase global quality by generating a synergetic effect (Ebioka et al., 2007).

Another example is (iii) the next-generation Human Digital Pipeline System, which enables a total linkage from design to manufacture through a digital pipeline for production preparation (Sakai and Amasaka, 2007b).

Furthermore, in order to realize the key to success in global production—uniform quality and simultaneous plant start-up worldwide—the authors introduce (iv) the how to build a linkage between a High Quality Assurance Production System and a Production Support Automated System (Sakai and Amasaka, 2014), (v) the Body Auto Fitting Model using NJ-GPM (Sakai and Amasaka, 2013), (vi) the Human-Integrated Assist System based on the Human Intelligence Production Operating System (Sakai and Amasaka, 2008), (vii) comparing experienced and inexperienced machine workers (Yanagisawa et al., 2013), and (viii) gaining mutual trust between logistics providers and shippers (Okihara et al., 2013).

Through these research cases, the departments related to manufacturing, production engineering, production management, purchasing management, and information systems are currently collaborating with one another, and results are being obtained in which the previously acquired results are being integrated and further developed (Amasaka et al., 2008).

8.1.4.3 Effectiveness of Advanced TMS

Some characteristic research cases that contributed to the establishment of Advanced TMS (Amasaka, 2007c), the core technology of departments related to business and sales, market survey, merchandise planning, purchasing management, publicity and advertisement, and promotions are as follows: (i) the implementation of the Intelligent Customer Information Marketing Model (ICIMM) to realize market creation (Amasaka, 2007c), (ii) the Scientific Mixed Media Model (SMMM) for boosting automobile dealer visits (Amasaka et al., 2013a), and (iii) the Total Direct Mail Model (TDMM) to attract customers (Ishiguro and Amasaka, 2012a,b).

The ICIMM recommends reform of the (dealer) shop front advertising and promotions, sales, and customer services, so that the expected results can be successfully achieved.

Through scientific verification of the customers' purchasing patterns, the combined effect of advertising promotion, consisting of TV commercials, newspaper ads, radio ads, flyers, train car ads, and direct mail/direct handing (DM/DH) employing SMMM and TDMM, was enhanced to raise the rate of customers' visits to automobile shops, as well as to realize "market creation."

All of these research cases are the results of strategic joint task team activities, in which the top management assumed a leadership role in finding the solution to the management technology issues utilizing Science SQC. Through company-wide partnering, QCD was simultaneously achieved and the strategic innovation of quality management was realized (Amasaka, 2008b).

8.1.5 Conclusion

This study discussed the proposal and effectiveness of a high-linkage model employing a structured triple management technology integrated system—Advanced TDS, Advanced TPS, and Advanced TMS—for expanding uniform quality worldwide and production at optimum locations. The model has been demonstrated based on the author's research at advanced car manufacturer A.

8.2 Automobile Exterior Color Design Development Model

8.2.1 Introduction

The idea of global quality competition in the auto industry has made developing body colors that best match the exterior design of vehicle models a critical factor in terms of product strategy, as color selection has the ability to affect consumer purchasing behavior.

The rapid global advancement currently underway brings with it increasingly diverse and complex personal values and subjective impressions, which are difficult to fully grasp. However, manufacturers that cannot accurately identify these consumer values and subjective impressions and incorporate the corresponding elements in their vehicle designs will find it difficult to remain competitive in the market.

Previously, the author (Amasaka, 2002, 2007) developed and verified the effectiveness of the Customer Science application system Customer Information Analysis and Navigation System (CS-CIANS), a business approach that scientifically indexes customer quality demands employing Advanced TDS through New JIT (refer to Sections 5.1, 7.1 and 8.1).

The authors have therefore concluded that traditional design processes, which are implicit and rely heavily on designer intuition and experience-based rules of thumb, must be reformed. To achieve this, the author has created the Automobile Exterior Color Design Development Model (AECD-DM) (Muto et al., 2011; Takebuchi et al., 2012) utilizing Science SQC (Amasaka, 2003a, refer to Section 5.4).

With a concrete target, the authors first use statistical modeling to visualize the factors involved in successfully meeting customer quality demands. Then, using customer values as a starting point, they create indexes that allow automakers to develop a business approach that will align the concepts and approaches of divisions involved in exterior design and color development. Finally, linkages are forged between the identified factors of success and the indexes.

The authors then apply their model to body color development at a specific company and obtain the given results.

8.2.2 Background

Prior research on this topic has been presented by Arima (2002), Kawamura (2003), Amasaka (2007, 2008), and others (Amasaka, 1995, Amasaka et al., 1999; Fujieda, et al., 2007; Yamaji and Amasaka, 2009b). However, the work of these authors focused on just a portion of the tasks associated with developing vehicle body colors. Research that utilizes a comprehensive business approach involving both paint manufacturers and automakers has yet to be carried out. In contrast, this paper strategically applies the customer science application system CS-CIANS constructed by Amasaka (2005a).

CS-CIANS is a method for generating customer value in a way that recognizes the necessity of reforming design work processes with product planning departments, with the aim of making customer intentions scientific. It is a networking system that includes a navigation system for analyzing customer information (via multivariate analysis) and analysis case examples. The authors expanded this model in order to create a business approach model from a perspective that includes both paint manufacturers and automakers.

8.2.3 Research Method

In order to understand how vehicle body color development processes were currently being carried out, the authors conducted an on-site investigation of color designers at three automakers and three paint manufacturers (Takebuchi et al., 2012b; Muto et al., 2011).

8.2.3.1 On-Site Investigations

The investigation revealed that vehicle body color development was carried out in six steps. First, a market survey was conducted. Next, the product planning department came up with a product concept.

An exterior designer at the automaker then came up with the exterior design and shape of the vehicle. In the fourth step, the automaker had a color designer come up with a color concept. The automaker's color designer then sent a request to the color designer at the paint manufacturer asking them to develop the paint. This step involved repeated discussion about specific colors between the two color designers. Finally, the designers communicated their color concept to a technician (a paint mixologist) who actually mixed and developed the paint.

However, there were problems with this process. Sometimes the designers could not come to a subjective agreement about the paint, or there were variations in work flow because they were basing their decisions on intuition and rules of thumb. The reality is that designers find it quite difficult to gain a concrete understanding of customer needs.

8.2.3.2 Visualizing Problem Areas and Work Processes

Problems in the current work process identified through the on-site investigations were converted into binary data, using problems as samples and keywords as variables, and then distributed using cluster analysis. The same data were also subjected to the quantification method of the third type of analysis, where the distribution was plotted and grouped into clusters on the same scatter diagram as shown in Figure 8.2. Data on the color development process were also grouped as shown in Figure 8.3, making the relationship between problem areas and the work process clear.

A closer look at the groups of relevant data in Figures 8.2 and 8.3 reveals that problems with information (6) occur during the market survey stage [1]. The figures also show heavy overlap between the vehicle color designer [2] and paint designer [3] groups, indicating that both designers are grappling with the same problems.

Figure 8.2 reveals that those problems have to do with indicators and standards (3), instincts, tricks of the trade, and personal experience (7), and gaps between the roles of different departments (8). At the same time, it is clear that among color designers [4], there are problems associated with shared aims (1), general means of

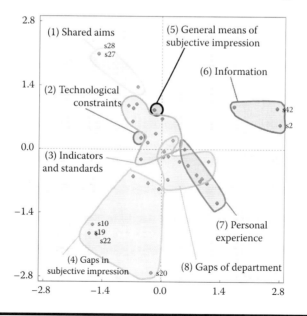

Figure 8.2 Group into clusters.

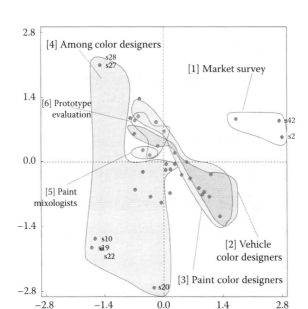

Figure 8.3 Group into the color development process.

subjective impression and gaps in subjective impression (4). For paint mixologists [5] and during the prototype evaluation stage [6], problems occur with technological constraints (2) and indicators and standards (3).

8.2.3.3 Visualizing Success Factors and Elements

8.2.3.3.1 Fact-Finding Research at Manufacturers

A series of shared question items were developed for automakers and paint manufacturers based on each question. There were 7 valid responses collected from paint manufacturers and 12 collected from automakers.

8.2.3.3.2 Identifying Success Factors and Elements at Different Manufacturers

The results of the fact-finding survey conducted in Section 8.2.3.3.1 were converted into binary data using the same method outlined in Section 8.2.3.2: success factors were used as samples and keywords as variables and then the data were distributed using a cluster analysis.

The analysis revealed that the five factors that allowed paint manufacturers to successfully create the product quality their customers demanded were (1) market surveys, (2) technological constraints, (3) communication, (4) scientific analysis, and (5) collaboration with other departments. Automakers identified six success factors: (1) market surveys, (2) technological constraints, (3) communication, (4) scientific analysis, (5) collaboration with other departments, and (6) evaluations and standards.

Text mining was then used to get a clearer picture of the core elements involved in each success factor. A word network diagram was then used to identify the connections among different elements, as shown in Figure 8.4, and the strength of these different connections was calculated using partial correlation coefficients between the elements.

A partial correlation coefficient of 0.7 or above was assigned the indicator +++, 0.6–0.7 was assigned ++, and 0.5–0.6 was assigned +. Figure 8.5 shows the results of this analysis for paint manufacturers.

8.2.3.3.3 Visualizing Success Factors and Elements at Both Manufacturers

The success factors and elements identified at the different types of manufacturers were visualized using a single figure. They were then combined into one using all of the partial coefficients of 0.7 or greater. Using the example of the scientific analysis shown in Figure 8.5 for instance, partial correlation coefficients for "quantifying color concepts," "quantifying color," and "scientific data analysis" are 0.7 or more.

These are thus combined into a single success factor: "quantifying color concepts and colors through scientific analysis." Next, the elements identified during this process are visualized in the six separate categories (refer to Figure 8.7 inside): shared elements at both manufacturers, elements only at automakers, and elements only at paint manufacturers.

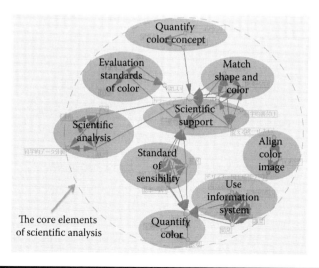

Figure 8.4   Extraction example of text mining.

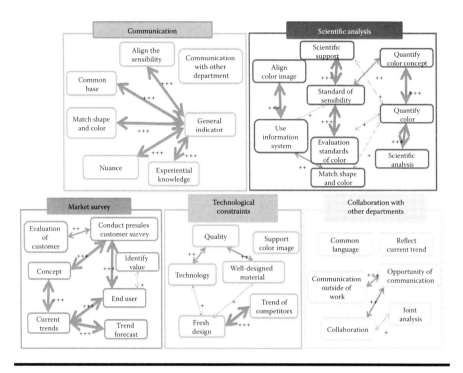

Figure 8.5 Success factors and elements at paint manufacturers.

8.2.3.4 Constructing the Business Approach Model

8.2.3.4.1 Indexing Customer Values

In this section, customer values were used as a basis for an indexing procedure aimed at developing a business approach that aligned the concepts and approaches used by departments involved in vehicle exterior design and body color development. In order to do this, survey data were collected on-site at the AMLUX consultation space owned by Toyota Motor Corporation. A total of 417 data items (including free response data) related to vehicle body color were taken from about 6000 items. The analysis was narrowed down to the 134 free response items related to the most common vehicle type (a sedan).

Text mining was used to identify characteristic colors for each age group. An analysis was then conducted for each group. The customer opinion results for participants in their 40s fell into three general groups: (1) opinions on existing colors, (2) color requests, and (3) awareness of color when purchasing a vehicle. An analysis of data from other age groups showed that all customer opinions could be divided into these three categories.

These were converted into three indicators demonstrating customer values: (1) color evaluation, (2) color preference, and (3) focal points during purchase.

Organizing the opinions of huge numbers of customers into these three indicators should allow all departments to align their concepts and approaches. Figure 8.6 shows the results of this analysis for the 40s age group.

8.2.3.4.2 Constructing an Automobile Body Color Development Approach Model

Using the success factors and elements identified in Figure 8.7 inside and linking them using the three indicators developed in Section 8.2.3.4.1, the Automobile Body Color Development Approach Model in Figure 8.7 was developed with the aim of generating the quality that customers demand.

This business approach model adds three new indicators to the current vehicle body color development work process shown in the center of the figure and then links them with the success factors and elements from Figures 8.6 through 8.11. The result is a set of guidelines that manufacturers can use to meet the quality needs of their customers.

8.2.4 Application Example

Developing the above-mentioned AECD-DM, the authors conducted a preference survey on young people's lifestyles and determined the elements most desired by young people when choosing automobile exterior colors. The authors then determined the color elements that have the most influence on customers

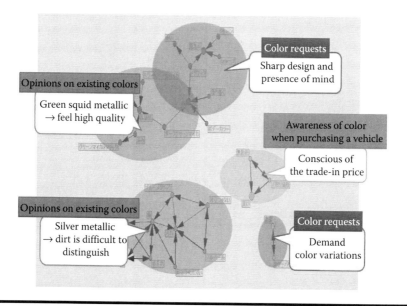

Figure 8.6 Example of analysis of customers in their 40s.

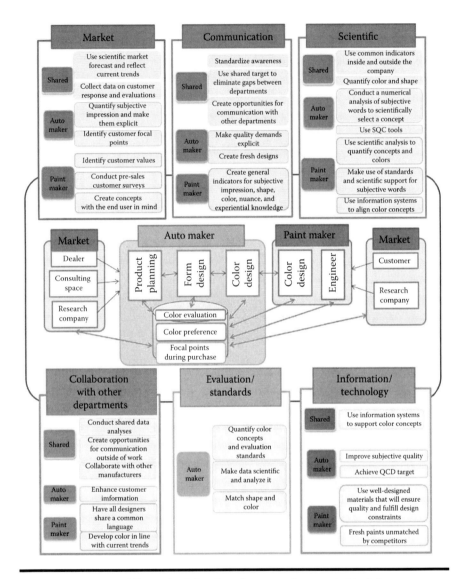

Figure 8.7 Automobile Body Color Development Approach Model.

aesthetically and conducted correspondence analysis on the preferences and color elements using Science SQC. The findings were subsequently used in the development of exterior colors.

8.2.4.1 Importance of Textual Expressions of Exterior Colors

Color can be described in terms of three elements: hue, luminosity, and intensity. However, automobile exterior colors are recognized not only in terms of these three

Global Manufacturing Strategy ■ 241

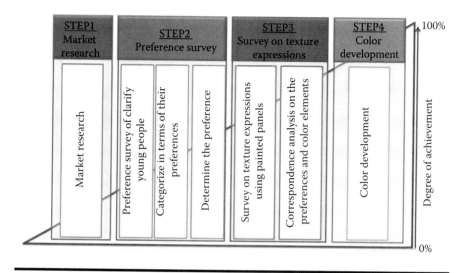

Figure 8.8 Automobile Exterior Color Design Approach Model.

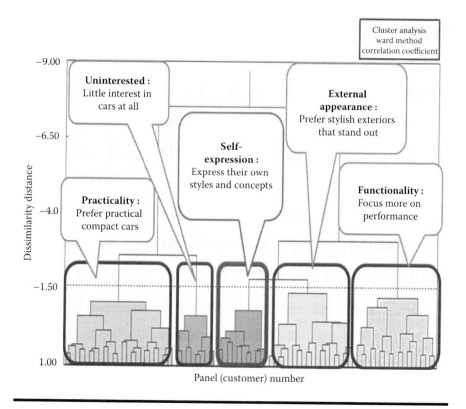

Figure 8.9 Categorizing preferences using cluster analysis.

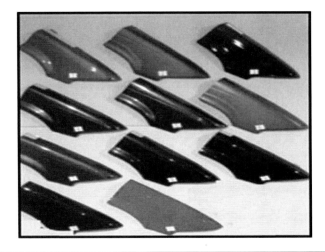

Figure 8.10 Painted panels used for surveys.

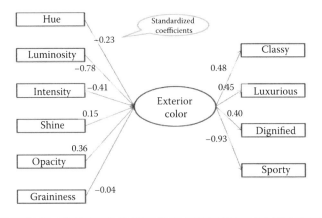

Figure 8.11 Path from color elements to exterior color.

elements, but also in terms of other qualities, such as the elements of intensity of reflected light, cityscape reflected in the colors, metallic appearance, and depth. Qualities such as depth and metallic appearance are textural expressions that are unique to automobiles and are difficult to quantify. Two black cars with different textural expressions will give different impressions.

Automobiles have a three-dimensional form, and textural expressions are important in making this unique form stand out. The development of new paint technologies has enabled paints to be designed so their brightness and hue differ when viewed from different angles through use of reflective materials in the paints to vary their luminosity. Such developments in paint technology make it possible to achieve a variety of textural expressions and are becoming increasingly important for automobiles.

Automobile manufacturers seek to create appealing automobiles by strategically varying the textural expressions of each vehicle's design concept. In other words, the textural expressions in automobiles constitute a significant and influential factor in the car-buying process (Koga et al., 2003; Fujieda et al., 2007; Kawasumi and Suzuki, 2008; Park et al., 2009; Tokinaga and Utsunomiya, 2009; Watanabe and Jo, 2009; Yomo and Kato, 2010).

8.2.4.2 Proposal of the AECD-AM

The authors believe it is vital to visualize customer preferences in order to create exterior colors that are appealing to customers; therefore, this study seeks to firmly establish the Automobile Exterior Color Design Approach Model (AECD-AM) throughout the development process, as shown in Figure 8.8. This approach/model is described below together with actual examples.

8.2.4.2.1 Market Research (Step 1)

In Step 1, market research was conducted to determine young people's preferences and find out what they consider important in a car. First, observations were carried out in two areas of Tokyo, Japan: between Harajuku and Shibuya, where there are many young people; and in Hiroo, where there are fewer young people.

Comparing the cars being driven in these areas, it was found that in Harajuku, the cars gave the impression that young people were asserting themselves, preferring vivid colors that stand out. In contrast, the cars in Hiroo gave a more luxurious impression. The cars had shiny surfaces displaying opaque, grainy colors with depth, and there was a preference for basic colors such as black and silver. In both cases, it was found that color preferences varied according to both cityscape and lifestyle.

8.2.4.2.2 Preference Survey (Step 2)

In Step 2, the authors conducted a survey to clarify young people's preferences and their perspectives concerning automobiles. First, a preference survey sheet was created to determine the preferences of young people targeted by the survey. Participants were asked to evaluate the following items on a seven-point scale: (1) 4 items concerning interest/enthusiasm for automobiles and (2) 15 items concerning preferred automobile designs (preferences) (Appendix 8.1). They were then asked to comment freely on (3) hobbies/interests, (4) most-read fashion magazines, and (5) favorite colors.

Four collage panels each were made for men and women based on fashion trends, and participants were asked which trends they most identified with. This survey was aimed at men and women in their twenties, and answers were obtained from a total of 74 men and women.

Based on the results of the preference survey, the following analysis was conducted to clarify car-related preferences. First, cluster analysis was conducted using the data on preferences gained from the seven-point evaluation as shown in Figure 8.9. The differences in respondents' preferences were categorized into five groups and the groups were labeled as follows according to the respondents' particular interests and preferences: practicality, uninterested, self-expression, external appearance, and functionality.

Based on the free opinions given, the characteristics of each group showed that respondents in the "practicality" group prefer practical compact cars, and those in the "uninterested" group show little interest in cars at all. Those in the "external appearance" group prefer stylish exteriors that stand out, while those in the "functionality" group focus more on performance. Respondents in the "self-expression" group show a strong tendency to prefer cars that help to express their own status or image, and members of this group tend to have the most interest in cars. This study focuses on developing exterior colors for people who value self-expression.

Principal component analysis was conducted to clarify which aspects of automobile design are desired by each group. Having clarified each group's car-related preferences in this way, we were able to determine the four preferences most important to people who value self-expression (classy, luxurious, dignified, and sporty). The authors also found that respondents who said they value self-expression tend to have a boyish or casual style and prefer red, black, and brown.

Accordingly, it is important to narrow down the colors to red, black, and brown and focus on the above four preferences when developing exterior colors for people who value self-expression. In order to successfully achieve this, it is necessary to determine the extent to which color influences customers aesthetically; therefore, a survey was conducted on textural expressions in Step 3.

8.2.4.2.3 Survey on Textural Expressions (Step 3)

A survey was conducted on textural expressions in order to quantify the extent to which automobile exterior colors influence customers aesthetically. First, the authors identified which of the many elements of color have the most influence on customers aesthetically. In addition to the three elements of color (hue, luminosity, and intensity), the authors also quantitatively identified three textural expressions unique to automobile colors (shine, opacity, and graininess).

Shine is the intensity of reflected light, opacity is the variance in brightness between areas where light does or does not fall, and graininess is the size of particles giving the color its grainy appearance. Thus, a total of six elements of color (hue, luminosity, intensity, shine, opacity, and graininess) were identified.

Next, the authors conducted a survey using 11 painted panels with varying color elements to determine how customers are influenced aesthetically by color variations including the above textural expressions. For this study, the panels were shaped and painted like car panels to show the relevant textural expressions. First, the painted panels were selected. Three panels were initially prepared for the survey, using three colors (red, black, and brown) with minimal opacity and graininess.

Then, a further 8 panels were prepared, varying the effect of the opacity and graininess equally to produce a total of 11 panels as shown in Figure 8.10. The six elements of color (hue, luminosity, intensity, shine, opacity, and graininess) of each panel were measured. The results are shown in Table 8.1. Then, participants were asked to look at each painted panel and evaluate, according to a seven-point scale, the extent to which each panel gave the desired impression for the four preferences (classy, luxurious, dignified, and sporty).

Aesthetic evaluation data were obtained from the results. The survey was aimed at men and women in their twenties, and answers were obtained from a total of 94 men and women, as shown in Table 8.2.

Table 8.1   Six Elements of Color of Each Panel

Panel No.	Hue (H45°)	Luminosity (L45°)	Intensity (C45°)	Shine (60°)	Opacity (L15–110°)	Graininess (HG)
1	35.00	83.89	41.17	1.79	0	89.90
2	16.79	39.00	32.00	33.90	75.70	91.50
3	23.54	60.10	33.00	4.01	53.60	90.40
4	27.49	56.55	28.06	35.70	59.20	89.10
5	6.19	2.10	342.00	34.79	89.70	96.50
6	3.44	2.80	31.00	19.94	17.90	90.80
7	30.45	10.96	44.15	59.85	29.20	95.90
8	9.90	8.04	10.56	0.59	0	94.50
9	5.48	1.10	255.50	19.05	51.20	93.00
10	2.77	0.70	260.00	11.90	4.00	89.60
11	1.66	0.12	93.07	3.64	0	92.60

Table 8.2   Aesthetic Evaluation Data (Preference)

Participants	Panel No.	Classy	Luxurious	Dignified	Sporty
1	1	4	3	2	4
2	1	7	4	4	5
3	1	2	3	6	7
4	1	3	1	5	7
:	:	:	:	:	:
94	11	5	7	4	3

The results of the survey on textural expressions were used to determine the influence of variations in each color element in terms of the preferences. Covariance structure analysis was conducted to determine the correspondence relationship between color elements and preferences. The results were then used to provide information to contribute to the development of exterior colors appealing to people who value self-expression.

In the model diagram used for this covariance structure analysis, the "exterior color" latent variable was chosen as the latent variable that would satisfy the four vehicle design preferences of those who value self-expression. A path analysis using AMOS 6.0 was then conducted in order to determine the cause-and-effect relationship between the six color elements and exterior color.

Further paths were then drawn between the four preferences and a latent variable representing the exterior color. This exterior color variable consolidates the preferences of people in the self-expression group when evaluating automobiles and the paths enable the influences to be understood. The coefficients of each path are standardized coefficients.

The absolute values of these standardized coefficients indicate the degree of influence, while the direction of influence is indicated by positive or negative numbers as shown in Figure 8.11.

The model was subjected to a goodness-of-fit test once it was corrected using a modification index, resulting in reasonable scores of 0.979 goodness of fit index (GFI), 0.895 adjusted GFI (AGFI), 0.038 standardized root mean square residual (SRMR), and 0.097 root mean square error of approximation (RMSEA). The results of the t-test for each path coefficient showed that the path coefficient from graininess to exterior color was not significant, but that the remaining path coefficients were significant.

The standardized coefficients of paths shown on the model diagram can be used to determine which color elements are required and how they need to be altered to bring the color closer to the target preferences. For example, in developing a more classy color, the influence of luminosity is greatest at -0.37 ($=-0.78 \times 0.48$). This shows that the degree of luminosity should be lowered to create a more classy color.

Table 8.3 shows the standardized coefficients for the effect of color elements desired for exterior colors. The highest values are for luminosity (-0.78), intensity (-0.41), and opacity (0.36), showing that these elements have the greatest influence. The coefficients also show that the desired color can be achieved by lowering the degree of luminosity and intensity and raising the level of opacity. Colors can then be developed based on these findings.

8.2.4.2.4 Color Development (Step 4)

In Step 4, the authors developed colors aimed at the self-expression group, based on the findings from Step 3. Table 8.3 shows that colors preferred by people in the self-expression group can be achieved by lowering the degree of intensity and

Table 8.3 Path from Color Elements to Exterior Color

Color Elements	Standardized Coefficients
Hue	−0.23
Luminosity	−0.78
Intensity	−0.41
Shine	0.15
Opacity	0.36
Graininess	−0.04

luminosity and raising the level of opacity. Lowering the degree of intensity and luminosity results in subdued colors, giving a composed impression. Raising the level of opacity accentuates the color, giving a dignified impression.

Based on these findings, three painted panels were produced using the colors from the survey (red, black, and brown). When developing the colors for the panels, the authors focused primarily on luminosity, intensity, and opacity to produce colors to suit the preferences of people in the self-expression group, as shown in Figure 8.12.

8.2.4.2.5 Effectiveness of the AECD-AM

Respondents in the self-expression group were asked to freely evaluate the three developed colors. The respondents gave positive evaluations, saying they liked how classy the colors were and that they would like a car with colors such as these. This shows that the target users were satisfied with the exterior colors developed using this model.

Feedback was then obtained from designers at automobile and paint manufacturers to validate the effectiveness of the model. The feedback confirmed that

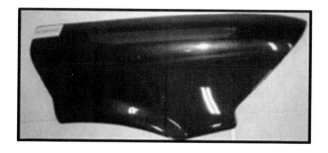

Figure 8.12 Exterior color designed for those pursuing self-expression.

quantifying the findings in this way makes it possible to gain an unbiased view of aesthetic influences on customers, thereby validating the model's effectiveness.

8.2.5 Verifying the Automobile Exterior Color Design Approach Model

In order to confirm the effectiveness of the new business approach model, the author requested that an automaker and a paint manufacturer evaluate it. The model was favorably received, meriting the following responses: "this process is something I intuitively understand, but it's good to see all of these elements finally combined into a single model;" "this model would be valuable as a tool for designing a checklist for each work process;" and "the model is a good tool for developing a shared awareness of work approaches." The effectiveness of the model was thus confirmed.

8.2.6 Conclusions

In this study, the authors used statistical science in the form of the customer science application system CS-CIANS (a method for generating customer value, which they had previously created and verified) to construct the AECD-DM. This represents a strategic expansion of customer science.

Specifically, data collected from companies were statistically analyzed in order to visualize success factors in meeting customer quality demands. Indicators reflecting customer values were then constructed. By linking these success factors and indicators, AECD-DM was created and its effectiveness confirmed. In the near future, the author will apply this research to other industries and promote further expansion of this business approach model.

8.3 Highly Reliable Development Design CAE Model

8.3.1 Introduction

In recent years, the automotive industry has been engaged in a global production strategy for simultaneous achievement of quality, cost, and delivery (QCD), aiming to achieve worldwide uniform quality and production at optimum locations in an effort to prevail and survive in global quality competition (Amasaka, 2007b). One of the specific measures taken was an urgent improvement of intelligent productivity in the advanced manufacturing processes of planning and development, design, prototyping evaluation, mass production preparation, and mass production for the purpose of offering highly reliable products to create customer value in a short period of time.

Among other things, a close look at the development design and production process stage reveals an excessive repetition of "experiment, prototyping, and evaluation" that prevents the "scale-up effect" generated in the bridging stage between prototyping, experiment, evaluation, and mass production. Therefore, innovation of the development and production method, as well as reduction of the development period, is a top priority (Amasaka, 2007f, 2008a).

Against this background, the author focuses on the technical requirements for high assurance CAE for establishing a development design quality assurance system that is indispensable for CAE in the automotive industry. A prerequisite for automotive development and design is to derive highly reliable CAE analysis results that show no gap between the actual machine lab tests and the analysis results. The rational integration of overall optimality and partial optimality needs to be achieved through the process of problem–theory–algorithm–modeling–computer as a technical requirement to be included in highly reliable CAE software (Amasaka, 2008a).

For this reason, innovation to promote the advance from the conventional evaluation-based development, which uses the prototyping and experiment process (a method based on the confirmation of real goods for improvement) and has long supported highly reliable designing, to a CAE prediction-based design process is urgently needed. To accomplish, this a new development design technique was established: the Highly Reliable Development Design CAE Model. This is the key to strategic promotion of the Advanced TDS, Strategic Development Design Model, which utilizes the core element TDS of New JIT (Amasaka, 2002).

In an effort to realize this, the author has proposed the high quality assurance model for super-short-period development design, a Highly Reliable Development Design CAE Model, and demonstrated its effectiveness (Amasaka, 2007b). The author explains a "highly reliable numerical simulation through the use of a concrete target. As an application example, a simulation technology was developed for molding urethane foam and it was implemented as Production of CAE Software for Molding Automotive Seat Pads: Urethane Foam Molding Simulator (Amasaka, 2007b).

8.3.2 Need for a New Global Development Design Model

At present, however, advanced companies in the world, including Japan, are shifting to global production to realize uniform quality worldwide and production at optimum locations for survival amid fierce competition (Amasaka, 2007b,d).

To attain successful global production, technical administration, production control, purchasing control, sales administration, information systems, and other administrative departments should maintain close cooperation with clerical and indirect departments while establishing strategic cooperative and creative business linkages with individual development, production, and sales departments, as well as with outside manufacturers (suppliers).

Today, when consumers have quick access to the latest information in the worldwide market thanks to the development of information technology (IT), strengthening of a development design that utilizes TDS has become increasingly important. Simultaneous attainment of QCD requirements is the most important mission for developing highly reliable new products ahead of competitors, as shown in Figure 5.2 (refer to Section 5.1).

As shown in the figure, the main elements are composed of (a) collection and analysis of both internal and external information focusing on the importance of design thought, (b) design process development and its management, (c) design technology to create a general solution, and (d) design guidelines (theory, action, decision making) to train designers.

Recently, the author has been verifying the effectiveness of proposed TDS at advanced corporations, including the Toyota Motor Corporation. In order to manufacture attractive products, therefore, the author requires the urgent establishment of a new global development design model for the next generation.

8.3.3 Advanced TDS, Strategic Development Design Model

Currently, to continuously offer attractive, customer-oriented products, it is important to establish a *new development design model* that predicts customer needs (Amasaka, 2004b, 2005b). In order to do so, it is crucial to reform the business process for development design (Amasaka, 2007b). Manufacturing is a battle against irregularities, and it is imperative to renovate the business process in the development design system and to create a technology so that serious market quality problems can be prevented in advance by means of accurate prediction/control.

For example, as a solution to technical problems, approaches taken by design engineers, who tend to unreasonably rely on their own past experience, must be clearly corrected. In the business process from development design to production, the development cost is high and the time period is prolonged due to the "scale-up effect" between the stages of experiments (tests and prototypes) and mass production. In order to tackle this problem, it is urgently necessary to reform the conventional development design process. Focusing on the successful case mentioned above, the author deems it a requisite for leading manufacturing corporations to balance high-quality development design with lower cost and shorter development time by incorporating the latest simulation CAE and Science SQC (Kuriyama, 2002; Magoshi et al., 2003; Hashimoto et al., 2005; Amasaka, 2008a).

Against this background, it is vital not to stick to the conventional product development method, but to expedite the next-generation development design business process in response to a movement toward digitizing design methods. Having said the above, the author proposes the Advanced TDS, Strategic Development Design Model, as described in Figure 8.13, and further updates

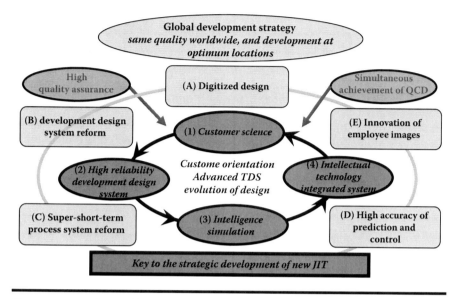

Figure 8.13 Advanced TDS, Strategic Development Design Model.

the TDS, a core technology of New JIT (Amasaka, 2008a). New JIT is aimed at the simultaneous achievement of QCD by high-quality manufacturing, which is essential to realize customer satisfaction (CS), employee satisfaction (ES), and social satisfaction (SS).

For realization, (1) customers' orientation (subjective implicit information) must be scientifically interpreted by means of Customer Science (Amasaka, 2002, 2005a), namely, converting the implicit information to explicit information by objectifying the subjective information using Science SQC (Amasaka, 2003a) so as to (2) create a *Highly Reliable Development Design System*," thereby (3) eliminating prototypes with accurate prediction and control by means of *intelligence simulation*.

To this end, it is important to (4) introduce the *Intellectual Technology Integrated System*, which enables sharing of knowledge and the latest technical information possessed by all related divisions.

8.3.4 Highly Reliable Development Design CAE Model

Investigation into the management technology issues concerning managers and administrators of advanced corporations indicates that development design puts primary emphasis on technical problems in the process of finding a solution, and resources are concentrated on the pressing issue of developing new models and products on a proposal basis, as shown in Figure 2.2 (refer to Chapter 2).

8.3.4.1 Revolution in Manufacturing Development Design and the Evolution of CAE

The author takes the automobile industry as a representative example of the manufacturing industry, as shown in Figure 2.1 (refer to Chapter 2) (Takeoka and Amasaka, 2006; Amasaka, 2008a). The conventional process of automobile development/production (from planning to mass production) was carried out through the first and the second cycles of "experiment–prototyping–evaluation." As a result, it took approximately 40 months from the start of development to the beginning of mass production.

Recently, however, the development production period of automobile development/production (from planning to mass production) has been further shortened from 2 years to 1 year, which includes the process of designing–prototyping–experimental evaluation–production preparation–mass production trial. This process is now anticipated by means of (1) simultaneous engineering (SE) activities, (2) advancement of computer-aided design/manufacturing (CAD/CAM; 2D→3D solid) and IT, (3) introduction of and a wider range of applications of CAE, and (6) the advancement of knowledge integration, cutting down the number of prototypes and overlapping stages in the experiment–prototyping–evaluation cycles required (Amasaka, 2007b; Amasaka et al., 2005).

Now, amid fierce competition and demand for time reduction in product development, what is normally called "rework" resulting from various production/quality-related problems is virtually impossible. This intensifying competition, coupled with a shortage of development and design specialists, has been addressed by increasing CAE investment and bringing in an outsourced workforce. It has been observed, however, that because of insufficient development of training programs to foster highly skilled CAE engineers, the effectiveness of CAE has been weakened and the authors' development/design process aimed at simultaneous achievement has been hindered.

Through a case study of a leading corporation, the author has also grasped the effectiveness of CAE utilization and the importance of CAE education (Sakai and Amasaka, 2006a; Amasaka, 2007b). Also, studies by the author have demonstrated the effectiveness of incorporating SQC, which expands the effectiveness of CAE and its range of application (Kusune et al., 1992; Amasaka et al., 1996; Amasaka, 2003c). Based on the above knowledge, the impact of CAE and obstacles to be overcome are plotted in the relation diagram from the standpoint of CAE management and simultaneous achievement of QCD, which realizes the high quality assurance of automobiles, as shown in Figure 8.14 (Amasaka, 2007b).

By summarizing the diagram, it becomes clear that one of the problems in applying CAE for the realization of simultaneous achievement of QCD is the failure to understand the mechanism of the technical problems encountered and apply it to a CAE model (Amasaka, 2005b, 2007d,f). The second point observed is

Global Manufacturing Strategy ■ 253

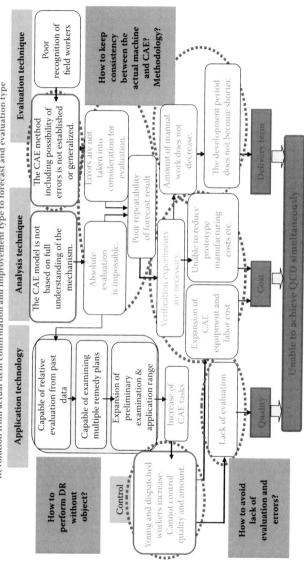

Figure 8.14 Issues when applying CAE to development design reform.

that, as a substitute for prototypes and experimental evaluation, this CAE analysis proves to be insufficient for reliable prediction and control.

The gap (analysis error) between the analysis and the experimental evaluation data can be as much as a few percent, and at present, the establishment of CAE software and its usage taking error into account is not at a satisfactory level. Therefore, despite its expansion, CAE cannot be regarded as making a sufficient contribution to the simultaneous achievement of QCD and reduction in development time (Amasaka, 2005b; Amasaka et al., 2005).

8.3.4.2 Highly Reliable Development Design CAE Model

For this reason, innovation to promote the advance from the conventional evaluation-based development, which uses the prototyping and experiment process (a method based on the confirmation of real goods for improvement) that has long supported highly reliable design, to a CAE prediction-based design process is urgently needed. To accomplish this, a new development design technique was established: the Highly Reliable Development Design CAE model utilizing Advanced TDS (Amasaka, 2007b).

Therefore, the author discusses the *extraction of issues* with a view to creating the highly reliable CAE model. Next, the developers engaging in CAE (car body manufacturers: Hino Motors, Ltd. and Toyota Motor Corporation; parts manufacturers: NHK Spring Co., Ltd.; software makers: Mizuho Information & Research Institute, Inc.; system developers: Mathematical Systems Inc., Tsukuba University, and Aoyama Gakuin University, etc.) have jointly sorted out the free responses to the questions given and summarized the required technical requests, as shown in Figure 8.15 (Amasaka, 2007f; Tanabe et al., 2006).

This model illustrates the techniques belonging to the domains of (1) problem setting, (2) modeling, (3) algorithm, (4) theory, and (5) computing (calculation technology). These techniques are being used to realize the systemization or formulation of working level problems, development of the kind of algorithms that utilize calculation resources more efficiently, logical analysis of the algorithms, and improvement in hardware and software technology for accelerating the calculation speed.

These are the development targets for all kinds of new and old technologies related to computer science that have been actively promoted throughout the world. Far from intending to thoroughly cover the field, this figure simply lists some names of the main techniques in each domain, but it helps us to see the large number of options available for elemental technologies involved in CAE as we try to improve it. However, from the standpoint of implementing CAE as a problem-solving method on a working level, the sheer number of, and a wide selection of, these elemental technologies is not sufficient (Whaley et al., 2000; Nonobe and Ibaraki, 2001; Enrique and Parallel, 2005).

This is because CAE is thought to be a process consisting of multiple elemental technologies. The process of CAE starts with (1) setting problems to be solved and

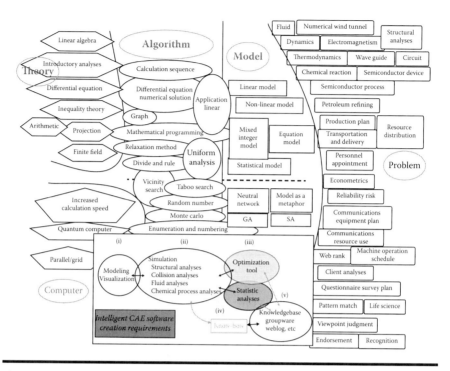

Figure 8.15 Highly Reliable Development Design CAE Model utilizing Advanced TDS (problem–model–algorithm–theory–computer).

(2) modeling these problems as some type of mathematical formula. In CAE, when using calculators as a means to analyze the model, this means of analysis needs to be provided in the form of a calculation procedure, namely, (3) algorithms, so that the software can perform the calculation. The validity, applicable range, and performance or expected precision of such algorithms themselves can be deduced from (4) some kind of theory.

Needless to say, the technology related to the computer itself, functioning as hardware to realize the algorithms, undoubtedly has a large effect on the success of CAE. In addition, the elemental technologies composing the process need to be those that cohere with one another and complement any weaknesses contained therein for realization of highly reliable CAE. The intelligence CAE software creation requirements are shown in Figure 8.15 (illustration: from (i) to (iv)) (Amasaka et al., 2005). Though algorithms themselves might be excellent in theory, if they are not properly and efficiently implemented in the calculator, favorable results cannot be expected. As the performance of algorithms largely depends on the compatibility with the modeling, errors in the modeling hinder the efficient performance of the algorithms even if the problem setting is correct.

Compatibility between the algorithms and calculator cannot be overlooked since algorithms which can draw out the best performance from the calculator are able to produce the desired results. In short, when appropriate combinations among the elemental technologies are not selected, the entire process of CAE does not function. In other words, success in CAE depends on the "collective strengths" of the elemental technologies, and this is what is asserted here.

Skilled CAE engineers are not experts in all the fields of the elemental technologies, but they understand their characteristics and interactions as implicit knowledge and thus conduct selection and combination to obtain favorable interactions and consequently the desired results. The formulation of such implicit knowledge confined to the personal know-how of the engineers is an indispensable step to be taken for sophistication of CAE as a problem-solving method and therefore it is positioned as a major theme in authors' work (Amasaka, 2005b; Amasaka et al., 2005; Takeoka and Amasaka, 2006).

8.3.5 Application Example: Highly Reliable Numerical Simulation—Production of CAE Software for Molding Automotive Seat Pads

In this section, a highly reliable numerical simulation is explained that applies Advanced TDS. As an application example, "the production of CAE software for molding automotive seat pads" is presented (Amasaka, 2007f; Enrique and Parallel, 2005). For the purpose of reducing the trial production period and improving the precision of automotive seat pad molding, a simulation technology was developed for molding urethane foam and it was implemented as the technical element analysis of the urethane foam molding simulator, as shown in Figure 8.16 (Amasaka, 2007b).

In the development process, particular effort was made to remove the empirical rules and to implement universal equations with a view to responding to original design shapes and complex composition. Consideration was also given to practical use at production sites by enabling simulation in a short period of time.

8.3.5.1 Grasping the Problematic Phenomena

The phenomena associated with urethane foaming in a mold are characterized by its expansion due to its changing composition through chemical reactions and also considerable changes in viscosity. These are the major characteristics of the simulation.

These phenomena can be roughly divided into the following: chemical reactions, flow of a mixture of raw materials and urethane foam, transition of the chemical types inside the mixture, rise in airflow and air pressure, and rise in temperature.

These in turn can be sorted out as follows:

1. Multiphase flow of gas, liquid, and solid
2. Volume increase and free liquid surface behavior due to foaming

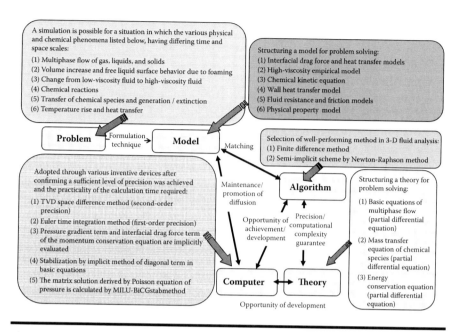

Figure 8.16 Technical element analysis for urethane foam modeling simulator.

3. Change from low-viscosity fluid to high-viscosity fluid
4. Rise in temperature caused by chemical reactions and heat generation, as well as heat transfer
5. Transition of chemical species and generation/extinction

8.3.5.2 Theoretical Analysis Model

The phenomena roughly grouped in Section 8.3.5.1 can be expressed by chemical and physical equations or analytic models as shown below:

1. Basic equation of multiphase flow
2. Mass transfer equation of chemical species
3. Biphasic interface model
4. Physical properties calculation model such as the viscosity coefficient
5. Chemical kinetic equation
6. Wall heat transfer and friction model

The basic equation of multiphase flow or mass transfer equation of chemical species above can be expressed by time development–based partial differential equations representing transfer and diffusion of the fluid energy or chemical species concentration. Other items representing the phenomena of heat transfer, wall friction, generation/extinction by chemical reactions, and so on can be added to those equations as needed. Analytic models such as the biphasic interface model, chemical kinetic equation, and wall heat transfer and friction model listed above are additional models, and they can be associated with time development–based partial differential equations such as basic equation of multiphase flow and mass transfer equation of chemical species.

These additional analytic models are also called constitutive equations, and they are indispensable for completing time development–based partial differential equations. Generally speaking, constitutive equations are used to model phenomena wherein temporal and spatial scales are largely different from the phenomena expressed in partial differential equations. They are also used in cases where the phenomena are difficult to express in equations or theoretical formulas dealing with physical properties. Many of these are often based on the data obtained through experiments, and the physical properties employed are usually calculated by polynomials or exponentials for physical values such as temperature or pressure.

8.3.5.3 Implementation of Intelligence and High Precision

In order to conduct simulation while linking the basic equations and constitutive equations in Section 8.3.5.2 on a computer, numerical modeling is needed. In the urethane foam molding simulator, the basic equations were digitized by using the finite difference method as a numerical analysis method, and the 3D space inside the mold was segmented by rectangular grids for computer simulation.

Setting physical chemical values such as pressure, temperature, gas volume fraction, chemical type concentration, and so on as unknown, digitized equations are solved to find the spatial distribution of these physical chemical values along with the progress of time. Sufficient consideration must be given to calculation errors and stability in the numerical solution approach. This was adopted in the urethane foam molding simulator along with various inventive devices after confirming a sufficient level of precision was achieved and the practicality of the calculation time required.

However, substantial calculation precision depends on the selection and structuring of the analytic models for the interfacial drag force and heat transfer models, chemical reaction velocity equation, and heat transfer and wall friction model rather than on the numerical solution method. Due caution is needed for the fluctuations, as well as measurement range and conditions, of the actual measured data when it comes to structuring models for simulation.

A completely wrong solution can be obtained instead of deviation in calculation precision, especially if a model fitted with polynomials is used beyond the measurement range. An important requirement for simulation is the capability to conduct calculation in a virtual condition beyond reality, and this point was also taken into consideration for the modeling process.

8.3.5.4 Development of Urethane Foam Molding Simulator

For the development of the urethane foam molding simulator, the following factors were taken into consideration in an effort to structure a realistic model: examination of the precision of the actual measured data, examination of models appropriate for the capabilities, development period and cost of the assumed calculator, avoidance of calculation instability stemming from temporal and spatial scales, efficient calculation methods leading to better stability, examination of the precision compared to the experimental testing results, and so on.

8.3.6 Conclusion

With a view to achieving worldwide quality competition and simultaneous achievement of QCD, the author has promoted the Advanced TDS, Strategic Development Design Model, which is an advanced form of the core New JIT technology, TDS. With a view to establishing a development design quality assurance system necessary for CAE in the advanced manufacturing industry, attention was given to the indispensable Highly Reliable Development Design CAE Model.

When this model was applied to a concrete target, it was demonstrated that a rational arrangement of partial as well as overall optimality is needed for the required technical elements consisting of problem–model–algorithm–theory–computer to be implemented in the highly reliable CAE numerical simulation.

The guidelines for the implementation of intelligence and high precision into CAE analytic software were presented, and their effectiveness was verified through the application example highly reliable numerical simulation for the production of CAE software for molding automotive seat pads: urethane foam molding simulator.

8.4 New Japanese Global Production Model
8.4.1 Introduction

The leading Japanese administrative management technology that contributed most to worldwide manufacturing from the second half of the twentieth century was the Japanese Production System, which is typified by the Toyota Production System. However, the Toyota Production System has been further developed and spread in the form of internationally shared global production systems such as just in time (JIT), and therefore it is no longer a proprietary technology of Japan.

Digital engineering is bringing about radical changes in the way manufacturing is carried out at manufacturing sites. This means that it is now necessary to reconstruct the principles of management technology and Japan's unique world-leading management technologies so that they will be viable even for the next generation of manufacturing.

With this in mind, this section focuses on the strategic deployment of the Advanced TPS, Strategic Production Management Model (Amasaka, 2002, 2007a,b). A core technology of this model is the TPS of New JIT, an innovative management technology principle that surpasses conventional JIT practices (refer to Section 8.1). With a concrete target, Amasaka and Sakai (2009a) have proposed a New Japan Global Production Model (NJ-GPM) to enable the strategic deployment of Advanced TPS. The aim of this model is to realize a highly reliable production system suitable for global production by reviewing the production process from production planning and preparation through production itself and process management.

The core technologies that constitute the model are the TPS Layout Analysis System (TPS-LAS), Human Intelligence Production Operating System (HI-POS), TPS Intelligent Production Operating System (TPS-IPOS), TPS Quality Assurance System (TPS-QAS), Human Digital Pipeline (HDP), and Virtual Maintenance Innovated Computer System (V-MICS). The model has proved to be effective at Toyota, a leading automobile manufacturer (Amasaka and Sakai, 2009a).

8.4.2 Advanced TPS, Strategic Production Management Model

8.4.2.1 Demand for New Management Technologies That Surpass JIT

Dramatic changes are occurring in today's manufacturing industry. It is vital for Japanese manufacturing not to fall behind in the advancement of management technologies. In order for a manufacturer to succeed in the future world market, it needs to continue to create products that will leave a strong impression on customers and to offer them in a timely fashion.

At present, however, the TPS, which is representative of Japanese manufacturing, has been further developed and spread in the form of internationally shared global production systems such as JIT and the Lean system; therefore, it is no longer a proprietary technology of Japan. It is not an exaggeration to say that the realization of competitive manufacturing—the simultaneous achievement of QCD—ahead of their competitors is what will ensure Japanese manufacturers' success in global marketing.

8.4.2.2 Basic Principle of Total Production System

The urgent mission for Japanese manufacturers is to reconstruct world-leading, uniquely Japanese principles of management technology and administrative management technology, which will be viable even for next-generation manufacturing (Amasaka, 2007h). Dramatic changes are taking place in today's manufacturing industry. In order to prevail in today's competitive manufacturing industry, which is often referred to as worldwide quality competition, the

Global Manufacturing Strategy ▪ 261

pressing management issue is to realize the kind of global production that can achieve so-called worldwide uniform quality and production at optimal locations (Amasaka et al., 2008).

Given this, the author (Amasaka, 2007h; Amasaka et al., 2008) has proposed the basic principle of New Manufacturing Theory, which is positioned as part of the evolution system of the TPS, as shown in Figure 5.3 (refer to Section 5.1).

This basic principle entails the core principles of New JIT, the next-generation management technology established by the authors: TDS, TPS, and TMS.

8.4.2.3 Advanced TPS: A New Japanese Global Production Model

Amasaka (2007b, 2008) and Amasaka and Sakai (2009a) have proposed the Advanced TPS, New Manufacturing Theory called the New Japanese Production Model, as introduced in Figure 8.17, in order to enable the strategic deployment of this new manufacturing theory TPS. The mission of Advanced TPS is to contribute to worldwide uniform quality and production at optimal locations as a strategic deployment of global production and to realize CS, ES, and SS through high quality assurance manufacturing.

With a concrete target, this model is the systemization of a new, next-generation Japanese production management system and involves the high cyclization of the production process for realizing the simultaneous achievement of QCD requirements. In order to make this model into a reality, it will be necessary to adapt it to handle digitalized production and reform it to realize an advanced production management system. Other prerequisites for realizing this include the need to create an attractive working environment that can accommodate the increasing number of older and female workers at the production sites and to

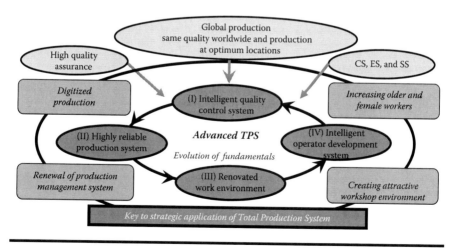

Figure 8.17 Advanced TPS, Strategic Production Management Model.

cultivate intelligent production operators. These measures need to be organically combined and spiraled up in order to make the simultaneous achievement of QCD possible.

In order to make this system into a reality, it will be necessary to (i) adapt it to handle digitalized production and (ii) reform it to realize an advanced production management system. Other prerequisites for realizing this include the need (iii) to create an attractive workplace environment that accommodates the increasing number of older and female workers at the production sites and (iv) to cultivate intelligent production operators. These four measures need to be organically combined and spiraled up in order to make the simultaneous achievement of QCD possible.

One of the technical elements necessary for fulfilling these requirements is the reinforcement of maintenance and improvement of process capabilities by establishing an intelligent quality control system. Second, a highly reliable production system needs to be established for high quality assurance. Third, reform is needed for the creation of a next-generation working environment system that enhances intelligent productivity. Fourth, intelligent production operators need to be cultivated who are capable of handling the advanced production system and an intelligent production operating development system needs to be established. Through strategic management of these elements, worldwide uniform quality and simultaneous launch (production at optimal locations) will be possible.

In order to offer customers high-value-added products and prevail in worldwide quality competition, it is necessary to establish an advanced production system that can intellectualize the production technology and production management system. This will in turn produce high performance and highly functional new products. The authors believe that what determines the success of global production strategies is the advancement of technologies and skills that are capable of fully utilizing the above-mentioned advanced production system in order to realize reliable manufacturing at production sites.

8.4.2.4 High-Cycle System for the Production Business Process Using Advanced TPS

Considering the points mentioned in Section 8.4.2.3, the authors have recognized the need to upgrade the intelligence of the production sites for global production in order to successfully carry out the strategic deployment of the new manufacturing system, Advanced TPS. Amasaka et al. (2006) and Amasaka and Sakai (2008) have proposed a high-cycle system for the production business process, as shown in Figure 8.18.

The objective of this system is effective management of the advanced production process in order to improve the intelligent productivity of production operators and to consolidate the information about highly cultivated skills and operating skills for advanced facilities into a commonly shared system.

Global Manufacturing Strategy ■ 263

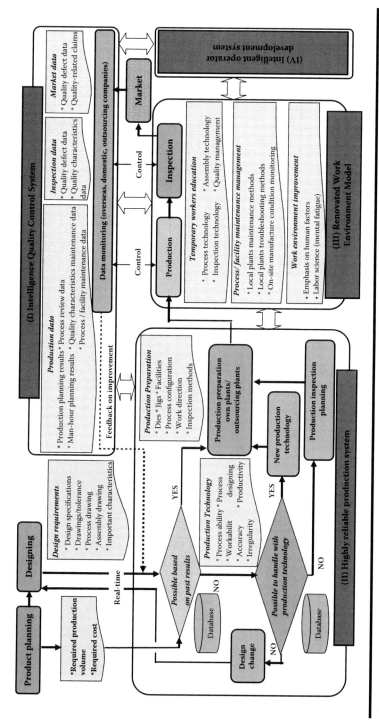

Figure 8.18 Upgrading intelligence of production: high-cycle system.

In this way, the production operators can upgrade their simple labor work to intelligent production operation. To accomplish this, the four key technologies depicted in Figure 8.18 (I–IV) are to be reformed for intelligent operation of the production sites:

(I) The *Intelligent Quality Control System* aims to achieve high quality assurance through digital engineering, reinforcement of quality incorporation focusing on intelligent control charts, and ensuring the process capability (Cp) and machine capability (Cm) through innovation of the operating and maintenance systems of production facilities.
(II) The *Highly Reliable Production System*, that enhances intelligent productivity with highly skilled workers, aims to construct a global production network system through incorporation of the latest technologies, such as CAE, CAD, robots, and the use of computer graphics (CG).
(III) The *Renovated Work Environment Model* aims to improve the value of labor and bring about a comfortable workplace environment that can accommodate the increasing number of older and female workers in the labor force.
(IV) The *Intelligent Operators Development System* aims to realize the early cultivation of highly skilled workers through utilization of visual manuals supported by the latest IT and virtual technology.

8.4.2.5 Creation of New Japanese Global Production Model Employing Advanced TPS

Global production must be deployed in order to establish the kind of manufacturing that is required to gain the trust of customers around the world by achieving a high level of quality assurance and efficiency and shortening lead times to reinforce the simultaneous achievement of QCD requirements. The vital key to achieving this is the introduction of a production system that incorporates production machinery automated with robots, skilled and experienced workers (production operators) to operate the machinery, and production information to organically combine them.

Thus, having recognized the need for a new production system suitable for global production, Amasaka and Sakai (2009a) have created a New Global Production Model (NJ-GPM), shown in Figure 8.19, to realize the strategic deployment of the Advanced TPS. The purpose of this model is to eradicate ambiguities at each stage of the production process from production planning and preparation through production itself and process management, and between the processes, in order to achieve a highly reliable production system for global production that will improve the reliability of manufacturing through the clarification and complete coordination of these processes.

More specifically, the model is intended to (i) employ numeric simulation CAE and CG right from the production planning stage to resolve technical issues before

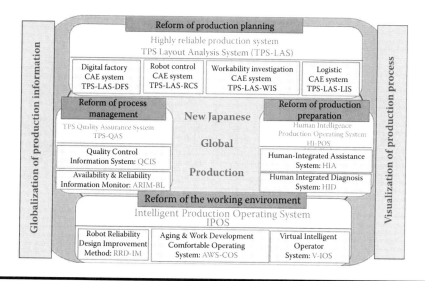

Figure 8.19 New Japanese Global Production Model (NJ-GPM).

they occur, (ii) reinforce production operators' high-tech machine operating skills and manufacturing capabilities, and (iii) visualize the above using IT in order to reform production information systems to create a global network of production sites around the world.

The six core technologies that constitute this model and their characteristics are described below.

1. Reform of production planning: The TPS-LAS is a production optimization system intended to realize a highly reliable production system by optimizing the layout of both the production site as a whole and each production process with regard to production lines (logistics and transportation), robots (positioning), and production operators (allocation and workability) through the use of numeric simulation (Sakai and Amasaka, 2006a). The TPS-LAS is made up of four subsystems: the Digital Factory CAE System (LAS-DFS), Robot Control CAE System (LAS-RCS), Workability Investigation CAE System (LAS-WIS), and Logistic Investigation CAE System (LAS-LIS).
2. Reform of production preparation: The HI-POS is an intelligent operator development system intended to enable the establishment of a new people-oriented production system whereby training is conducted to ensure that operators develop the required skills to a uniform level, and diagnosis is then carried out to ensure that the right people are assigned to the right jobs (Sakai and Amasaka, 2006b). HI-POS is made up of three subsystems: the Human Intelligence Diagnosis System (HID), Human Integrated Assist System (HIA), and Human Digital Pipeline System (HDP), as shown in Figure 8.20. (Details of HDP and HIA are given in Sections 10.3 and 10.4).

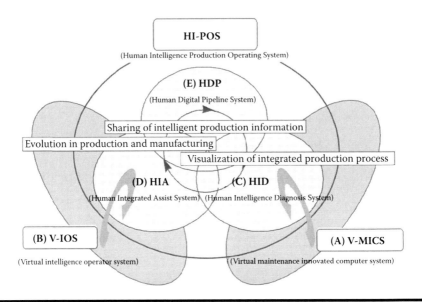

Figure 8.20 HI-POS architecture.

3. Reform of the working environment: The Intelligent Production Operating System (IPOS) is intended to lead to a fundamental reform of the work involved in production operations by raising the technical skill level of production operators and further improving the reliability of their skills for operating advanced production equipment within an optimized working environment. TPS-IPOS is made up of three subsystems: the Virtual Intelligent Operator System (V-IOS) (Sakai and Amasaka, 2003), Aging and Work Development Comfortable Operating System (AWD-COS) (Amasaka, 2007e), and Robot Reliability Design Improvement Method (RRD-IM) (Sakai and Amasaka, 2007a).
4. Reform of process management: The TPS Quality Assurance System (TPS-QAS) is an integrated quality control system intended to ensure that quality is built into production processes through scientific process management that employs statistical science to secure process capability (Cp) and machine capability (Cm) (Amasaka and Sakai, 2009b). TPS-QAS is made up of two subsystems: the Quality Control Information System (QCIS) and Availability and Reliability Information Monitor System (ARIM) (Amasaka and Sakai, 1998).
5. Visualization of production processes: The HDP (Sakai and Amasaka, 2007b) ensures that top priority is given to customers through manufacturing with a high level of quality assurance. This involves the visualization of intelligent production information throughout product design, production planning and preparation, and production processes, thereby facilitating the complete

coordination of these processes. This system enables the high cyclization of business processes within manufacturing.
6. Globalization of production information: The Virtual Maintenance Innovated Computer System (V-MICS) (Sakai and Amasaka, 2005) is a global network system for the systemization of production management technology necessary to realize a highly reliable production system, which is required to achieve worldwide uniform quality and production at optimal locations.

The NJ-GPM created is fundamental to the strategic deployment of Advanced TPS. Through the operation of a dual system involving both V-MICS and HDP, this new model integrates the core technologies from production planning and preparation through working environments and process management. In the next section, the author verifies the effectiveness of this research through some examples illustrating the deployment of NJ-GPM.

8.4.3 Example Applications

In this section, the authors (Amasaka and Sakai, 2009a) introduce some research examples of NJ-GPM, which has contributed to the advancement of management technology at advanced car manufacturer A.

8.4.3.1 TPS Layout Analysis System

A simulation of main body conveyance using TPS-LAS (and its four constituent subsystems) is shown in Figures 7.21 and 7.22 (refer to Section 7.3) to illustrate a highly reliable production system that has contributed to the reform of production planning (Sakai and Amasaka, 2006a).

Firstly, the necessary production machinery is modeled, and a hypothetical production line is set up within a "digital factory" on a computer. TPS-LAS-DFS is then used to reproduce the flow of people and parts within the production site. This enables any interference between production machinery and production cycle times to be checked in advance using simulations. One type of advance simulation uses TPS-LAS-RCS for the optimum placement of welding robots for the main body to ensure that no interference occurs.

Next, advance verification is performed using TPS-LAS-WIS to ensure that the predetermined work (standardized work) is carried out within the predetermined cycle time with no waste (*muda*) or overburdening (*muri*). Then, TPS-LAS-LIS is used to establish optimized conveyance routes between processes and determine optimum buffer allocations.

On body production lines in particular, where hangers are used to suspend the bodies for each vehicle model, simulations can be used to determine in advance whether there are too many or too few hangers at entrances and exits. Similarly, by predicting downtime when conveyance equipment collisions occur, it is possible to verify in advance whether buffer levels and availability will fall below target levels. Additionally, by determining the optimum number of body hangers and adjusting the main body conveyance routes, these simulations enable reductions in operating efficiency on the lines to be predicted.

TPS-LAS is currently being deployed as part of global production strategies and is proving to be effective both in Japan and overseas.

8.4.3.2 Human Intelligence-Production Operating System

The authors have implemented HI-POS by using its two constituent subsystems: HID and HIA (Sakai and Amasaka, 2006b). Figure 8.21 shows an example of a Total Link System Chart (TLSC) that represents the combined application of HID and HIA and illustrates the following points: (a) improved clarity and accuracy of analysis, (b) clearly structured production process evaluation criteria, (c) clearly indicated administrative links among organizations, (d) a bird's-eye view of work and information flows, (e) clarity of knowledge and know-how, (f) confirmation of available resources, and (g) issue detection and resolution.

A TLSC such as the one shown here is used to flush out any hidden problems. The problems found at various levels are clarified and categorized according to the KJ method (the creative problem solution technique using brainstorming; Kawakira, 1967). Logical reasoning is applied to trace the root causes of the problems, and the appropriate evidence is gathered and organized. This is followed by the formulation and evaluation of countermeasures. Items taken up (problems) are analyzed to evaluate the extent of improvement and the costs involved.

These systems and the TLSC used to represent them are currently being employed to promote proactive *kaizen* (improvements), which is proving to be effective both in Japan and overseas.

8.4.3.3 TPS Intelligent Production Operating System

Sakai and Amasaka (2007a) are implementing the Intelligent Production Operating System (TPS-IPOS) by using three subsystems. Firstly, the Virtual Intelligent Operator System (V-IOS) is intended to improve the skills of new (inexperienced) production operators both in Japan and overseas. For example, at special training centers with simulations of actual assembly lines, as shown in Figure 8.22, both (a) training processes for assembly work and (b) work training systems for assembly work are employed in the training of operators. Then, once a certain level of skills has been mastered, operators progress to actual assembly lines where they are promptly and methodically developed as highly skilled and experienced technicians using (c) standard work sheets extracted from the aforementioned HID.

Secondly, the Aging and Work Development Comfortable Operating System (AWD-COS) constitutes a fundamental reform of work and labor. Therefore, the author (Amasaka, 2007e) has initiated a company-wide project called Aging and

Global Manufacturing Strategy ■ 269

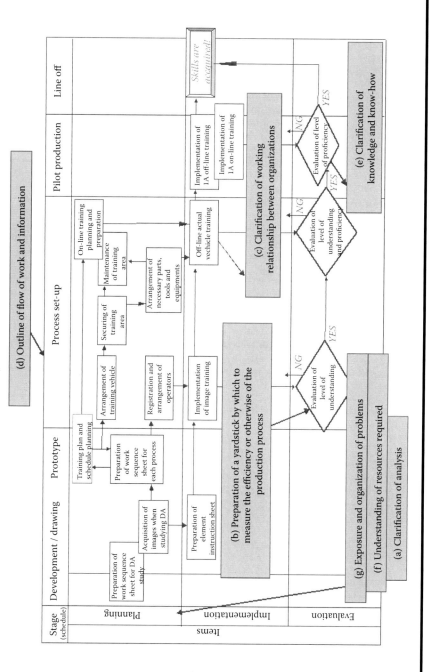

Figure 8.21 Total Link System Chart (TLSC) for HI-POS (using both HID and HIA).

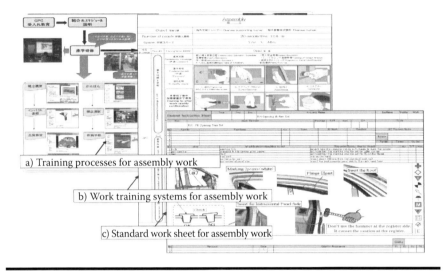

Figure 8.22 Virtual Intelligent Operator System (V-IOS).

Work Development 6 Programs Project (AWD6P/J) in order to combat the effects of aging, as shown in Figure 6.9 (refer to Section 6.2).

- Project I: Arousing motivation in workers
- Project II: Reviewing working styles to reduce fatigue
- Project IV: Improving heavy work with user-friendly tools and equipment
- Project V: Creating thermal environments suited to the characteristics of assembly work
- Project VI: Reinforcing illness and injury prevention

Thirdly, the Robot Reliability Design Improvement Method (RRD-IM) is intended to improve the reliability of robots from development, production, introduction, and operation, right up until they wear out and are replaced (Sakai and Amasaka, 2007a). The body assembly line is a series model with multiple robots positioned, as shown in Figure 7.30 (refer to Section 7.4), and so the line's availability is determined by the number of robots introduced. Figure 7.34 (left) (refer to Section 7.4) shows a calculation used to obtain the relation between the number of robots (N) and robot MTBF (t), where monthly operating hours (T) = 400 h; significance level (α) = 0.05, failure repair time (r) = 1.2 h, and the line's required availability (A) = 98%. This shows that if 300 robots are introduced on a body assembly line, the necessary MTBF is 30,000 h, and, therefore, a tenfold improvement is required in the existing robot MTBF of 3,000 h.

The use of TPS-IPOS is proving to be very effective at new factories overseas, for example, where the target operating efficiency (QCD effect and safety) from start of production is being achieved at the same level and within the same timescales as factories in Japan.

8.4.3.4 TPS Quality Assurance System

This system enables the deployment of manufacturing with superior quality and productivity by integrating two high-precision quality control systems suitable for global production (Amasaka and Sakai, 2009b). Firstly, in order to analyze process management status in real time and enable diagnosis of process management abnormalities, the Quality Control Information System (QCIS), as shown in Figure 8.23, automatically creates control charts using process analysis functions such as (1) scroll function, (2) display of grouped and raw data, (3) innovative factorial analysis by layer, (4) *kaizen* history database, (5) abnormality diagnosis function, and (6) data links with other application software.

Secondly, the Availability and Reliability Information Monitor System (ARIM) gathers information on operating efficiency and failures for *andon* systems and clusters of machinery and equipment on each production line at factories in Japan and overseas, as shown in Figure 8.24. This information is used to carry out Weibull analysis of equipment failures and real-time reliability analysis in order to maintain a high level of machine reliability and maintainability, enabling increased operational efficiency on production lines. This TPS-QAS system enables fast and accurate process management on a global network, and it has been deployed with considerable effect.

8.4.3.5 Human Digital Pipeline System

The HDP system, as shown in Figure 8.25, has the following features (Sakai and Amasaka, 2007b): (i) It creates and supplies, in advance, Standard Work Sheets, on which production operators record each task in the correct order for jobs such as assembly work, by using design data for new products and facilities prepared

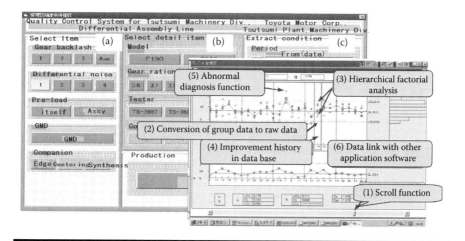

Figure 8.23 Outline of the control chart utilizing the software system T-QCIS.

272 ■ *New JIT, New Management Technology Principle*

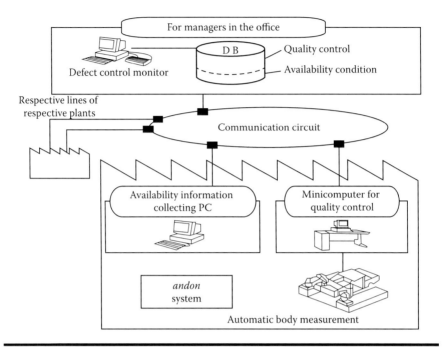

Figure 8.24 Outline of reliable ARIM.

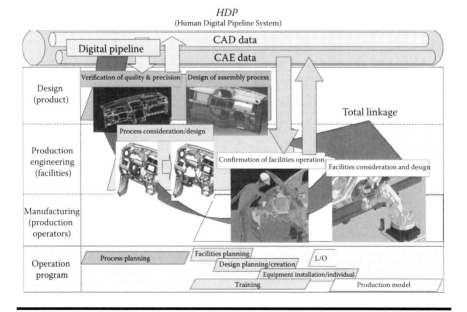

Figure 8.25 Outline of Human Digital Pipeline (HDP) System.

from design through to production technology, even if there are no production prototypes. (ii) Next, the HDP system enables visualization training for machining processes step-by-step in the order that parts are built up, even if the actual product does not yet exist. The system is proving to be very effective in raising the level of proficiency for processes requiring skills and capabilities at the production preparation stage.

8.4.3.6 Virtual-Maintenance Innovated Computer System

The V-MICS, as shown in Figure 8.26, takes a server and client system configuration, with a server specially set up for each production site (Sakai and Amasaka, 2005). Production operators are able to browse information using databases (DB) and CG whenever necessary from the client computers at each maintenance station via the network and can also input any special items as necessary.

The servers at each site are synchronized with the central server (V-MICS server) so that any new information is simultaneously recorded and sent out to each server. This enables knowledge and information for each process to be shared and experienced virtually on computers among sites within and outside different countries. Coordination with the aforementioned TPS-LAS, HI-POS, TPS-IPOS, TPS-QAS, and HDP has enabled the strategic operation of a global production system with considerable effect.

8.4.4 Conclusion

In order to reconstruct the principles of management technology and Japan's unique world-leading management technologies so that they will be viable even

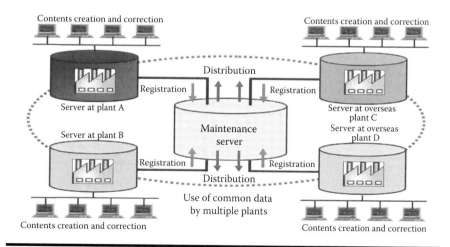

Figure 8.26 Outline of Virtual Maintenance Innovated Computer System (V-MICS).

for the next generation of manufacturing, the author has created Advanced TPS, a New Global Production Model. The author has also created a New Japanese Global Production Model (NJ-GPM) to enable the strategic deployment of Advanced TPS, and its effectiveness has been verified at Toyota.

8.5 Intelligent Customer Information Marketing System

8.5.1 Introduction

The author (Amasaka, 2002) has touched on the development of the principle of New JIT and its validity as a new management technology for twenty-first-century manufacturing. New JIT innovates the business process of each division, which encompasses sales, development, and production (refer to Section 5.1).

This section discusses the validity of the Advanced TMS, Strategic Marketing Development Model as a core technology of the TMS of New JIT, which contributes to the construction of a Strategic Marketing Development Model (SMDM) (Amasaka, 2007a,c; Yamaji and Amasaka, 2009a). In concrete terms, a model that enables the sales, marketing, and service divisions nearest the customers to systematically determine their tastes and desires is necessary.

At present, however, the system for applying scientific analytical methods to customer data has not been satisfactorily established. In some cases, its importance has not even been recognized. This section aims to create a Scientific Customer Creative Model (SCCM), which is a form of strategic marketing utilizing Advanced TMS.

This section also introduces the effectiveness of Advanced TMS that reflects latent customer needs through scientific marketing application examples via the Intelligent Customer Information Marketing System (ICIMS), based on the author's experience at advanced car manufacturer A (Amasaka, 2007c).

8.5.2 Need for a Marketing Strategy That Considers Market Trends

Today's marketing activities require more than just short-term strategies, such as product, price, place, and promotion (4P) activities in the business and sales divisions. After the collapse of the bubble economy, the competitive environment in the market changed drastically. Since then, only companies that have implemented strategic marketing quickly and aggressively have enjoyed continued growth (Okada et al., 2001). After close examination, it was said that strategic marketing activities must be conducted as company-wide, core corporate management activities that involved interactions between all divisions inside and outside the company (Jeffrey and Bernard, 2005).

Therefore, a marketing management model needs to be established so that the business, sales, and service divisions, which carry out development and design for appealing product projects and are closest to customers, can organizationally learn customers' tastes and desires by means of the continued application of objective data and scientific methodology (James and Mona, 2004; Amasaka, 2005a). However, at present, the organizational system has not yet been fully established in these divisions; in some cases, even the importance of this system has not been realized (Gary and Arvind, 2003).

8.5.3 Importance of Innovating Dealers' Sales Activities

Considering recent changes in the marketing environment, it is now necessary to implement innovation of business and sales activities to accurately grasp the characteristics and changes of customer preferences independently from convention. Contact with customers has never called for more careful attention and practice (Amasaka, 2011).

For example, it is now more important to construct and develop an intelligent customer information network, CS-CIANS, using customer science, that systematically improves know-how related to the application of customer information software with respect to users of various vehicle makes (refer to Sections 5.1, 6.4, and 7.1). This information network turns customer management and service into a science by utilizing Science SQC according to customers' involvement with their vehicles in daily life (Amasaka, 2002, 2003a).

To realize the innovation of business and sales activities, as shown in Figure 8.27, the following three factors are important: merchandise, shop and selling power, and shop appearance and operation. Of these three factors, (1) innovation in building ties with customers is particularly important in shop appearance and operation. It constitutes the basis for the innovation of (2) business negotiation, (3) employee image, and (4) after-sales service (Amasaka et al., 1998; Amasaka, 2011).

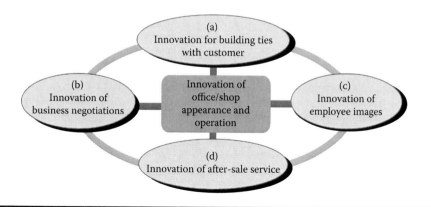

Figure 8.27 Innovation of business and sales activities.

8.5.4 Advanced TMS, Strategic Development Marketing Model

In light of recent changes in the marketing environment, what is needed now is to develop innovative business and sales activities that are unconventional and correctly grasp the characteristics and changes of customer tastes. Contact with customers has never called for more careful attention and practice. To devise an appealing, customer-oriented marketing strategy, it is important to evolve current market creation activities (Nikkei Business, 1999; Amasaka, 2001, 2002).

Therefore, the author proposes the Advanced TMS, Strategic Development Marketing Model, as described in Figure 8.28, which further updates TMS (for details, refer to Sections 5.1 and 8.1).

8.5.5 Creation of Scientific Customer Creative Model Utilizing Advanced TMS

One vital point of the strategic marketing structure is its definition; marketing activities should be defined from closed marketing activities, which are limited to the business and sales division, to open marketing activities, which can be performed through steady linkage with all other divisions in a company-wide framework. The aim is evolution of market creation through high quality assurance and innovating dealers' sales activities by utilizing the scientific approach of Advanced TMS. The authors present a Scientific Customer Creative Model (SCCM), which takes the form of strategic marketing, as shown in Figure 8.29. In the figure, the entire structure consists of three domains: (1) marketing strategy, (2) manufacturing process, and (3) market and customers. In each domain, the key marketing

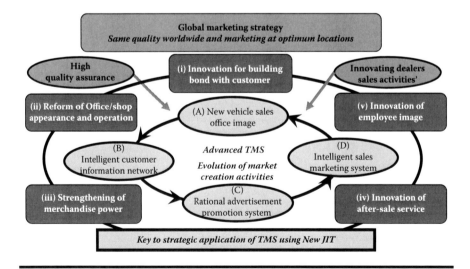

Figure 8.28 Advanced TMS, Strategic Development Marketing Model.

Global Manufacturing Strategy ■ 277

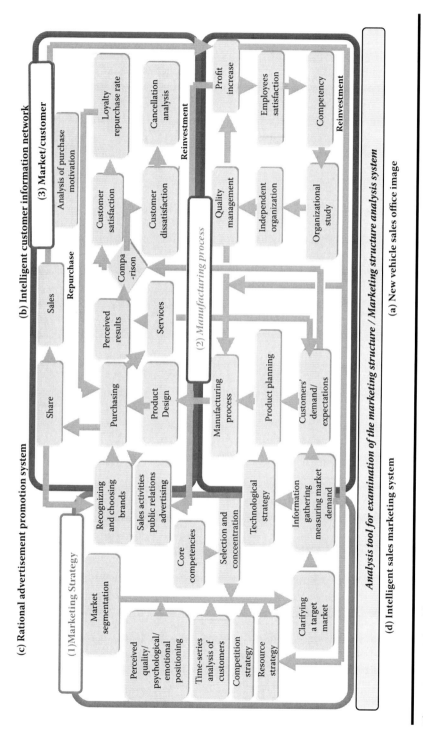

Figure 8.29 Scientific Customer Creative Model (SCCM).

items are linked by paths to show how they are associated (Amasaka et al., 2004; Shimakawa et al., 2006).

The outline of SCCM is shown in the following. First of all, in the (1) *marketing strategy* domain, the key point is how the market segment and the target market are determined. In general, the target market is determined based on the company's core competencies, competition strategy, and resource strategy over the medium and long terms. By introducing a scientific analysis approach that uses IT, it clarifies a potential target market from the changing market or the customer structure analysis. Secondly, in the (2) *manufacturing process* domain, the key point is to collect/analyze customers' demands and expectations precisely. At this time, it is important to consider what value the customers want. When implementing information collection/analysis, customer value is described in numerical form from many different viewpoints, and a new product that is aimed at enhancing customer value is implemented through the flow of planning→development →production.

Furthermore, in the (3) *market/customer* domain, the key point is to learn the structure of the customer's motivation to buy products, customer satisfaction (CS), and loyalty. Then, it is necessary to extract the elements for customer retention (CR) from this data and utilize them for specific *kaizen* activities such as reflecting it in future products. It is important to develop an analysis tool for close examination of the marketing structure and a marketing structure analysis system that will support marketing activities in these three domains from a strategic marketing viewpoint, as shown in Figure 8.29.

8.5.6 Application: Establishment of Intelligent Customer Information Marketing System

This section discusses the reality and effectiveness of the Intelligent Customer Information Marketing System, which the author established recently through the application of Advanced TMS. The typical execution example of ICIMS is the Japanese Sales Marketing System, which involves the Customer Purchasing Behavior Model of Advertisement Effect (CPBM-AE).

8.5.6.1 Customer Purchasing Behavior Model of Advertisement Effect (CPBM-AE) for Automotive Sales

Recently, the total amount of advertising expenditure by Japanese enterprises has increased almost in-line with increases in GDP (Nikkei Institute of Ad, 2006; Dentsu, 2006). The breakdown by medium is as follows: TV advertisements (TV-CM, 33.9%), newspaper ads (18.8%), flyers (8.0%), magazines (7.1%), and radio (3.2%). In the case of Toyota, which has topped Japanese advertising expenditure for 7 consecutive years, the rate of advertising expenditure to operating profit is 7.3% (average).

This ratio is constantly high for a number of enterprises. According to automotive dealers' empirical knowledge, number one in rank for major mass effect is TV-CM, but as far as the author knows, there are no studies on the quantitative effects of TV-CM, such as the proposed CPBM-AE (Niiya and Matsuoka, 2001). The author is interested in quantitatively turning the purchasing behavior of customers who visit dealers with intentions to buy after watching TV-CM into explicit knowledge. Moreover, the visiting ratio of customers will rise and the effect of marketing and sale activities will be drastically enhanced due to an improved understanding of the effects of TV-CM and the effects of a mix of media such as newspaper ads and flyers.

8.5.6.1.1 Proposing Customer Purchasing Behavior Model of Advertisement Effect

Figure 8.30 shows the CPBM-AE prepared on the basis of research of the authors (Amasaka, 2003b, 2009) at the time of introduction of new cars. As background for the influential factors in the figure, the authors established the following *CPBM-AE1*: recognition of the vehicle name (R)→interest in the vehicle (I)→desire to visit a dealer to see the vehicle (D-1)→consideration of purchasing (C-1)→visit to a dealer for purchasing contract (P-1). This was influenced by a TV-CM, newspaper ad, radio, flyer, DM/DH, or presence/business talk over approximately 2 months. From this, the authors found that a *CPBM-AE2* exists: consideration of purchasing the vehicle (C-2)→desire to visit a dealer (D-2)→impression from actually seeing the vehicle at the dealer (I-2)→purchasing contract (P-2).

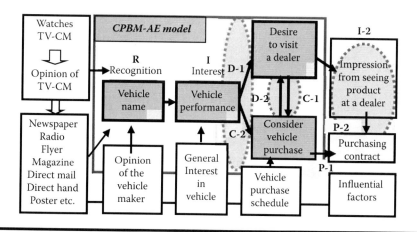

Figure 8.30 Customer Purchasing Behavior Model of Advertisement Effect (CPBM-AE).

8.5.6.1.2 TV Advertisement Effect Introducing Two Newly Released Cars

For example, supposing that customers have their purchasing desire aroused by watching a TV-CM for a newly released car, what percentage of customers would visit their dealers? It is important for future marketing strategies to conduct a dynamic survey of customers' purchasing behavior. However, because of the enormous scale of market surveys and the limited opportunities for them, there have been no actual examples of factorial analysis up to the present (Niiya and Matsuoka, 2001; Ikeo, 2006; Amasaka, 2009). With the help of advanced car manufacturer A, the author (Amasaka, 2003) conducted a dynamic survey of customers' purchasing behavior resulting from a TV-CM introducing two new passenger vehicles called the Funcargo and Platz in Japan (refer to reference data, Amasaka, 2003b). As a result of the investigation analysis, the author could confirm a CPBM-AE-1/2 for customer purchasing.

Figure 8.31 shows the results of analysis that applied the CPBM-AE-1 at the time of the introduction of the new vehicle, Funcargo. The total mean curve indicates that 1/3 (34.6%) of customers recognized the new vehicle name (R) from the TV-CM alone. The number of customers drops by half (18.4%) at the interest stage (I), dropping by a further half (8.1%) at the desire to visit a dealer stage (D-1). At the considering purchase stage (C-1), the figure drops to 9.6% at best, even with the addition of D-2→I-2, as stated above. Moreover, at the visiting stage in the figure, the ratio of customers that visited a dealer fails to reach 2%. This implies the need to establish an effective mixed-media model in the future. It has been verified that this analysis has a similar dynamic trend for the Platz and does not vary for sex of purchasers, age, or area.

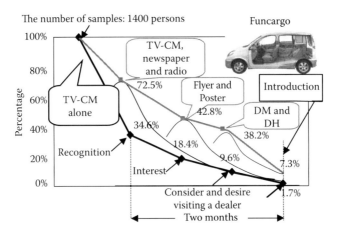

Figure 8.31 Reality of CPBM-AE of customer actions.

8.5.6.1.3 Establishment of an Effective Mixed-Media Model

Amasaka (2003b, 2009) presents the most effective mixed-media model (CPBM-AE-2) for increasing the rate of dealer visits, as shown in Figure 8.31 (the upper line graph in the middle of the figure). Compared to the effect of TV-CM alone, the mixed-media effect of TV-CM, newspaper ads, and radio improved vehicle name recognition (R) to 72.5%.

Similarly, the use of flyers and posters increased interest in the vehicle (I) to 42.8%. The use of DM/DH increased the desire to visit the dealer and vehicle purchase consideration (D-2, C-2, I-2) to 38.2%. The cumulative effect produced an end result where the rate of dealer visits (P-2) increased greatly to 7.3%. However, the challenge is to strengthen new initiatives that will lead to an even greater increase in the rate of dealer visits.

8.5.6.2 Japanese Sales Marketing System

Amasaka (2001, 2011) has constructed the Japanese Sales Marketing System (JSMS) based on his experience at advanced car manufacturer A, as a way to aid sales marketing through innovative bond building with the customer, as shown in Figure 8.32 (refer to reference data, Amasaka, 2001).

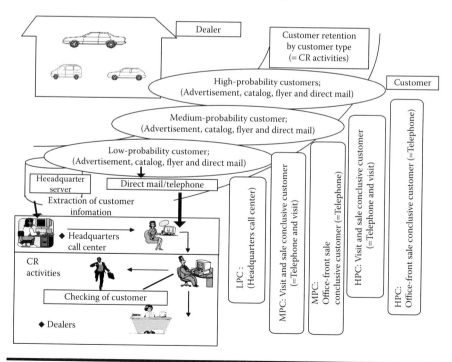

Figure 8.32 Outline of Japanese Sales Marketing System (JSMS).

This system combines IT and statistical science to make practical use of customer data in order to (i) increase the rate of customer retention and (ii) acquire new customers. In order to increase the rate of dealer visits and vehicle purchases by current loyal users, they are stratified into high-probability customers (HPCs), medium-probability customers (MPCs), and low-probability customers (LPCs), and then a sample is taken of the marketing, sales, and service items that the customers demanded, and customer satisfaction is taken as follows:

1. The CR activities based on customer type are adopted by classifying HPCs and MPCs into those who visit the shop and those who must be visited by our staff, taking characteristics at new car purchase into account. A system is established so that the shop manager directly receives MPCs when they visit the shop without fail in order to promote visits to the shop by HPCs. Thus, the frequency of contact with customers is increased. Further, sales and service activities focus on telephone calls for customers who visit the shop, and telephone calls and home visits for those who require visits by our staff.
2. As for LPCs, who have less contact with the sales staff, a telephone call center is established within the dealer, as shown in Figure 8.32, to accumulate know-how related to the effective use of customer information software. The two-step approach is adopted as the practical sales policy where telephone calls are used to follow up on the effect of publications, advertisements, catalogs, fliers, and direct mail. As expected, excellent results have been reported at Netz Chiba and other dealers who applied the Toyota sales marketing system described here.

The system then uses these data to enhance the daily marketing, sales, and service activities. The system can also be made use of when visiting customers and to help acquire new customers at the time they visit a dealer. Due to the application of JSMS, which involves the previously introduced CPBM-AE, T-ICIMS operation has recently contributed to an increase in the sales share of vehicles at advanced car manufacturer A in Japan (40%/1998 to 46%/2005) (Nikkei Business, 1999).

8.5.7 Conclusion

In this section, the author proposed the Advanced TMS, Strategic Development Marketing Model as a core technology TMS of New JIT, which contributes to constructing a Strategic Development Marketing System. The aim is to bring about an evolution of market creation through high quality assurance and innovating dealers' sales activities utilizing a scientific approach.

So, the authors created a SCCM), which has the structure of strategic marketing. Further, the author introduced the effectiveness of application examples via the ICIMS.

Appendix 8.1

Q1. To what extent do you consider the following characteristics of automobile design to be important?

No.	Characteristic of Automobile Design	Very Important						Not Important at all	Comments
1	Sophisticated	7	6	5	4	3	2	1	
2	Traditional	7	6	5	4	3	2	1	
3	Chic	7	6	5	4	3	2	1	
4	Orthodox	7	6	5	4	3	2	1	
5	Original	7	6	5	4	3	2	1	
6	Curving	7	6	5	4	3	2	1	
7	Advanced	7	6	5	4	3	2	1	
8	Sporty	7	6	5	4	3	2	1	
9	Classy	7	6	5	4	3	2	1	
10	Luxurious	7	6	5	4	3	2	1	
11	Fashionable	7	6	5	4	3	2	1	
12	Compact	7	6	5	4	3	2	1	
13	Practical	7	6	5	4	3	2	1	
14	Dignified	7	6	5	4	3	2	1	
15	Youthful	7	6	5	4	3	2	1	

Q2. To what extent do you agree with the following statements?

No.	Item	Strongly Agree						Strongly Disagree	Comments
1	An automobile is merely a transportation device	7	6	5	4	3	2	1	

No.	Item	Strongly Agree						Strongly Disagree	Comments
2	The economic burden accompanying possession and use of an automobile is large	7	6	5	4	3	2	1	
3	An automobile is an indispensable item in life	7	6	5	4	3	2	1	
4	Many color variations are necessary in an automobile	7	6	5	4	3	2	1	

References

Amasaka, K. 1995. A construction of SQC intelligence system for quick registration and retrieval library, *Lecture Notes in Economics and Mathematical Systems*, Springer, 445: 318–336.

Amasaka, K. 2000. A demonstrative study of a new SQC concept and procedure in the manufacturing industry, *An International Journal of Mathematical and Computer Modeling*, 3(10–12): 1–10.

Amasaka, K. 2001. Proposal of "Marketing SQC" to revolutionize dealers' sales activities, in *Proceedings of the 16th International Conference on Production Research*, pp. 1–9, July 31, 2001, Prague, Czech Public, (CD-ROM).

Amasaka, K. 2002. New JIT, a new management technology principle at Toyota, *International Journal of Production Economics*, 80: 135–144.

Amasaka, K. 2003a. Proposal and implementation of the "Science SQC" quality control principle, *International Journal of Mathematical and Computer Modelling*, 38(11–13): 1125–1136.

Amasaka, K. 2003b. A demonstrative study on the effectiveness of "Television Ad" for the automotive sales, in The 18th Annual Technical Conference of the Japan Society for Production Management, pp. 113–116, September 21, 2003, Nagasaki Institute of Applied Science, Nagasaki, Japan, [in Japanese].

Amasaka, K. 2003c. A "dual total task management team" involving both Toyota and NOK, in *Proceedings of the Group Technology/Cellular Manufacturing World Symposium*, pp. 265–270, Columbus, Ohio.

Amasaka, K. 2004a. Applying New JIT—A management technology strategy model at Toyota—Strategic QCD studies with affiliated and non-affiliated suppliers, in *Proceedings of the 2nd World Conference on Production and Operations Management Society*, pp. 1–11, May 1, 2004, Cancun, Mexico, (CD-ROM).

Amasaka, K. 2004b. Development of "Science TQM", a new principle of quality management: Effectiveness of strategic stratified task team at Toyota, *International Journal of Production Research*, 42(17): 3691–3706.

Amasaka, K. 2005a. Constructing a customer science application system "CS-CIANS"—Development of a global strategic vehicle "Lexus" utilizing New JIT, *WSEAS Transactions on Business and Economics*, 3(2): 135–142.

Amasaka, K. 2005b. Interim report of WG4's studies in JSQC research on simulation and SQC (1)—A study of the high quality assurance CAE model for car development design, in *1st Technical Conference of Transdisciplinary Federation of Science and Technology*, pp. 93–98, November 25, 2005, Nagano, Japan [in Japanese].

Amasaka, K. 2007a. High linkage model "Advanced TDS, TPS & TMS": Strategic development of "New JIT" at Toyota, *International Journal of Operations and Quantitative Management*, 13(3): 101–121.

Amasaka, K. 2007b. New Japan Production Model, an advanced production management principle: Key to strategic implementation of New JIT, *International of Business and Economics Research Journal*, 6(7): 67–79.

Amasaka, K. 2007c. The validity of advanced TMS, a strategic development marketing system utilizing New JIT, *The International Business and Economics Research Journal*, 6(8): 35–42.

Amasaka, K. 2007d. Highly reliable CAE model, the key to strategic development of Advanced TDS, *Journal of Advanced Manufacturing Systems*, 6(2): 159–176.

Amasaka, K. 2007e. Applying New JIT—Toyota's global production strategy: Epoch-making innovation in the work environment, *Robotics and Computer-Integrated Manufacturing*, 23(3): 285–293.

Amasaka, K. 2007f. Highly reliable CAE model, the key to strategic development of the New JIT, in *Proceedings of the 18th Annual Conference of the Production and Operations Management Society*, pp. 1–20, Dallas, Texas, (CD-ROM).

Amasaka, K. 2007g. The validity of "TDS-DTM": A strategic methodology of merchandise development of *New JIT*—Key to the excellence design "*LEXUS*", *The International Business and Economics Research Journal*, 6(11): 105–115.

Amasaka, K. (ed.). 2007h. *New Japan Model: Science TQM—Theory and Practice for Strategic Quality Management*, Maruzen, Tokyo, Japan [in Japanese].

Amasaka, K. 2008a. An integrated intelligence development design CAE model utilizing New JIT: Application to automotive high reliability assurance, *Journal of Advanced Manufacturing Systems*, 7(2): 221–241.

Amasaka, K. 2008b. Strategic QCD studies with affiliated and non-affiliated suppliers utilizing New JIT, in Putnik, G.D. and Cruz-Cunha, M.M. (eds), *Encyclopedia of Networked and Virtual Organizations*, Information Science Reference, Hershey, PA, pp. 1516–1527.

Amasaka, K. 2009. The effectiveness of flyer advertising employing TMS: Key to scientific automobile sales innovation at Toyota, *The Academic Journal of China-USA Business Review*, 8(3): 1–12.

Amasaka, K. 2011. Changes in marketing process management employing TMS: Establishment of Toyota sales marketing system, *China–USA Business Review*, 10(7): 539–550.

Amasaka, K. 2012. Constructing optimal design approach model: Application on the Advanced TDS, *Journal of Communication and Computer*, 9(7): 774–786.

Amasaka, K., Baba, J., Kanuma, Y., Sakai, H., and Okada, S. 2006. The evolution of technology and skill the key to success: Latest implementation of New Japan Model "Science TQM", in *The 23th Annual Technical Conference of the Japan Society for Production Management*, pp. 81–84. March 19, 2006. Osaka City University Osaka, Japan [in Japanese].

Amasaka, K., Ishii, T., Mitsuhashi, T., Tanabe, T., and Tsubaki, H. 2005. A study of the high quality assurance CAE model of the car development design (2), in *The 35th Annual Technical Conference of the Japanese Society for Quality Control*, pp. 147–150, November 12, 2005, Kansai University, Osaka, Japan [in Japanese].

Amasaka, K., Kishimoto, M., Murayama, T., and Ando, Y. 1998. The development of "marketing SQC" for dealers' sales operating system, in *The 58th Technical Conference of Journal of the Japanese Society for Quality Control*, pp. 76–79, May 31, 1999, Tokyo, Japan [in Japanese].

Amasaka, K., Kurosu, S., and Morita, M. 2008. *New Manufacturing Principle: Surpassing JIT—Evolution of Just In Time*, Morikita-Shuppan, Tokyo [in Japanese].

Amasaka, K., Nagaya, A., and Matsubara, W. 1999. Studies on design SQC with the application of Science SQC—Improving of business process method for automotive profile design, *Japanese Journal of Sensory Evaluations*, 3(1): 21–29 [in Japanese].

Amasaka, K., Ogura, M., and Ishiguro, H. 2013a. Constructing a Scientific Mixed Media Model for boosting automobile dealer visits: Evolution of market creation employing TMS, *International Journal of Business Research and Development*, 3(4): 1377–1391.

Amasaka, K., Onodera, T., and Kozaki, T. 2013b. Developing a Higher-Cycled Product Design CAE Model: The evolution of automotive product design and CAE, *Journal of Communication and Computer*, 10(10): 1292–1306.

Amasaka, K. and Sakai, H. 1998. Availability and Reliability Information Administration System "ARIM-BL" by methodology in "inline-online SQC", *International Journal of Reliability, Quality and Safety Engineering*, 5(1): 55–63.

Amasaka, K. and Sakai, H. 2008. Evolution of TPS fundamentals utilizing New JIT strategy: Proposal and validity of Advanced TPS at Toyota, *Journal of Advanced Manufacturing Systems*, 9(2): 85–99.

Amasaka, K. and Sakai, H. 2009a. Proposal and validity of New Japan global production model "G-NJPM" utilizing New JIT: Strategic development of advanced TPS surpassing TPS, in *The 1st Annual Conference of the Japanese Operations Management and Association*, pp. 15–30, June 20, 2009, Aoyama Gakuin University, Tokyo, Japan [in Japanese].

Amasaka, K. and Sakai, H. 2009b. TPS-QAS, new production quality management model: Key to New JIT—Toyota's global production strategy, *International Journal of Manufacturing Technology and Management*, 18(4): 409–426.

Amasaka, K. and Sakai, H. 2011. The new Japan global production model "NJ-GPM": Strategic development of Advanced TPS, *The Journal of Japanese Operations Management and Strategy*, 2(1): 1–15.

Amasaka, K., Nakaya, H., Oda, K., Oohashi, T., and Osaki, S., 1996. A study on estimating vehicle aerodynamics of lift, *Journal of the Institute of Systems Control and Information Engineers*, 9(5): 229–237 [in Japanese].

Amasaka, K., Watanabe, M., Ohfuji, T., Matsuyuki, A., and Ebine, 2004. Toward the establishment of the quality management technology of the client extreme priority, in *The 74th Technical Conference of the Japan Society for Quality Control*, pp. 141–144, May 29, 2004, Tokyo, Japan [in Japanese].

Amasaka, K. and Yamaji, M. 2006. Advanced TDS, total development design management model, in *Proceedings of the IABR Conference*, pp. 1–15, May 3, 2006, Cancun, Mexico, (CD-ROM).

Ando, T., Yamaji, M., and Amasaka, K. 2008. A study on construction of automobile design concept support methods—Visualization of form and customer sensibility utilizing design CAD, in *The 38th Annual Conference of the Japanese Society for Quality Control*, pp. 177–180, November 8, 2008, Tokyo, Japan [in Japanese].

Arima, M. 2002. Kaisetsu Color Design Tosyoku Sekkei Readings: Techno Cosmos, Dai 15 Gou, pp. 53–58 [in Japanese].

Dentsu. 2006. http://www.dentsu.co.jp/(2006).

Doos, D., Womack, J. P., and Jones, D. T. 1991. *The Machine That Changed the World—The Story of Lean Production*, Rawson/Harper Perennial, New York.

Ebioka, K., Sakai, H., and Amasaka, K. 2007. A New Global Partnering Production Model "NGP-PM" utilizing "Advanced TPS", *Journal of Business and Economics Research*, 5(9): 1–8.

Enrique, A. and Parallel, M. 2005. *A New Class of Algorithms*, Addison Wiley.

Fujieda, S., Masuda, Y., and Nakahata, T. 2007. Development of automotive color designing process, *Journal of Society of Automotive Engineers of Japan*, 61(6): 79–84 [in Japanese].

Gary, L. L. and Arvind, R. 2003. *Marketing Engineering: Computer-Assisted Marketing Analysis and Planning*, Pearson Education.

Hashimoto, H. et al. 2005. *Automotive Technological Handbook (3), Design and Body*, Chapter 6, *CAE*, pp. 313–319. Society of Automotive Engineers of Japan, Tosho Shuppan-sha Tokyo [in Japanese].

Hayes, R. H. and Wheelwright, S. C. 1984. *Restoring Our Competitive Edge: Competing Through Manufacturing*, Wiley, New York.

Ishiguro, H. and Amasaka, K. 2012a. Proposal and effectiveness of a highly compelling direct mail method: Establishment and deployment of PMOS-DM, *International Journal of Management and Information Systems*, 16(1): 1–10.

Ishiguro, H. and Amasaka, K. 2012b. Establishment of a strategic Total Direct Mail Model to bring customers into auto dealerships, *Journal of Business and Economics Research*, 10(8): 493–500.

James, A. F. and Mona, J. F. 2004. *Service Management: Operations, Strategy, and Information Technology*, McGraw-Hill, New York.

Jeffrey, F. R. and Bernard, J. J. 2005. Best face forward, *Diamond Harvard Business Review*, August: 62–77.

Kawakita, J. 1967. *The Conception Method—For the Development of Creativity*, Chuukou-Shinsho, Tokyo, Japan [in Japanese].

Kawamura, M. 2003. Zidai wo Utsusu Ryukousyoku to Color Design no Yakuwari Readings: Jidosya gizyutu Dai 57 kan Dai 5 Gou, pp. 15–20 [in Japanese].

Kawasumi, M. and Suzuki, S. 2008. Measurement of color discrimination ellipse for surface color with texture, *Journal of the Color Science Association of Japan*, 32(3): 166–174 [in Japanese].

Koga, K., Takahashi, T., Yamane, T., and Abe, K. 2003. Automotive basecoat with high flip-flop, *Journal of Society of Automotive Engineers of Japan*, 57(5): 31–35 [in Japanese].

Kuriyama, T. 2002. The requests to marketing CFD software in the automotive industry, *Journal of Japan Society of Computational Fluid Dynamics*, 10(2): 208–213.

Kusune, K., Suzuki, Y., Nishimura, S., and Amasaka, K. 1992. The statistical analysis of the spring-back for stamping parts with longitudinal curvature, quality, Japanese. *Society for Quality Control*, 22(4): 24–30 [in Japanese]

Magoshi, K. et al. 2003. Simulation technology applied to vehicle development, *Journal of Society of Automotive Engineers of Japan*, 57(3): 95–100 [in Japanese].

Muto, M., Miyake, R., and Amasaka, K. 2011. Constructing an Automobile Body Color Development Approach Model, *Journal of Management Science*, 2: 175–183.

Nakamura, M., Enta, Y., and Amasaka, K. 2011. Establishment of a model to assess the success of information sharing between customers and vendors in software development, *Journal of Management Science*, 2: 165–173.

Nezu, K. 1995. *Scenario of the Jump of US—Manufacturing Industry Based on CALS*, Industrial Research Institute, Tokyo [in Japanese].

Niiya, T. and Matsuoka, F. (eds.). 2001. *Foundation Lecture of the New Advertising Business*, Senden-kaigi, Tokyo [in Japanese].

Nikkei Business. 1999. Renovation of shop, product and selling method-targeting young customers by "NETS", pp. 46–50, March 15, 1999, Tokyo, Japan [in Japanese].

Nonohe, K. and Ibaraki, T. 2001, An improved tabu search method for the weighted constraint satisfaction problem, INTOR, 39(2), 131–151.

Okada, A., Kijima, M., and Moriguchi, T. (eds.). 2001. *The Mathematical Model of Marketing*, Asakura-shoten, Tokyo [in Japanese].

Okihara, D., Takada, A., Murakami, K., and Amasaka, K. 2013. Constructing a model for selecting kaizen actions to gain mutual trust between logistics providers and shippers, in *Proceedings of the 14th Asia Pacific Industrial Engineering and Management System*, pp. 1–8 (CD-ROM), December 6, 2013, Cebu, Philippines.

Park, S., Kamaike, M., and Nagao, T. 2009. Single picture impression structure and compound picture impression harmonization of user, environment and car, *Design Science in Information Systems Research*, 5(6): 59–66 [in Japanese].

Sakai, H. and Amasaka, K. 2003. Construction of "V-IOS" for promoting intelligence operator—Development and effectiveness for "visual manual format", in *The 18th Annual Conference of the Japan Society for Production Management*, pp. 173–176. September 21, 2003, Nagasaki Institute of Applied Science, Nagasaki, Japan [in Japanese].

Sakai, H. and Amasaka, K. 2005. V-MICS, advanced TPS for strategic production administration: Innovative maintenance combining DB and CG, *Journal of Advanced Manufacturing Systems*, 4(6): 5–20.

Sakai, H. and Amasaka, K. 2006a. TPS-LAS model using process layout CAE system at Toyota, advanced TPS: Key to global production strategy New JIT, *Journal of Advanced Manufacturing Systems*, 5(2): 1–14.

Sakai, H. and Amasaka, K. 2006b. Strategic HI-POS, intelligence production operating system: Applying advanced TPS to Toyota's global production strategy, *WSEAS Transactions on Advances in Engineering Education*, 3(3): 223–230.

Sakai, H. and Amasaka, K. 2007a. The Robot Reliability Design and Improvement Method and the advanced Toyota Production System, *Industrial Robot: International Journal of Industrial and Service Robotics*, 34(4): 310–316.

Sakai, H. and Amasaka, K. 2007b. Human Digital Pipeline method using total linkage through design to manufacturing, *Journal of Advanced Manufacturing Systems*, 6(2): 101–113.

Sakai, H. and Amasaka, K. 2008. Human-integrated assist system for intelligence operators, *Encyclopedia of Networked and Virtual Organization*, II(G-Pr): 678–687.

Sakai, H. and Amasaka, K. 2013. Establishment of Body Auto Fitting Model "BAFM" using "NJ-GPM" at Toyota, *Journal of Japanese Operations Management and Strategy*, 4(1): 38–54.

Sakai, H. and Amasaka, K. 2014. How to build a linkage between high quality assurance production system and production support automated system, *Journal of Japanese Operations Management and Strategy*, 4(2) (forthcoming).

Shimakawa, K., Katayama, K., Oshima, Y., and Amasaka, K. 2006. Proposal of strategic marketing model for customer value maximization, in *The 23th Annual Technical Conference of the Japan Society for Production Management*, pp. 161–164, March 19, 2006, Osaka, Japan [in Japanese].

Takebuchi, S., Asami, H., and Amasaka, K. 2012a. The Automobile Exterior Color Design Approach Model: Linking form and color, *China-USA Business Review*, 11(8): 1113–1123.

Takebuchi, S., Nakamura, T., Asami, H., and Amasaka, K. 2012b. The Automobile Exterior Color Design Approach Model. *Journal of Japan Industrial Management Association*, 62(6E): 303–310.

Takeoka, S. and Amasaka, K. 2006. The present condition and the foresight of CAE in the field of the development of automobiles—The applications of CAE for the super-short-term development, in *The 80th Technical Conference of Japanese Society for Quality Control*, pp. 103–106, May 26, 2006, Tokyo, Japan [in Japanese].

Tanabe, T., Tsubaki, H., and Amasaka, K. 2006. Fusion and evolution of requisite technologies for CAE—Scheduling applications as an example, in *The 80th Technical Conference of Japanese Society for Quality Control*, pp. 111–114, May 26, 2006, Tokyo, Japan [in Japanese].

Taylor, D. and Brunt, D. 2001. *Manufacturing Operations and Supply Chain Management: Lean Approach*, Thomson Learning, High Holbon, London.

Tokinaga, I. and Utsunomiya, C. 2009. Advertising strategy about color variation, *Design Science in Information Systems Research*, 56(4): 85–90 [in Japanese].

Watanabe, A. and Jo, K. 2009. Consideration about the correlation nature of the cluster of a fashion, and the taste color of clothes, *Journal of the Color Science Association of Japan*, 28: 42–43 [in Japanese].

Whaley, R. C., Petitet, A., and Dongarra, J. J. 2000. Automated empirical optimization of software and the ATLAS project, Technical report, University of Tennessee, Knoxville, TN, Department of Computer Science, University of TN, Knoxville, TN 37996.

Womack, J. P. and Jones, D. T. 1994. From lean production to the lean enterprise, *Harvard Business Review*, March–April: 93–103.

Yamaji, M. and Amasaka, K. 2009a, Proposal and validity of intelligent customer information marketing model: Strategic development of *Advanced TMS*, *The Academic Journal of China-USA Business Review*, 8(8): Serial No.74, 53–62.

Yamaji, M. and Amasaka, K. 2009b. Intelligence design concept method utilizing customer science, *The Open Industrial and Manufacturing Engineering Journal*, 2: 21–25.

Yanagisawa, K., Yamazaki, M., Yoshioka, K., and Amasaka, K. 2013. Comparison of experienced and inexperienced machine workers. *International Journal of Operations and Quantitative Management*, 19(4): 259–274.

Yomo, A. and Kato, T. 2010. Grouping tastes in color by personal preference and lifestyles, *The Institute of Image Information and Television Engineers Technical Report*, 34(18): 59–61 [in Japanese].

Chapter 9

Innovative Deployment of an Advanced Total Development System, Total Production System, and Total Marketing System

9.1 Automobile Form and Color Design Approach Model

9.1.1 Introduction

The idea of global quality competition in the auto industry has made developing exterior colors that best match the exterior design of vehicle models a critical factor in terms of product strategy, as color selection has the ability to affect consumer purchasing behavior. The rapid global advancement currently underway brings with it increasingly diverse and complex personal values and subjective impressions, which are difficult to fully grasp.

However, manufacturers that cannot accurately identify these consumer values and subjective impressions and incorporate the corresponding elements in their vehicle designs will find it difficult to remain competitive in the market. The author has therefore concluded that traditional design processes, which are implicit and rely heavily on designer intuition and experience-based rules of thumb, must be reformed employing an advanced Total Development System (TDS) through New Just in Time (New JIT) (Amasaka, 2002, 2007b,c; Amasaka et al., 2008).

To address these issues, the author has created the Automotive Form and Color Design Approach Model, which is designed to optimally match form and body color (the key elements of automotive exterior design) in a way that customers find attractive (Muto et al., 2013; Takebuchi et al., 2012a). In order to identify unspoken subjective customer impressions (preferences), the model objectifies (quantifies) form and body color qualities and outlines related to cause-and-effect relationships utilizing customer science through Science Statistical Quality Control (SQC) (Amasaka et al., 1999; Amasaka, 2003b; Takebuchi et al., 2010, 2012b; Muto et al., 2011, 2013).

The following list presents a detailed outline of the steps taken in this research:

1. SQC Technical Methods are used to identify the main elements that younger buyers are looking for in automotive body colors.
2. A survey is conducted using painted panels to find out what color elements generate subjective customer impressions.
3. Three-dimensional computer-aided design (3-D CAD) software is used to assign numerical values to form and color.
4. The results are used to generate research-oriented CAD models, which are used along with biometric devices to quantify the impact that form and color have on subjective customer impressions.
5. The relationship between survey data assessing subjective impressions and the qualities of form and body color are identified.
6. The author applies the newly developed Automotive Exterior Design Approach Model to optimally match form and body color in a way that customers find attractive, and the desired results are obtained.

9.1.2 Research Background

One of the most significant challenges facing Japanese industry in the twenty-first century is producing appealing products. Conversely, the challenge faced by industry in the twentieth century was achieving product consistency. It is clearly not enough to focus on price when developing products, especially nowadays when preferences are so varied and customers demand products designed to suit individuals. Similarly, the trial-and-error approach to development that was seen during the period of high economic growth is no longer viable.

In the automobile industry also, it is vital to properly take into account customer preferences and values in the planning and development of new products. This means it is important to verbalize (imagery/conceptualization) customers' feelings (implicit knowledge) and reflect this information on design drawings in the process of designing (explicit knowledge) through the application of Customer Science, which scientifically analyzes customer preferences (Amasaka et al., 1999; Okabe et al., 2006; Amasaka, 2007c).

Much research has been conducted on the subject of automobile exterior design (Satake et al., 2004; Asami et al., 2010; Takimoto et al., 2010). Conventional research has focused on the form of automobiles, but there has been little research on the relationship with color, which has a profound effect on purchasing.

Some research was conducted to identify customers' subjective impressions by changing the color while fixing the form (Fujieda et al., 2007; Asami et al., 2011; Takebuchi et al., 2010). As far as the author knows, however, there has never been any research to identify customers' subjective impressions of vehicle form and color when they are simultaneously changed (i.e., research in which both form and color are treated quantitatively).

This research study aimed to reveal how the relationship between form and color can affect customers' subjective impressions. Specifically, the author used an eye camera and an electroencephalogram (EEG; a device that measures brain waves) to clarify which aspects of form attract the most attention of customers. The author also revealed the relationship between form and color that can affect customers' subjective impressions by reproducing form and color on a CAD design and employing statistical inference.

9.1.2.1 Necessity of Numerical Representation of Form Using CAD

In order to visualize the influence rate of profile changes on customers' subjective impressions, it is important to quantify them. This research makes use of 3-D CAD as a means to quantify profiles.

There are three advantages to using 3-D CAD: (1) It is possible to perform numerical conversions of profiles (parameterization) using CAD software that allows numeric definitions of form. (2) Because exterior design can be viewed from different angles (front, side, rear, etc.), it is possible to compare different views with customers' subjective impressions. (3) 3-D CAD is also the primary modeling tool in the actual design process.

9.1.2.2 Research on Automobile Form Design Support Method

Similar research was done by Asami et al. (2010, 2011). They constructed an Automobile Form Design Support Method, which used biometric devices along with visualization technology and statistics to establish the relationship between vehicle form and color and customers' subjective responses.

First, they analyzed lines of sight using an eye camera to understand the perspectives from which customers observe overall automobile design. The subjective data collected during this process were then used to create a numerical model that measures form and color according to an objective scale. Finally, this information was used to construct and assess an evaluation model, and the given results were obtained. The effectiveness of this model was then verified.

However, a possible research subject is the creation of an automobile design approach that matches the form with the color (Amasaka, 2005a; Okabe et al., 2006; Yamaji et al., 2010; Takimoto et al., 2010). This research goes further by taking into account the unique qualities of automotive body color, quantifying form and color, and identifying how these elements relate to customers.

9.1.3 Creating an Automobile Form and Color Design Approach Model

To optimally match form and color in a way that customers find attractive, the author has created a new Automotive Form and Color Design Approach Model utilizing Science SQC as shown in Figure 9.1.

The author believes that it is important to visualize customer preference and to make use of this information in product development; thus, the author created the model shown in Figure 9.1, which plays an important role in product development. Steps 1–4 show an outline of this model.

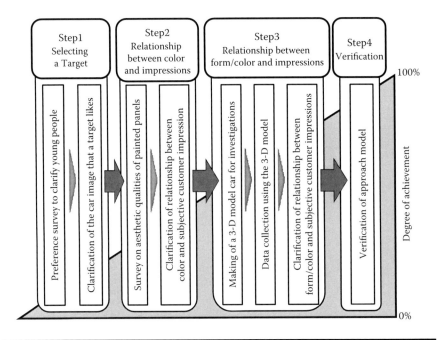

Figure 9.1 An automotive form and color design approach model.

9.1.3.1 Selecting a Target (Step 1)

In selecting a target, the author grouped customers by preference and then identified what the target group wanted from automotive design. A preference study was conducted using a fashion collage and the results were subjected to cluster and principal component analyses. The results revealed that younger customers could be divided into five groups according to their preference—practically oriented, disinterested, self-expressive, appearance oriented, and function oriented.

The characteristics of the self-expressive group were strong preferences and a desire to express their personal status and image through their vehicle. This group was also the most interested in cars, so the author decided to use the self-expressive group as the subject of this study. The analysis results also indicated three basic colors preferred by the target group—black, brown, and red. Four preference elements for this group were then identified: classy, luxurious, dignified, and sporty. In this way, the author arrived at his target group, four preference elements, and three preferred base colors.

9.1.3.2 Identifying the Relationship between Color and Subjective Customer Impressions (Step 2)

In order to grasp the relationship between the four subjective automobile qualities that were prized by members of the target group (classy, luxurious, dignified, and sporty), a survey was conducted using painted panels to identify the effect of color on subjective customer impressions.

This research study used three basic color characteristics (hue, brightness, and saturation) plus three additional qualities unique to automotive coloration, for a total of six automotive color parameters—hue, brightness, saturation, luminosity, shading, and graininess. Luminosity indicates the degree of light reflected, while shading indicates the difference in brightness between areas of light and shadow. Graininess refers to the size of the sparkling materials in the paint that reflect light when it hits the vehicle (Fujieda et al., 2007; Satake et al., 2004).

The author then integrated these color parameters with the three automotive colors preferred by the target group (red, brown, and black) to come up with a total of 11 colored panels. Test subjects were asked to look at the panels and rate how well each one evoked the four subjective qualities (classy, luxurious, dignified, and sporty). These data were then subjected to a covariance structure analysis.

In the model diagram used for this covariance structure analysis, exterior color was designated as the latent variable satisfying the four subjective qualities (classy, luxurious, dignified, and sporty).

The author then conducted a path analysis to figure out the cause-and-effect relationships between the six color parameters and exterior colors and then calculated standardized estimates. Looking at the standardized estimates in the model diagram shown in Figure 9.2, we can see the level of impact of each parameter as well as the directionality of that effect.

The results tell us that in order to produce exterior colors that match the preferences of the self-expressive group, we need to increase brightness and saturation while strengthening shading characteristics. The next section will deal specifically with the quality of shading.

9.1.3.3 Identifying the Relationship between Form/Color and Subjective Color Customer Impressions (Step 3)

In this section, a CAD model was used to conduct a subjective impression survey and clarify the relationship between automotive form/design and subjective customer preferences. In order to build the CAD model used in the survey, the author focused particularly on designing the four priority areas and distinctive factors known to contribute significantly to automotive form.

As shown in Figure 9.3, these were identified in prior research (Takebuchi et al., 2010, 2012a; Muto et al., 2011, 2013) as: (1) the edge of the hood, (2) the beltline, (3) the front pillar (A pillar), and (4) the rear pillar (C pillar).

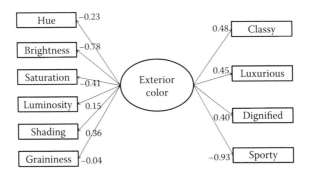

Figure 9.2 The model diagram of covariance structure analysis and standardized estimates.

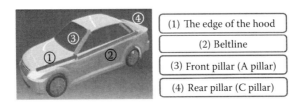

Figure 9.3 Four priority areas of the automotive model.

With these four priority areas defined as key factors, the degree to which form was modified for each factor was used as one standard to define the model vehicles created for the survey. Here, the author considered the overall balance of the vehicle design, setting up the following two core elements in order to maintain integrity of form: (1) depth of the edge of the hood and angle of the beltline (Form Element A) and (2) the front and rear pillars (Form Element B).

By keeping the modifications to these standards consistent in terms of direction, the author was able to come up with four vehicle models for the subjective impression survey that maintained design balance. With these four vehicle models (vehicle model: I–IV) as a base, three color standards (exterior color: red, brown, and black) and two shading standards (quality: big and small) were applied to create a total of 24 vehicle options as shown in Figure 9.4.

Two kinds of data were then collected using the vehicle models created for the subjective impression survey—subjective evaluation data and biometric measurement data (using an eye-tracking camera). For the first kind of data, the subjects were asked to look at the vehicle models prepared for the subjective impression survey and rate the degree to which they felt that the subjective automobile qualities applied to each vehicle model on a seven-point scale. For the second kind of data, the subjects looked at the same vehicle models while wearing an eye-tracking camera to measure line-of-sight information.

Next, the subjective evaluation data and the vehicle model parameters were subject to a Quantitative Theory Type I analysis in order to pinpoint the degree to which aspects of the vehicle—such as form and color qualities—impact subjective customer impressions. The Quantitative Theory Type I analysis was conducted using the 12 subjective qualities as external criteria and form and color—Form Element A, Form Element B, color, quality (shading)—as items. The author looked at partial regression coefficient values from the items to the external criteria to identify the level of impact for each variable as well as the direction of that impact.

One example from the analysis follows. When looking at the external criterion "sporty," the author found a determination coefficient of 0.628. For the individual items—Form Element A—it was found that raising the angle of both the beltline and the hood edge gave the vehicle a more "sporty" form. For Form Element B,

	Form Element A		Form Element B	
	Edge of the hood	Beltline	A pillar	C pillar
Type A	15°	0°	30°	20°
Type B	45°	5°	30°	20°
Type C	15°	0°	60°	60°
Type D	45°	5°	60°	60°

×

Exterior color	Quality (shading)
Three colors (red, brown, black)	Big, small

Figure 9.4　The parameters of the automotive model.

it was found that if brown was used as the standard color, changing the color to black made the vehicle seem less sporty, while changing it to red made it seem more sporty. Finally, the author found that eliminating shading effects caused the vehicle to lose its sportiness. These results made it possible to identify both the degree and the direction of impact that form and color have on creating the desired vehicle image.

Next, the author defined the interplay between aspects of the vehicle—such as form and color—by clarifying the interactions between the different variables with an analysis of variance. This was done in order to identify the effects that occur between the variables themselves. Here, the example of "sporty" is used as the objective variable in the analysis of variance as shown in Table 9.1. Factor A is Form Element A, factor B is Form Element B, factor C is exterior color, and factor D is color quality.

The results showed an interaction between factors A and B, A and C, and A and D, where a modification in each of these variables had a powerful impact on the other variable. When a cause-and-effect diagram is created with "sporty" as the objective variable, the results indicate that raising the angle of Form Element A, increasing the angles in Form Element B, making the color red, and intensifying the color quality create the most "sporty" impression.

As can be observed, more shading led viewers to concentrate their gaze on the edge of the hood as shown in Figure 9.5a, while less shading caused them to focus more attention along the beltline as shown in Figure 9.5b.

Using the eye-tracking camera, the author also collected biometric measurement data on line-of-sight movements as well as the length of time that subjects held their gaze at any one place. The line-of-sight heat map in Figure 9.5 compares two models with different shading characteristics.

This suggests that the level of shading changes the way that viewers look at the edge of the hood and the beltline by impacting the viewers' line of sight. At the same time, the author was able to see the difference in how the eye tracks with and without shading by tracing the amount of time that viewers spent gazing at particular points when shading was used as the color quality factor.

The conclusion is that the difference in how the beltline and the character line are seen is a function of the level of shading, since shading affects customers' line of sight. Therefore, variables such as form and color do not exist independently, but have a mutual effect on one another. This insight about mutual interaction was found to be true for each variable used in the analysis of variance study.

9.1.3.4 Verification (Step 4)

In order to check the validity of the foregoing analysis, the author conducted a survey on the CAD models that had the "sporty" and "luxurious" automotive design qualities favored by the self-expressive group, to see whether those models actually matched customers' images of them (Figure 9.6).

Innovative Deployment ■ 299

Table 9.1 Analysis Example of "Sporty" Used as the Objective Variable

	Factor	Sum of Squares	Degree of Freedom	Variance	Variance Ratio	Statistical Test	P Value (Upside)
1	A	310.537	1	310.537	276.402	**	.000
2	B	0.104	1	0.104	0.093		.761
3	AB	11.704	1	11.704	10.418	**	.001
4	C	54.808	2	27.404	24.392	**	.000
5	AC	11.325	2	5.663	5.040	**	.007
6	BC	1.108	2	0.554	0.493		.611
7	D	168.337	1	168.337	149.833	**	.000
8	AD	34.504	1	34.504	30.711	**	.000
9	BD	0.504	1	0.504	0.449		.504
10	CD	3.675	2	1.838	1.636		.197
11	Error ABCD	252.788	225	1.124			
12	Total	849.396	239				

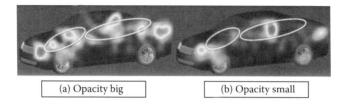

Figure 9.5 The line-of-sight heat map.

Figure 9.6 The vehicle model for inspection.

The sporty model received an average verification score of 6.6 on the verification questionnaire, while the luxurious model received an average verification score of 6. These results confirm the validity of the analysis results obtained in Section 9.1.3.3, since these average scores were higher than the scores that other models received on those particular subjective quality evaluations.

Next, the author used a device that measures brain waves to quantitatively evaluate customer preferences and provide another indicator to evaluate subjective impressions. The equipment allows researchers to separate brain wave frequency into four bands—delta waves (<4 Hz), theta waves (4–8 Hz), alpha waves (8–13 Hz), and beta waves (>13 Hz). It is known that the brain emits more alpha waves when experiencing pleasure and more beta waves when experiencing displeasure. As an indicator for evaluating customer preference, the author provided images for the subjects to view and then recorded their brain waves, focusing on the value calculated for alpha waves over the calculation value for beta waves, or α/β (Kawano et al., 2004; Takebuchi et al., 2012b).

The author showed the test subjects two CAD images of different vehicle models and measured their brain waves as they viewed the images. By extracting the alpha and beta waves measured as the subjects looked at the CAD models, and then calculating the ratio of alpha to beta for each one, the author found that the size of the alpha–beta ratio was consistent with the degree to which the subjects liked a given vehicle model. The ratio of alpha to beta waves used in this study can therefore be considered an effective way to grasp customer preference. With this, the validity of the analysis results obtained in Section 9.1.3.3 was considered verified.

9.1.4 Conclusions

The author created an Automobile Form and Color Design Approach Model. Form and body color qualities were objectified (quantified) in order to grasp unspoken subjective customer impressions (preferences). Related cause-and-effect relationships were then clarified and the desired results obtained.

Specifically, the author selected younger customers as the target of the study; he then collected both subjective evaluation data and biometric measurement data on them and used statistical methods to analyze these data. The results indicated the relationship between subjective customer impressions and changes in vehicle form, color, and color qualities. The author was thus able to find an ideal form and body color match that caused customers to find a vehicle attractive.

This approach model can be used in the trial production stage of vehicle models as a way to quantitatively capture the subjective characteristics that a particular model evokes. In addition, designers can use this model to determine which design parameters they can modify to bring a model closer to having the subjective characteristics that they want to achieve.

Further research is needed to verify the effectiveness of this approach model in additional analysis conditions, such as those that use vehicle models other than sedans, and other color qualities such as depth and transparency.

Currently, researchers are conducting research on automobile exterior color and interior color matching, reflecting on the research results of this model (Koizumi et al., forthcoming).

9.2 Optimal Computer-Aided Engineering (CAE) Design Approach Model

9.2.1 Introduction

In recent years, the advanced manufacturing industry has been faced with the urgent task of drastically reducing their development design times in order to respond quickly to changing consumer needs.

One of the most important challenges for manufacturers is how to strengthen and enhance the CAE of analysis in order to achieve fast and high-quality development design processes (Amasaka, 2007c, 2008, 2010a,b).

To address these issues, the author conducted research on an Optimal CAE Design Approach Model utilizing the Advanced TDS, Strategic Development Design Model (See Section 8.3.3 and 8.3.4).

In this section, the author addresses the technological problem of oil seal leakage in automotive drivetrains as a way to construct an Optimal CAE Design Approach Model for quality assurance (QA; Amasaka, 2012; Amasaka et al., 2012; Nozawa et al., 2012, 2013). The model is used to explain the cavitation caused by metal particles (foreign matter) that is generated through transaxle wear, a pressing issue in the automobile industry.

9.2.2 CAE in Product Design: Application and Issues

9.2.2.1 CAE in the Product Design Process

The time between product design and production has been significantly reduced in recent years with the rapid spread of global production. QA has become increasingly critical, making it essential that the product design process—a critical component of QA—be reformed to ensure quality (Kume, 1990; Amasaka, 2010a).

Figure 9.7 shows the typical product design process that is currently used by many companies (Amasaka, 2010a). The figure shows that companies first create product design instructions based on market research and planning. They then use these instructions to make specific product design specifications (drawings) that are promptly converted to digital format so that they can be suitably processed and applied. The data are primarily used in numerical simulations known as computer-aided engineering.

CAE analysis enables engineers to determine whether the product design specifications result in sufficient product quality—if not, the specifications are promptly optimized. Manufacturers then carry out effective preproduction, testing, and evaluation of prototypes (prototype testing). A final evaluation of the

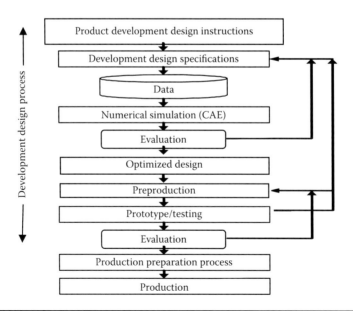

Figure 9.7   CAE in the development design.

development design is conducted, and if there are no problems, companies move to production preparation and process design in order to get ready for full-scale production.

9.2.2.2 CAE Analysis: Current Status and Issues

In recent years, CAE and other numerical simulations have been applied to a wide variety of business processes, including research and development; design, preproduction, and testing/evaluation; production technology; production preparation; and manufacturing. These and other applications are expected to have effective results (Magoshi et al., 2003; Yoo et al., 2010; Amasaka, 2010a).

The product design process, for example, is typically one that is guided by unspoken experiential knowledge and rules of thumb, leading to prototype testing that is guided by repeated trial-and-error efforts. In this age of global quality competition, using CAE as a predictive evaluation method in design work is expected to contribute greatly to shortening the development design time and improving quality (Amasaka, 2007c, 2008, 2010b).

Despite these high expectations, conventional forms of CAE analysis have resulted in figures that have deviated as much as 10%–20% from those obtained through prototype testing evaluations. This means that many companies are now stuck with applying CAE only to the monitoring task of comparative evaluations of old and new products—despite the enormous amount of funds that companies have invested in CAE development.

There are two absolute requirements for precise (highly reliable) CAE analysis methods that can both prevent the critical technical problems plaguing manufacturers from recurring and contribute to new product designs. The first is reducing the deviation from prototype testing evaluation figures to 5% or less, and the second is evaluating the absolute values needed for tolerance designs.

9.2.3 Application of the Advanced TDS, Strategic Development Design Model

Currently, to continuously offer attractive, customer-oriented products, it is important to establish a new development design model that predicts customer needs. In order to do so, it is crucial to reform the business process for development design using the core principle of TDS through New JIT (Amasaka, 2002, 2007c, 2008, 2010b). Manufacturing is a battle against irregularities; therefore, it is imperative to renovate the business process in the development design system and to create a technology so that serious market quality problems can be prevented in advance by means of accurate prediction/control.

For example, in solving technical problems, the approaches taken by design engineers, who tend to rely unreasonably on their own past experience, must be clearly corrected. In the business process from development design to production, the development design cost is high and the time period is prolonged due to the "scale-up effect" between the stages of experiments (tests and prototypes) and mass production. In order to tackle this problem, it is urgently necessary to reform the conventional development design process.

Focusing on the successful case mentioned above Section 8.3, the author deems it a requisite for leading manufacturing corporations to balance high-quality development design with lower cost and shorter development time by incorporating the latest simulation CAE and Science SQC (Amasaka, 2003, 2004, 2008). Against this background, it is vital not to stick to the conventional product development method, but to expedite the next-generation development design business process in response to a movement toward digitizing design methods.

Having said this, the author established the Advanced TDS, Strategic Development Design Model as the key to the high cyclization of development and design processes, the so-called New Japanese Development Design Model as described in Figure 8.13 (Amasaka, 2007c,g). This model is aimed at simultaneously achieving quality, cost, and delivery (QCD) through high-quality manufacturing, which is essential to realize customer satisfaction (CS), employee satisfaction (ES), and social satisfaction (SS; refer to Section 8.3).

To realize these aims, customers' orientation (subjective implicit information) must be scientifically interpreted by means of Customer Science (Amasaka, 2002, 2005a), namely, converting the implicit information to explicit information by objectifying the subjective information using Science SQC to create a Highly Reliable Development Design System, thereby reducing prototypes through accurate prediction and control by means of intelligence numerical simulation.

To this end, it is important to introduce the Intellectual Technology Integrated System, which enables sharing of knowledge and the latest technical information possessed by all related divisions.

9.2.4 Optimal CAE Design Approach Model

The Optimal CAE Design Approach Model in Figure 9.8 shows the procedural elements in Steps 1–5 of the business process that is used in automotive product design when applied to drivetrain oil seal leaks (Amasaka, 2012; Amasaka et al., 2012).

9.2.4.1 Step 1: Define the Problem

Step 1, defining the current situation, means explaining problematic technological issues regarding the number of functional breakdowns to clarify why the breakdowns are occurring as well as the mechanism that is generating them. Experts inside and outside the company use their collective wisdom in collaborative activities, applying the latest statistical methods and investigating and analyzing complex cause-and-effect relationships to define the problem in minute detail and reason out the faulty mechanism.

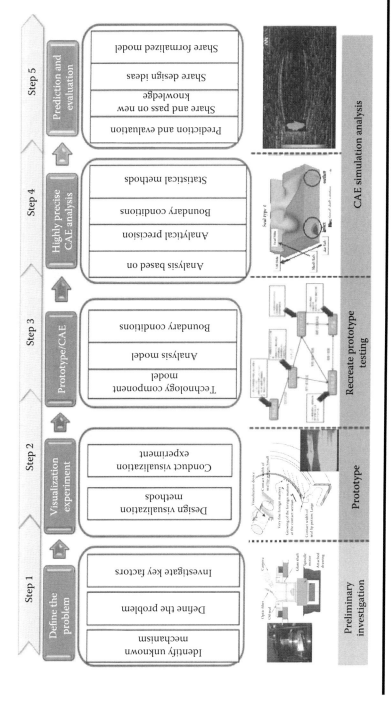

Figure 9.8 Optimal CAE design approach model.

9.2.4.2 Step 2: Conduct a Visualization Experiment

In Step 2, the visualization experiment, prototype testing is conducted in order to visualize the mechanism (dynamic behavior) that is generating the defect. This is how the faulty mechanism is further defined by utilizing Science SQC (Refer to Section 5.4.3 and Section 7.2.4). Specifically, using the Science SQC approach, SQC Technical Methods are applied to accurately explain the fault and conduct a factor analysis (Amasaka, 2008).

These methods use new seven (N7) QC tools, general SQC methods, reliability analysis (RE), multivariate analysis (MA), and design of experiment (DE) (Amasaka, 2004). The use of these methods allows previously overlooked and unidentified latent factors to be discovered and the faulty mechanism to be demonstrated via a logical thought process.

9.2.4.3 Step 3: Aligning Prototype Testing and CAE

In Step 3, the key factors (technology components) identified in Steps 1 and 2 that are generating the fault are subjected to a numerical simulation (CAE). The numerical simulation makes it possible to match the results obtained through prototype testing and CAE analysis (two-dimensional [2-D] and 3-D model) using absolute values, for which there is no discrepancy between the prototype testing and CAE analysis.

At this point, all business processes are scientifically and comprehensively optimized using the following steps, which must be part of a highly precise CAE analysis: clarify the phenomenon (defining the problem); modeling; calculation methods (algorithm); theory; and computer (calculation technology) using the Highly Reliable CAE Analysis Technology Component Model as shown in Figure 9.9 (Amasaka, 2007c, 2010a).

In using these steps, it is absolutely essential to clearly model the cause-and-effect relationships in unexplained mechanisms identified during prototype testing

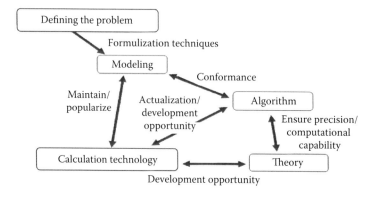

Figure 9.9 Highly Reliable CAE Analysis Technology Component Model.

(visualization) (Amasaka, 2007g; Yamada and Amasaka, 2011). To conduct a precise numerical simulation, there must be both an accurate theory and an experimental model that can logically define the impact of the latent factors identified during the experiment. Selecting a model with logical calculation procedures, analytical modeling, and algorithms is a must, with the goal of qualitatively modeling the fault (mechanism) (Figure 9.10).

9.2.4.4 Step 4: Conduct a Highly Precise CAE Analysis

In Step 4, conducting a highly precise CAE analysis, a highly reliable numerical simulation (quantitative modeling/highly precise CAE analysis) is carried out. This makes it possible to predict and control the absolute values needed for the CAE analysis based on the knowledge gained in Step 3 (Figure 9.9).

9.2.4.5 Step 5: Predict and Evaluate

Highly reliable CAE analysis makes CAE analysis for predictive evaluation possible when carrying out the business processes in Steps 1–4 in minute detail. In the past, CAE modeling was often conducted in an illogical way, with conventional CAE analysis and an undefined faulty mechanism. The result was a significant discrepancy (10%–30%) between prototype testing and CAE analysis values, often confining the use of CAE analysis to the auxiliary monitoring task of comparative evaluations of old and new products.

CAE analysis for predictive evaluation (Steps 1–4) makes precise prediction and evaluation possible by pinpointing key factors, accurately identifying the development design factors that must be optimally controlled, and making them explicit by incorporating them into drawings and manufacturing techniques. This contributes to swift design development that includes generating new knowledge and new

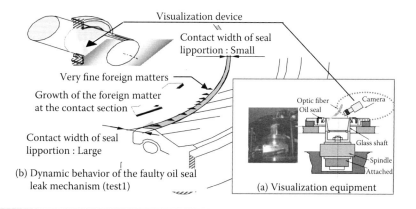

Figure 9.10 Faulty mechanism (oil leaks due to foreign matter).

design ideas, sharing explicit models, and creating assets out of new technology that are passed on and developed.

9.2.5 Application: Drivetrain Oil Seal Leaks: Cavitation Caused by Metal Particles in the Transaxle

In this chapter, the author uses both prototype testing and CAE, applying the Optimal CAE Design Approach Model to explain undiscovered technological mechanisms; the author then develops a model based on his investigative process (Amasaka, 2008, 2010b, 2012; Ito et al., 2010; Amasaka et al., 2012; Nozawa et al., 2012, 2013). The model is used to analyze "cavitation" caused by metal particles in the transaxle oil seal leaks.

9.2.5.1 Oil Seal Function

An oil seal on an automobile's transaxle prevents the oil lubricant within the drive system from leaking around the driveshaft. It is comprised of a rubber lip molded onto a round metal casing. The rubber lip grips the surface of the shaft around its entire circumference, thus creating a physical oil barrier. In this case, the sealing ability of microscopic roughness on the rubber surface is of primary importance (Amasaka, 2004).

The design parameters for the sealing condition of the oil film involve not only the design of the seal itself, but also external factors such as the shaft surface conditions, shaft eccentricity, and so on (Amasaka, 2010b; Amasaka, et al., 2012). Contamination of the oil by minute particles was found to be of particular importance to this problem since these are technical issues that involve not only the seal, but also the entire drivetrain of the vehicle.

9.2.5.2 Understanding the Oil Seal Leakage Mechanism through Visualization

Oil leaks and similar problems can result in immediate and critical vehicle defects. One of the primary causes of oil seal leaks is wear to the convex areas of the oil seal (o/s) where it comes into contact with the surface of the driveshaft, which is rotating at high speeds (Refer to Section 7.2.4). The author is applying the Optimal CAE Design Approach Model to this issue in order to resolve it (Ito et al., 2010; Amasaka, 2012; Nozawa et al, 2013).

9.2.5.2.1 Defining the Problem and Using a Visualization Device (Step 1)

This section addresses a second unexplained problem: metal particles (foreign matter) generated from rotation wear in drivetrain gears. The dynamic behavior of the faulty oil seal leak mechanism causing these metal particles to form is outlined using the developed visualization device (equipment) in Figure 9.10a, in order to turn this "unknown oil leak mechanism" into explicit knowledge (Refer to Section 7.2.4).

As shown in the figure, the oil seal was immersed in lubrication oil in the same manner as the transaxle, and the driveshaft was changed to a glass shaft that rotated eccentrically via a spindle motor to reproduce the operation that would occur in an actual vehicle. The sealing effect of the oil seal lip was then visualized using an optical fiber. It was conjectured that in an eccentric seal with one-sided wear, the foreign matter becomes entangled at the place where the contact width changes from small to large.

Three trial tests were carried out to ascertain if this was true or not. Based on an examination of faulty parts returned from the market and the results of the visualization experiment, it was observed that very fine foreign matter (which was previously thought not to impact the oil leakage problem) grew at the contact section, as shown in Figure 9.10b (test 1). It was also confirmed from the results of the component analysis that the fine foreign matter was a powder produced during gear engagement inside the transaxle gearbox.

9.2.5.2.2 Identifying the Oil Leakage Mechanism (Step 2)

The fine foreign matter in addition to microscopic irregularities on the lip sliding surface resulted in microscopic pressure distribution which eventually led to the degrading of the sealing performance as shown in Figure 7.14 (Test 2) (refer to Section 7.2). Also, the presence of this mechanism was confirmed from a separate observation that foreign matter had cut into the lip sliding surface, thereby causing aeration (cavitations) to be generated in the oil flow on the lip sliding surface.

This caused the deterioration of the sealing performance, as shown in Figure 7.15 (test 3) (refer to Section 7.2). This figure indicates that cavitations occur in the vicinity of the foreign matter as the speed of the spindle increases, even when the amount of foreign matter that has accumulated on the oil seal lip is relatively small.

As the size of the foreign matter increases, the oil sealing balance position of the oil seal lip moves more toward the atmospheric side and causes oil leaks at low speeds or even when the vehicle is at rest. This fact was unknown prior to this study, and therefore it was not incorporated into the original product design of the oil seals.

As a result of these efforts, the author was able to investigate the mechanism generating the oil seal leaks and use factor analysis to pinpoint the design elements in the oil seal and drivetrain gears that should have controlled the problem.

The mechanism involved cavitation occurring in rotating parts when foreign matter got wedged between sliding surfaces (on the lip surface). This happened in areas where there was variation in the size of the contact surface (from small to large) on the oil seal lip, caused by irregular wear and assembly variations.

The author used the knowledge obtained from the visualization experiment to logically outline the faulty mechanism in Figure 9.11. This was done in order

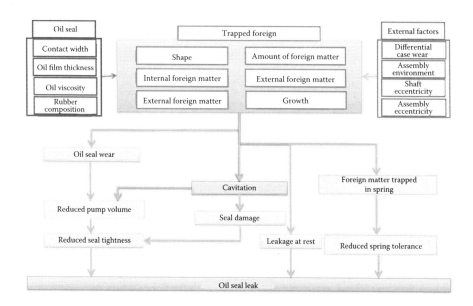

Figure 9.11 Oil seal visualization device and oil seal leak mechanism.

to capture the problem using the Highly Reliable CAE Analysis Technology Component Model. Using this process, the author was able to arrive at a hypothesis for why the cavitation was occurring, namely, factors such as low pump volume and seal damage had compromised the tightness of the seal and led to oil leaks.

9.2.5.3 Application of Optimal CAE Design Approach Model of Oil Simulator

9.2.5.3.1 Oil Simulator Using Highly Reliable CAE Analysis Technology Component Model (Step 3)

Using the Highly Reliable CAE Analysis Technology Component Model as shown in Figure 9.12, an oil seal simulator was created as an essential requirement for precise CAE analysis (Figure 9.9). As the figure indicates, the designs are optimized by integrating several aspects of the calculation process, including identifying the problem (root cause), conceptualizing the problem logically, and calculation methods (precision of calculators). Once the root causes of the problem are identified, it is critical that there is no discrepancy between the mechanism described and the results of prototype evaluations (Refer to Section 8.3.4).

The visualization experiment revealed that cavitation was occurring due to a weakening of the oil seal in areas (surfaces) that were in contact with the rotating driveshaft. This weakening was causing oil seal leaks. The Rayleigh–Plesset Model for controlling steam and condensation was used as a CAE analysis model that could explain the problem (Refer to Section 7.2.4).

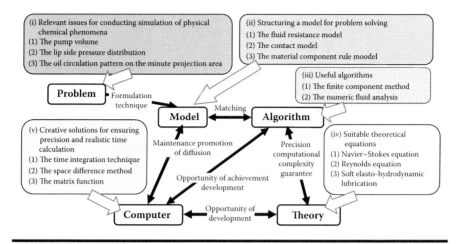

Figure 9.12 Oil seal simulator using the Highly Reliable CAE Analysis Technology Component Model.

The finite element method and nonstationary analyses were used as convenient algorithms. The Reynolds-averaged Navier–Stokes equation, Bernoulli's principle, and lubrication theory were appropriate theoretical formulas. Accuracy was ensured, and the time integration method was used to perform calculations in a realistic time frame. Each of the aforementioned elements was used to construct the oil seal simulator.

9.2.5.3.2 CAE Qualitative Model of the Basic Oil Seal Lip Structure (Step 4)

9.2.5.3.2.1 CAE Qualitative Model — The visualization experiment yielded the conditions on the sliding surface of the oil seal lip as a basic structural element. The author then used this element to construct the CAE Qualitative Model of the basic oil seal lip structure shown in Figure 9.13 to demonstrate sealing conditions. The model uses a statistical approximation of the slight roughness on the sliding surface to show the wedge effect created by minute projections.

In looking at the seal conditions on the sliding surfaces as a whole, the author concluded that the volume of inflow was greater at QAA′ than the outflow at QBB′, based on the fact that minute projections in section AA′ created a larger wedge effect than the minute projections in section BB′. These conditions also generated the oil circulation pattern on the minute projection area of the sliding surfaces, which meant that wear could be prevented by separating the two surfaces (Sato et al., 1999; Kameike et al., 2000).

The result obtained from incorporating the CAE Qualitative Model has helped to identify and refine the high-precision sealing mechanism of oil seals. The validity of the CAE Qualitative Model–sealing mechanism analysis was verified against

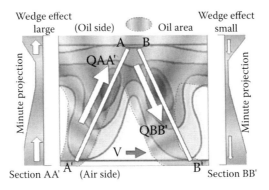

Figure 9.13 CAE qualitative model of the basic oil seal lip structure.

the results of actual vehicles and tests with a difference rate of 2%. The study conducted by the author has established this as a predictive engineering method for the functional designing of oil seal parts.

9.2.5.3.2.2 Two-Dimensional CAE Analysis — Using the technological elements mentioned above, a 2-D CAE analysis (2-D analysis) was used to conduct a numerical simulation that would accurately describe the behavior of the oil on the problematic minute projection areas. Figure 9.14 shows the results of this analysis. It shows the space between the driveshaft near minute projection AA′ and the seal where oil is getting trapped (The example of minute projection BB′ is omitted).

This 2-D analysis shows that shear stress is being generated by the fluid (oil) due to the rotation of the driveshaft and that the seal side flow direction is being reversed as the minute projections narrow the fluid channel.

9.2.5.3.2.3 Three-Dimensional CAE Analysis — Next, a 3-D analysis was conducted using a structural model of the sliding surfaces as a whole. This model took into account the direction of the oil flow in a third dimension (depth) based on the knowledge gained from the visualization experiment and the 2-D CAE analysis.

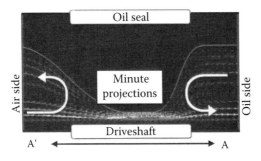

Figure 9.14 Two-dimensional analysis (example of minute projection AA′).

The model was used to perform a numerical simulation of the oil film present on the sliding surfaces. The analytical model shown in Figure 9.15 was constructed based on the CAE qualitative model of the basic oil seal lip structure. By imposing conditions such as shaft rotation speed, the amount of oil flow on the oil side and air side could be calculated. The oil flow to the seal side and to the air side was compared, producing similar results to the visualization experiment.

9.2.5.3.3 CAE Analysis (Step 5)

A cavitation is generated when oil collides with a foreign substance. The flow velocity near the foreign substance rises and there is a fall in pressure. This decreased pressure falls below the saturated vapor pressure, which results in a weakening of the oil flow and the generation of a cavitation.

9.2.5.3.3.1 Fluid Speed Analysis Example
— A fluid speed analysis similar to the one in Figure 9.16 was then conducted in order to look more closely at the mechanism causing cavitation. The analysis revealed that rapid changes in the fluid speed were occuring in the vicinity of foreign particles, and that the fluid speed dropped immediately before the oil collided with the foreign matter. This led to the conclusion that the presence of foreign particles was having an effect on oil flow.

9.2.5.3.3.2 Pressure Analysis Example
— Comparing cavitation and the fluid speed analysis results against the results of the pressure analysis shown in Figure 9.17 reveals that in areas of reduced pressure, oil was disappearing inside the cavities being formed—meaning that drops in pressure were likely being caused by these concave areas.

9.2.5.3.3.3 Cavitation Analysis Example
— Figure 9.18 shows the CAE analysis results at a rotation speed of 1100 rpm. This analysis confirmed the cavitation occurring around foreign matter, thus replicating the results of the visualization experiment. At the same time, the finding that cavitation becomes more significant as the rotation speed of the driveshaft increases was similarly replicated.

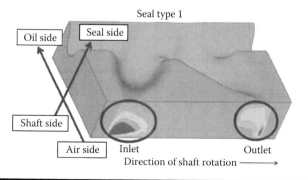

Figure 9.15 Three-dimensional analysis.

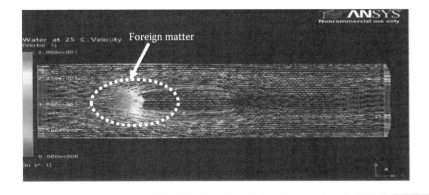

Figure 9.16 Fluid speed analysis around foreign matter.

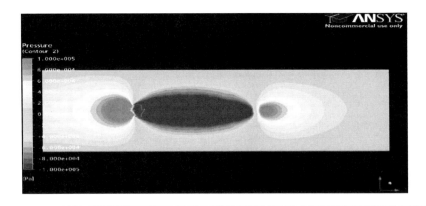

Figure 9.17 Pressure analysis around foreign matter.

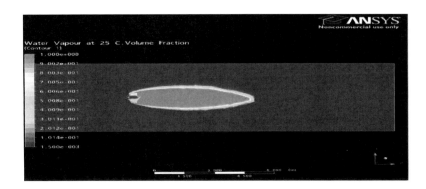

Figure. 9.18 Cavitation analysis around foreign matter.

9.2.5.3.3.4 Verification and Consideration — The above CAE analysis allowed the author to clarify the faulty mechanism causing cavitation; namely, that the presence of metal foreign particles was affecting the strength of the oil flow, causing drops in pressure in areas with faster oil flow and creating cavities. In addition, a similar analysis of changes in the shape and size of the foreign particles revealed that these changes were also causing changes in cavitation. These CAE analysis results indicate a close link between particle size/shape and cavitation.

The preproduction and testing/evaluation of prototypes add a significant amount of time and cost to the development process. By its renon, precise CAE allows manufacturers to eliminate preproduction (as well as prototype testing/evaluation) and still predict the mechanism causing cavitation and oil leaks. The CAE analysis allowed the author to re-create the changes in the flow speed and pressure around the foreign metal particles that were causing cavitation.

The deviation between the CAE analysis results and the results of the prototype testing was <5%, attesting to the usefulness of precise CAE analysis in certain cases.

9.2.5.3.3.5 Quality Improvement — These results led to two measures to improve design quality (shape and materials): (1) strengthen gear surfaces to prevent the occurrence of foreign matter even after the B10 life (L10 Bearing to mean time between failures [MTBF]) to over 400,000 km (improve quality of materials and heat treatments) and (2) formulate a design plan to scientifically ensure optimum lubrication of the surface layer of the oil seal lip where it rotates in contact with the driveshaft. The result of these countermeasures was a reduction in oil seal leaks (market complaints) to less than one-twentieth their original incidence as shown in Figure 7.19 (refer to Section 7.2).

9.2.5.3.3.6 Application to Similar Problem Solving — The author was able to apply the Optimal CAE Design Approach Model to critical development design technologies for automotive production, including predicting and controlling the special characteristics of automobile lifting power, antivibration design of door mirrors (Amasaka, 2010b), urethane seat foam molding (Amasaka, 2010b), and loosening bolts (Yamada and Amasaka, 2011; Onodera and Amasaka, 2012; Onodera et al., 2013; Amasaka et al., 2013).

In each of these cases as well, discrepancy was 3%–5% versus prototype testing. Based on the achieved results, the model is now being used as an intelligent support model for optimizing product design processes.

9.2.6 Conclusion

In this section, the author constructed an Optimal CAE Design Approach Model by utilizing the Advanced TDS, Strategic Development Design Model as a predictive evaluation method in design work, which is expected to contribute a great deal to shortening development design time and improving quality. The model primarily used numerical simulation to clarify the technological mechanism generating oil leaks as a result of metal particles entering oil seals in the transaxle.

9.3 Human Digital Pipeline System for Production Preparation

9.3.1 Introduction

Nowadays, Japanese manufacturers are establishing production bases in various countries, and are deploying global production based on the concept of "uniform quality worldwide and production at optimum locations." However, the author shares the sense of crisis about the current situation of Japan's exclusive patent, that is, competitiveness in manufacturing—high quality assurance.

The author has therefore established the Advanced TPS, Strategic Production Management Model for global production (Amasaka, 2007a,b; Amasaka and Sakai, 2008, 2009a, 2011). The mission of Advanced TPS in global deployment is to realize (a) an intelligent quality control system, (b) a high reliability production system, (c) the reformation of the work environment and (d) the development of intelligent operators (refer to Sections 8.1 and 8.3).

In an effort to realize the improvement in the intelligent productivity of production operators in the arena of global production, the author has grasped the necessity to create a new, human-centered production mechanism that offers creative and rewarding jobs. The author has therefore proposed the Human Intelligence Production Operating System (HI-POS) (Sakai and Amasaka, 2006b; Amasaka and Sakai, 2011).

As an implementation system of HI-POS for the global production strategy (refer to Section 8.4), the Human Digital Pipeline (HDP) System was established, enabling a total linkage of intellectual production information from designing to manufacturing through a digital pipeline (Amasaka et al., 2008; Sakai and Amasaka, 2007a).

More specifically, it is (1) a digital production system that prepares and offers the work procedure of production operators in advance, even without prototypes, by utilizing the design data of new products or facilities, and it is also (2) a production system that verifies a line process setup in advance. This accommodates the production planning needed for manufacturing a number of models on one line and completes the setting up of the line even before launching it.

The author has been studying its validity at advanced car manufacturer A. The use of this system has enhanced the skill level of production operators, competitive with that of Japan, and thus yielded the expected results.

9.3.2 Necessity of Cultivating Human Resources Compatible with Digital Production

Currently, Japanese manufacturers are endeavoring to realize customer-first manufacturing by rapidly deploying global production. The key to success in manufacturing lies in the ability to offer attractive merchandise to customers in a timely manner. As demonstrated in the cases of establishing overseas car manufacturing, the redoing of work processes in various stages from production preparation to mass production setup was repeated many times, and each time Japan provided support.

In this way, the start of production was realized through trial and error. Such a procedural difficulty mainly results from applying the Japanese method of production preparation along with the problem solutions derived from that method to the requirements of each case of overseas deployment (Nihon Keizai Shimbun, 2001; Asahi Shinbun, 2005).

Moreover, in the stage immediately prior to the start of mass production (production line off), each overseas manufacturing unit spends considerable person-hours preparing work process documentation similar to that prepared by other units; additionally, the training method and timing can be inconsistent between local manufacturing plants. These have caused problems related to productivity and quality at the beginning stage of mass production.

In order to overcome such issues, the author considered that the key to success was the study of a digital production system (DPS), that is, a production system that utilizes digital engineering. What is important in the utilization of the system is that the use of "DPS data" should not be limited to product development, design, and production preparation, but should be expanded to apply to the examination of workability, the preparation of work manuals and inspection methods, and even work training activities.

To realize the worldwide application of the DPS, it is necessary to acquire the skills related to information technology (IT) and the DPS, and it is deemed urgent to break away from the level of routine work operators, and to realize the cultivation of intelligence operators.

9.3.3 Proposing HI-POS of Intelligence Production System

While manufacturing in workshops is being transformed thanks to digital engineering, the engineering capability in manufacturing workshops often declines, thereby weakening the scientific production control that is used for quality incorporation in processes (Amasaka, 2003a). It is an urgent task to further advance the Toyota Production System strategically for higher-cycled, next-generation production business processes employing Total Production System (TPS) through New JIT (Amasaka, 2002), which are different from the conventional experiences of success (Intelligent Quality Control System, Highly Reliable Production System, Renovated Work Environment System and Intelligent Operator Development System) from the viewpoint of global production (refer to Section 5.1). The author therefore considered the necessity of including and organically integrating the four core elements with the strategic application of a conventional TPS in view of global production, and has proposed Advanced TPS as a global production strategy (Amasaka, 2007b).

The mission of Advanced TPS in global deployment is to realize definite terms as shown in Figure 8.17 (refer to Section 8.4), which (a) requires strengthening and improving the maintenance of process capability by establishing an intelligent quality control system; (b) requires establishing a highly reliable production system that achieves high quality assurance; (c) involves renovating the work environment to enhance intelligent productivity; and (d) requires developing intelligent operators (skill level improvement) and establishing an intelligent production operating system.

In recent years, in an effort to realize an improvement in the intelligent productivity of production operators in the arena of global production, the author has grasped the necessity of "the creation of a new, human-centered production mechanism" that offers creative and rewarding jobs. The author has therefore proposed HI-POS (Sakai and Amasaka, 2007), which is composed of three core systems—the HDP system, a Human Intelligence Assist (HIA) system, and a Human Intelligence Diagnosis (HID) system, as shown in Figures 8.19 and 8.20 (refer to Section 8.4). The basic requisites of the proposed HI-POS are the following three items:

1. A production system in which the reliability and maintainability of the entire production line are secured to prevent an operating rate decline and inferior quality, as well as to ensure an improvement in "operation skills of advanced production facilities."
2. In an integrated assembly industry, such as car manufacturing, there are many work areas where automation is difficult. Therefore, the acquisition of operators with the same technique/skill level as Japanese operators is necessary, particularly operators in developing countries who are assigned to work areas where low-productive manual line work is required.
3. An improvement in the intelligent diagnostic ability "highly skilled workmanship" of production operators themselves in the plants in developed countries that are equipped with advanced mass production facilities, particularly in the stage where quality is built up.

As IT and digitalization progress, more and more of the ever-changing "advanced production processes" are put into the so-called black box area. Before such processes are subject to individual discretion, the production operators need to cultivate more intelligence to realize the enhancement and homogenization of manufacturing on a worldwide scale.

9.3.4 Structuring "HDP" System and Its Components

Having said this, the author proposes, as an implementation system of HI-POS that is capable of realistically accommodating such needs, the HDP system (refer to Section 8.4), which enables the total linkage of "DPS data" from designing to manufacturing through a digital pipeline. More details about the HDP system as shown in Figure 8.25 (refer to Section 8.4.3.5) are discussed in the next section, but briefly, this system provides support ranging from tools to implementation methods for item (2) among the abovementioned three requirements for the global production strategy, which involves early acquisition of the same technique/skill level as Japanese operators (Sakai and Amasaka, 2007; Amasaka et al., 2008).

The HDP system proposed places emphasis on shifting the period of new model production to simultaneously realize the start of worldwide production. This system enables a real-time, total linkage of "DPS data" related to techniques and skills, from designing to manufacturing through a digital pipeline for the operators in both domestic and overseas production plants. By this means, training for highly intelligent work with intellectual productivity can be realized before long.

Figure 9.19 shows a conceptual schematic of the HDP system. This system is intended to prepare the work procedure of production operators in advance, outlining each of the operations in order, such as parts installation, by utilizing the "DPS data," even without prototypes. This system carries out the image training of production processes in the actual order of assembly operations. This is done at the preparatory stage before beginning mass production, and without having a real product to hand. Also, this system verifies the line process setup in advance to make sure that the production planning for manufacturing a number of models on one line can be carried out in the real production takt time.

This promotes the leveling of the operators' workload in each process, and completes the building up of the production line even before launching it. As a result, an improvement in the understanding and acquisition level of technique/skill processes in advance of mass production is possible.

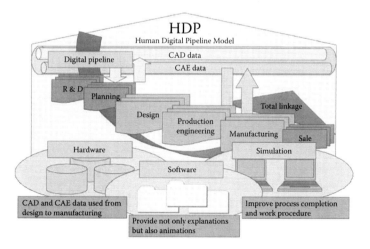

Figure 9.19 Schematic diagram of HDP.

9.3.4.1 Hardware Configuration of HDP System

The contents of such visual manuals should be revised as needed to reflect clearer ideas. The hardware configuring the HDP system is depicted in Figure 9.20. The conventional work procedure manuals that use handwritten letters and drawings are to be discontinued. Instead, intelligent, user-friendly operation manuals are prepared, which clearly present the items listed and instructions in an easily understood manner, and are provided to production operators.

More specifically, the CAD data as well as the CAE data used from development to production engineering are stored in (1) a product database (D/B) through the digital pipeline. Next, (2) a production information D/B, containing the production management information, such as production scale and volume as well as the parts arrangement information regarding the procurement status and locally procured parts, is connected to (3) a work procedure D/B, which accumulates typical examples of past work procedures, completing a total linkage.

In such a procedure, a work procedure manual is prepared from work data and parts data in advance and is provided to production operators. Concurrently, based on this work procedure manual, the routes that operators can take when moving in the production line are prepared. Then, from among these routes, the optimal combination of production operations is selected and arranged through simulation.

As a result, the process setup, which rearranges the work processes to correspond to multimodel production, will be verified before the start of mass production. In addition, the workload of each process is totaled for comparison, so that an uneven distribution of the workload (uneven distribution of the work amount

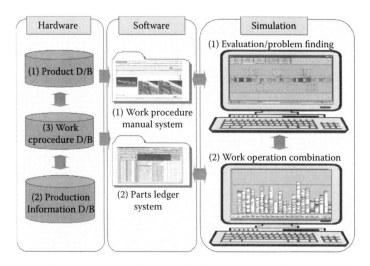

Figure 9.20 Hardware, software, and simulation configuration of HDP.

among differing vehicle models on the production line) is leveled out. The workload as well as the work posture of the operators is confirmed beforehand, and is then subjected to evaluation and fault finding.

9.3.4.2 Software Configuration of HDP System

The software configuration of the HDP system is depicted in Figure 9.20. The aforementioned work procedure manual is prepared using the Work Procedure Manual System. This system contains a wide range of information, such as the work data, which consist of work names, times, and locations; and the specification data, which consist of the specification, number, and quantity of parts, in addition to quality, work posture, instruments, safety, intuitive knacks and know-how, and so on. Based on all of this information, the work procedure manual is prepared.

Next, the visual data of parts are generated using the Parts Ledger System. In concrete terms, target parts are extracted using the information about their number, name, model, or quantity, and the associated 3-D shape data (CAD data) or verification data (CAE data) are searched.

Based on the work procedure manual system and the parts ledger system, the linkage between work and parts is made, and an elemental instruction sheet is prepared for each of the parts in the order of the steps that the parts are being assembled. For the types of work operations that cannot be fully instructed through explanations and photographs, video images are added to the visual manual to describe the acquired knacks and know-how, which ensure more accurate work operations by giving instructions on the procedure to be followed and the actions that should not be taken, using animation.

9.3.4.3 Simulation Algorithm of HDP System

9.3.4.3.1 Evaluation/Problem Finding

In order to allow for a production operator to move from his or her current work point to another work point in a straight line as shown in Figure 9.21, in addition to the speed of the conveyor, the following three factors need to be taken into consideration for the calculation of the coordinate locations below:

$$T = \left(-b + \sqrt{b^2 - ac}\right)\big/a$$

$$a = V^2 - V_{GX}^2 - V_{GY}^2 \quad b = -M_{GX} \times V_{GX} - M_{GY} \times V_{GY} \quad c = -M_{GX}^2 - M_{GY}^2$$

where:
  V = walking speed
  V_G = takt speed
  M_G = moving distance from the start point

V_X = X coordinate of the work point of V
V_Y = Y coordinate of the work point of V
M_{GX} = maximum moving time per unit time of V_X
M_{GY} = maximum moving time per unit time of V_Y
  T = walking time: Walking Time Calculation Formula

1. For simulating the avoidance route that is necessary to avoid contact between an operator and a body (vehicle), an advanced setting is programmed of the avoidance points for each move from one point to another, as shown in Figure 9.22. (For the work operation inside the vehicle body, four avoidance points, namely, the body front [engine compartment], rear [luggage], and each point on the right and left [center pillars] when viewing the whole body as a rectangle, are set. This program is designed to avoid any contact between a production operator and a vehicle body.)
2. Using the Dijkastra method (Dijkastra, 1959; Dijkastra and Feijen, 1991) (calculation method for the shortest route) for calculating the shortest route

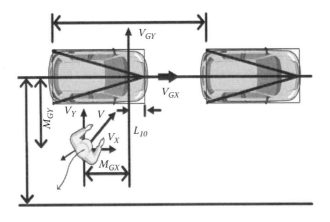

Figure 9.21 Calculation of coordinate locations.

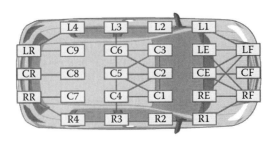

Figure 9.22 Advance setting for avoidance points.

and shortest distance, the next optimal work point for an operator to move to is calculated. (After determining the pattern of routes that a process operator can take for each vehicle model as shown in Figure 9.22, the points are conected with lines, for example, R3-C4-C2, and the shortest route of these patterns is calculated using the Dijkastra method.)
3. As part of the condition, the next movement coordinate is calculated using the time needed for each step. (The initial setting is 1 s.) The shorter this step time is, the more precisely the simulation can be carried out. (However, as the calculation amount increases, the load on the personal computer [PC] increases, too.)

Operator interference is verified through simulation to determine which combinations of vehicle models cause overlapping of operators in the time series. More specifically, the distance from the center coordinate of each operator is calculated, and if the results are smaller than the radius of the operators' movement, this will be defined as "operators' interference." Since it is possible to decide freely which models are put on the line in which order, simulation based on the aforementioned hardware information (production information D/B) is possible.

Through simulation, verification is used to determine whether an operator can complete the entire operation of one process within the set takt time, or if the line conveyor needs to be stopped. Based on the results of this simulation, advance verification is possible to determine which process on which vehicle model caused the operator to stop the line conveyor.

9.3.4.3.2 Work Operation Combination

Using the parameters of work hours and designated walking time, entered by means of the work procedure system mentioned earlier while discussing the software (Work Procedure Manual System), as well as the width and length of the vehicle body, an accumulative calculation is conducted. With regard to the walking time of an operator, the data computed through the abovementioned evaluation/problem finding procedure can be used as well. For the calculation of the average weight, the data imported by the software (Work Procedure Manual System) are used by setting as parameters in advance the production volume of a model, as well as the production volume of vehicle options.

An accumulative simulation will be carried out for the production workload for each vehicle model, which appears random according to the ratio of the production volume of each model. Based on this, the fluctuations in workload for each model can be verified beforehand, so that the process layout can be reexamined to realize the optimal load distribution. It is also capable of displaying graphs for each vehicle model, process by process, which enables the verification of the work hours for each vehicle model and, furthermore, singles out in advance the priority items for shortening the work hours in view of the process distribution.

9.3.5 Application Case: Deployment and Effect of HDP System

9.3.5.1 Assembly Procedure Manual Using Design Data

The following is an application case at an advanced manufacturer, Toyota Motor Corporation. Using the proposed HDP system, the work training of production operators was conducted in the shift to the production of new models at its preparatory stage for starting mass production. As an example, the work training for assembly operation can be mentioned.

The work training is carried out for the purpose of training production operators to be able to perform the work accurately according to the work procedure manual in connection with the start of new plants or the production of new models. Especially, the assembling operation requires the work to be done within the specified period of time with satisfactory precision, and a certain goal needs to be achieved in the early stage before the shift takes place.

Figure 9.23 shows an example of a detailed elemental instruction sheet (for the installation of window molding), as part of the assembly procedure manual prepared using the design data of assembly parts.

Such manuals can (a) give an easy explanation of installation methods by means of perspective visualization and color coding of the points requiring attention. For (b) the points requiring quality confirmation, 3-D data are provided for clearly presenting good and bad examples. Furthermore, real photographic images demonstrating good and bad examples are shown for difficult work operations, such as those dealing with locations not visible, along with clearer instruction.

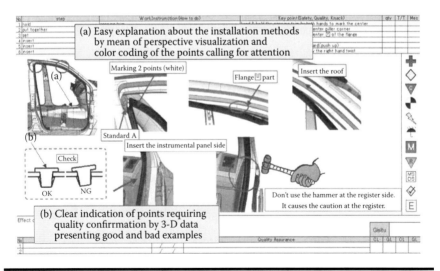

Figure 9.23 Example of element instruction sheet using work procedure manual.

9.3.5.2 Simulation Result of HDP System

9.3.5.2.1 Evaluation/Problem Finding

Figure 9.24 shows an example of verification of operators interfering with one another when determining which combinations of vehicle models cause operators to overlap in the time series. This diagram is an aerial view of the process in which a vehicle is carried on a conveyor at a fixed speed, and the walking routes of the operators performing assigned work operations are displayed in real time in straight lines.

When multiple operators performing different operations in one process overlap (that is when operators' interference occurs), this is displayed instantly, and the information on the combination of models is also output immediately.

Through simulation, verification is used to determine whether operators can complete their entire operation of one process in the set takt time. In the case that the line conveyor needs to be stopped because the assigned operation could not be completed in time, the conveyor on the screen is also stopped. Based on the simulation results, verification results are displayed showing which process, on which vehicle model, caused the operator to stop the conveyor.

9.3.5.2.2 Work Operation Combination

An example of the accumulated work operations for each process is shown in Figure 9.25 (left). The work hours are accumulated for each assembling location, specification, and priority. The walking time is automatically calculated from the traces connecting each part of a vehicle, and it is confirmed whether each operation is completed within the given takt time. In addition, an example of comprehending and confirming the net work ratio and the walking time is shown in

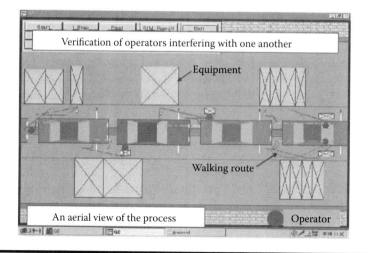

Figure 9.24 Example of verification of operators' interference with one another.

326 ■ *New JIT, New Management Technology Principle*

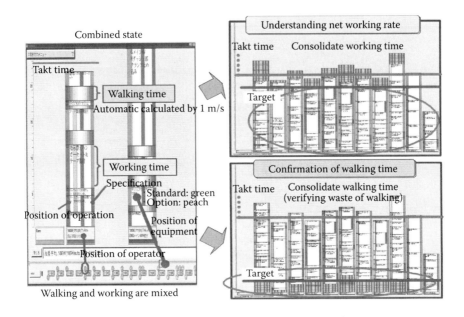

Figure 9.25 Example of accumulated work operations.

Figure 9.25 (left). By sorting out the working time and the walking time from the accumulated time results, the uneven distribution of the net working time and the nonworking time, such as time spent walking, are stratified to serve as a guideline for reviewing the process layout.

Figure 9.25 (right) shows the recalculation of the work accumulation after reshuffling the basic work operations between processes. Such changes can be easily made on the accumulative simulation screen by dragging and dropping with the mouse, immediately confirming the work points of previous and subsequent operations while automatically adjusting the walking time involved.

9.3.5.2.3 Contribution to Toyota's Global Production

As shown in Figure 9.26, the target of both productivity and quality was achieved at the stage of trial assembly before the start of mass production, demonstrating the accelerating effect. Also, the drop at transitional points, such as the cycle time change for increasing production capability, was avoided. The proposed system made a contribution to the global production strategy of advanced car manufacturer A, which requires achieving "uniform quality worldwide and production at optimum locations."

9.3.6 Conclusion

The author established the HDP system, which enables the total linkage of intellectual production information from designing to manufacturing through a

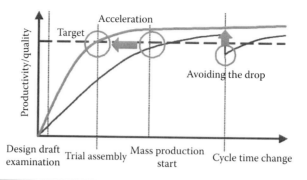

Figure 9.26 Productivity/quality evaluation.

digital pipeline. The HDP system was applied to the operations at an advanced manufacturing corporation, advanced car manufacturer A, and its validity was verified as an effective strategic move in the global production deployment of advanced car manufacturer A.

9.4 Body Auto Fitting Model Using New Japanese Global Production Model

9.4.1 Introduction

The leading Japanese management technology that has contributed most to worldwide manufacturing from the second half of the twentieth century is the Japanese Production System, which is typified by the Toyota Production System (Ohno, 1977). The Toyota Production System has been further developed and spread in the form of internationally shared global production systems.

With this in mind, the author has proposed the New Japanese Global Production Model (NJ-GPM) to enable strategic development based on the innovative deployment of the Advanced TPS, Strategic Production Management Model through New JIT (Amasaka, 2002, 2007b,d; Amasaka and Sakai, 2009b, 2011). The aim of this model is to realize a highly reliable production system suitable for global production by reviewing the production process from production planning and preparation through production itself and process management. The newly created NJ-GPM is fundamental to the strategic development of global production (refer to Sections 5.1 and 8.4).

Therefore, the author has established the Body Auto Fitting Model (BAFM) using the development of NJ-GPM and verified the effectiveness of this research through example application. Specifically, this study focuses on the fitting line by integrating the technologies using the BAFM at advanced car manufacturer A (a leading international corporation).

9.4.2 Background

9.4.2.1 The Current Condition and Problems of the Conventional Production System

Today's manufacturers in Japan are rapidly expanding their operations globally to be price competitive and they need to establish a new production system to suit their global strategies. Conventionally, well-experienced and highly skilled trainers go to local production sites and provide local production operators with individual, hands-on training.

The quality of training for production operators greatly relies on the personal capabilities of the highly skilled trainers. Different trainers give different training, which may confuse trainees and result in unevenness in the production operators' skill acquisition processes. Production operators in Japan have also experienced the same problems while passing on Japanese manufacturing technology.

9.4.2.2 Ideas for Three Important Points for Global Production

To ensure high quality and high efficiency, and to shorten lead time, the author has proposed three important components:

1. Production equipment using industrial robots
2. Skilled workers who operate the equipment (production operators)
3. Production systems including production data systems to activate the equipment and workers

The main factors for global production are to build a linkage among the individual processes in production planning, production preparation, actual production, and process control, to make those processes as perfect as possible, and to increase reliability in the manufacturing technology for global production (Sakai and Amasaka, 2007a,b).

9.4.3 Development of the New Global Production Model, Strategic Development of Advanced TPS

Global production must be developed in such a way as to establish the kind of manufacturing that is required to gain the trust of customers around the world. It must achieve a high level of QA and efficiency and shorten lead times to reinforce the simultaneous achievement of QCD requirements. The vital key to achieving these is the introduction of a production system that incorporates production machinery automated with robots, skilled and experienced workers (production operators) to operate the machinery, and production information to organically combine them.

Having recognized the need for a new production system suitable for global production, the author has created the NJ-GPM to realize the strategic development of the Advanced TPS, Strategic Production Management Model (Amasaka and Sakai, 2009b, 2011). This model eradicates ambiguities at each stage of the production process not only from production planning and preparation through production itself and process management, but also between the processes (refer to Section 8.3). The purpose is to achieve a highly reliable production system for global production that will improve the reliability of manufacturing through the clarification and complete coordination of these processes.

9.4.4 Establishment of Body Auto Fitting Model

The author has established the BAFM using the development of the NJ-GPM. The aim of the BAFM is to realize a highly reliable production system suitable for global production by reviewing the production process from production planning and preparation through production itself and process management.

The missions of the BAFM for global production are (i) to solve technical problems in advance in a production planning process by simulating technical problems with computer graphics created in CAE, (ii) to improve production operators' skills to operate automated equipment and the manufacturing technology, and (iii) to create production data that network and visualize the above objectives with the aid of IT.

To fulfill these missions, one requirement is that production operators must create a combination of processes in process planning and be well prepared so that they can easily do the processes. Digital engineering will lead to building a linkage among production processes to prevent technical problems, including:

1. Integrity between advanced equipment and its operators' skills.
2. Integrity between production operators' skills and their movements.
3. Integrity among the facility, its production operators' skills, and parts.
4. Another requirement is network systems that enable production operators to build the linkage. It is essential to build a linkage that disperses and integrates information globally while respecting local independence.

9.4.4.1 Four Essential Techniques to Build the Linkage among Production Processes in BAFM for Global Production

The idea in Figure 9.27 to build a linkage among production processes in the BAFM for global production consists of the following four essential techniques.

1. PSCS: Production Support CAE System
2. PSAS: Production Support Automated System

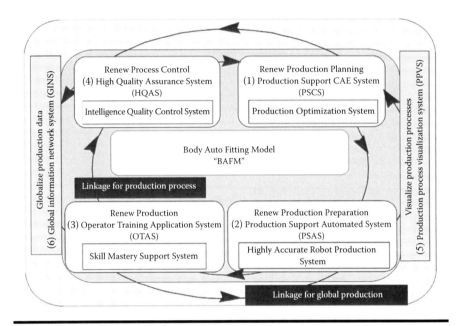

Figure 9.27 Outline of the Body Auto Fitting Model (BAFM).

3. OTAS: Operator Training Application System
4. HQAS: High Quality Assurance Production System

To use PSCS and PSAS, it is important to plan precisely from production planning stages and prevent problems by simulating. (i) Reforming production planning: CAE data simulation will optimize production requirements regarding production lines (logistics and transportation), robots (placing arrangement), and production operators (allocation and workability) for the entire plant. (ii) Reforming production preparation: A highly accurate robot production system will replace heavy-load processes that are currently done manually by workers.

For OTAS and HQAS, it is essential to improve, visualize, and systemize workers' skills (iii) in production processes. For those purposes, skill-mastering support systems will improve and judge workers' skills and place workers in the most appropriate positions in production processes. (iv) In process control, intelligent quality control systems will use IT control charts based on statistics and ensure process capacity (Cp) and machine capacity (Cm). It is vital to reform these stages in production planning, production preparation, mass production, and process control.

9.4.4.2 Two Essential Techniques to Build the Linkage to Globally Develop BAFM

The following two systems will support building the linkage to globally develop the BAFM among production operators.

In order to use the Production Process Visualization System (PPVS) and the Global Information Network System (GINS) in Figure 9.27, the following two systems are required.

1. Systems to visualize production processes in planning, preparation, and process control and gain better results on QCD activities
2. Servers and client systems to use production data globally

All of the above will implement the BAFM for global production to immediately establish the manufacturing technology that ensures high quality (Sakai and Amasaka, 2006).

In this section, the author verifies the effectiveness of a highly reliable robot production system for PSAS using the development of the BAFM.

9.4.5 Employing the Fitting Line Utilizing BAFM

9.4.5.1 Necessity of the Fitting and Fastening Automation

The recent labor shortage in Japan has imposed ever-increasing pressure on manufacturing sectors to reduce their dependence on manual labor. In body assembly plants, the automation of spot welding and similar operations is reaching nearly 100%, as shown in Figure 9.28. On the other hands, the automation of setting, fitting, and fastening of door, hood, and luggage compartment panels requiring highly skilled operators is the next important step in the complete automation of the body shop as shown in Figure 9.29.

Therefore, the author realized innovative unmanning of the fitting line by integrating technologies utilizing the BAFM, such as robotics (Sakai and Amasaka, 2007b,c), vision systems, bolt tightening, and product quality management; thus, the ability to automatically fit and fasten door, hood, and luggage compartment panels to the car body was achieved. This section shows the development of the

Operation	Automation level	Level of achievement	Quality importance
Parts setting S/A setting	30	△	—
Spot welding	88	◎	—
Brazing-finishing	10	○	○
Exterior component installation	8	×	○

Figure 9.28 Operations in body assembly plant and level of automation.

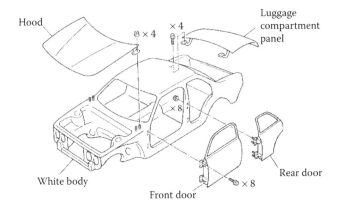

Figure 9.29 Schematic of exterior component installation.

following three items: (1) panel fitting accuracy, (2) automatic bolt tightening, and (3) integration into a flexible assembly line at Toyota, a leading international corporation.

9.4.5.2 Strategic Project Target

First, the strategic project target and the specific requirements were established at Toyota.

9.4.5.2.1 Fitting Quality Requirements

To ensure fitting accuracy, which provides a sleek exterior surface, strong emphasis was placed on the control of the gaps and the surface disparity between panels. Recent worldwide emphasis on quality required that high fitting accuracy requirements were set at $3\sigma: \pm 1.0$ mm (typical: $3\sigma: \pm 1.5$ mm).

9.4.5.2.2 Equipment Requirements (Bolt and Nut Supply and Tightening)

The most important basic technology regarding the installation equipment was the supply of bolts and nuts and their tightening processes. Achieving the target operational availability of 95% required the reliability of tightening (success rate per bolt) to be as high as 99.99%.

9.4.5.2.3 System Requirements

For the overall line side system, achieving high reliability and space efficiency was considered a must. Also, an overall control system combining all the individual technologies and high system flexibility was required.

9.4.5.3 Engineering Aspects

The engineering aspects to be cleared in order to meet the aforementioned requirements are shown in Table 9.2. The following sections will describe how the author overcame these hurdles.

9.4.6 Example Applications

In this section, the author introduces example application body auto fitting system of pioneering technology as applications of the BAFM, which has contributed to the advancement of management technology at advanced car manufacturer A (Amasaka and Sakai, 2009b).

9.4.6.1 A Fitting Accuracy

9.4.6.1.1 [A-1]: Setting the Fitting Quality Standards

Standards for fitting luggage compartment panels and doors were defined as described in the following sections.

9.4.6.1.1.1 Luggage Compartment Panel — The panel is fitted into an opening that is determined by several parts; right and left side member panels, upper back panel and lower back panel.

Table 9.2 Technical Goals in Development of Automated Body Fitting

	Requirement	Technical Goals
(A) Fitting accuracy	3σ|±|1.0 mm (variation in gap and surface disparity)	[A-1] Setting the fitting quality standards
		[A-2] Selecting the fitting adjustment method
		[A-3] Establishing the means of measuring the fitting position
(B) Assembly reliability	Tightening success rate: 99.99%	[B-1] Analysis of tightening mechanisms
		[B-2] Study of bolt and nut designs
		[B-3] Defining the equipment specifications

The configuration of the opening is not necessarily precise or consistent but has certain errors in the longitudinal and lateral dimensions. To achieve high fitting accuracy, the gaps (s_1–~s_4) and surface disparity (d_1–d_4) between the side member panels and the luggage compartment panel are measured with sensors attached to a robot, which holds and "fits" the luggage compartment panel. The gaps distributed between the left and right are equalized while the surface disparity is corrected to the target value; $s_1 = s_2$, $s_3 = s_4$ and $d_1 = d_2$, $d_3 = d_4 = d'$, respectively, as shown in Figure 9.30.

9.4.6.1.1.2 Doors — To determine the orientation of the body, initially the measurement datum was taken from an underbody panel. This was because the underbody pallet carries the white body and an underbody panel was assumed to be "held" in the most accurate position.

These measurements led to a deviation of $\sigma = 0.2$–0.4 mm in relation to the door opening. However, if two holes on the side member panel are used as the datum, the dimensional deviation is $\sigma = 0.1$–0.2 mm for the door opening as shown in Figure 9.31.

The two datum holes are measured via CCD cameras attached to the door panel jig. The jig is in turn attached to a robot via a tool changer and 2-D measurements are then possible through coordinate conversion algorithms.

9.4.6.1.2 [A-2]: Selecting the Fitting Adjustment Method (Gap, Surface Disparity, and Datum Hole Measurement)

9.4.6.1.2.1 Measurement System — The measurement of the gaps and the surface disparity between the luggage compartment panel and the opening essentially requires 3-D measurement.

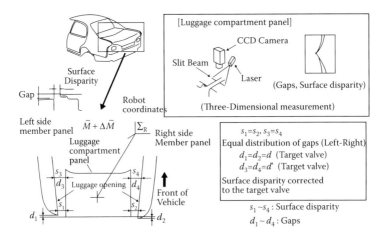

Figure 9.30 Luggage-fitting standards.

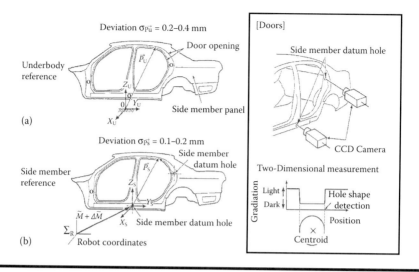

Figure 9.31 Standards for door fitting.

While there are several methods of 3-D measurement such as holography and interference fringe pattern analysis, we chose triangulation, which could be done with a simple instrument and is considered to be the most practical method.

In triangulation, a slit-shaped laser beam is applied to the object of measurement and the reflected light is monitored by a CCD camera. For the 2-D measurement of the door fitting datum holes, a CCD camera was used, which measures light intensity variation and then digitizes the image data to recognize the contour of the object (circle).

The digitized data were then used to calculate the object centered as shown in Figure 9.32.

9.4.6.1.2.2 Preliminary Study and Initial Problems

Luggage compartment panel: Figure 9.33a shows the result of preliminary gap and surface disparity measurements, indicating that a midpoint on the curvature is mistakenly recognized as the true end point.

This error was caused by the logic, which, as the surface was being scanned, recognized the point where the slit image disappeared. For example, on scanning line l1, a slight change in the gloss of the body surface could be misconceived as an end point. The software was revised to counteract these types of errors as shown in Figure 9.33b.

Doors: The result of a test operation with a 2-D sensor is shown in Figure 9.34, which shows (i) normal detection and (ii) a case where there was a surface flaw (surface scratch). The resulting hole centered offset caused an error, which lowered

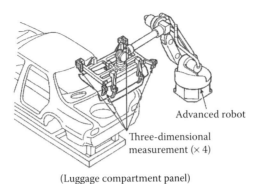

Figure 9.32 Outline of measurement system.

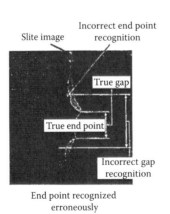
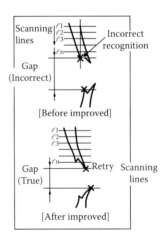

Figure 9.33 Measurement system for luggage compartment panel and doors: (a) measurement of gap and surface disparity; (b) end point detection logic.

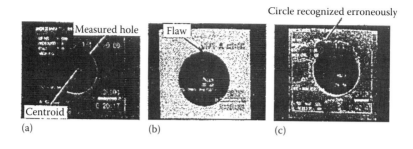

Figure 9.34 Measurement with two-dimensional CCD camera (photo 2): (a) normal detection, (b) presence of flaw, and (c) blurred image.

the measurement accuracy and (iii) a measurement failure where the measurement system stopped operation.

The prototype measurement system described above was improved by introducing the following ideas.

Luggage compartment panel:

1. Development of an electronic noise reduction program

Doors:

1. Statistical processing of the image recognition data
2. Development of centered location logic for imperfect circles

9.4.6.1.3 Problem Solutions

The following description uses the 2-D hole recognition sensor as an example.

1. The selected method quantitatively represents the characteristic properties $f(\mu, \sigma)$ of the image. These characteristic properties are then statistically confirmed using a pattern recognition process that utilizes 18 typical patterns stored in a database as shown in Figure 9.35.

The logic was developed by first assuming that the image pattern that corresponds to $\mu + n\sigma$ is image A' and the pattern that corresponds to $\mu - n\sigma$ is image A'' with regard to one characteristic property $f(\mu, \sigma)$ (where μ is the mean value of the characteristic property, σ is the deviation, and the threshold level is set at $\mu \pm n\sigma$). These two images are then used to determine a characteristic property C, which is determined by $\mu - n\sigma < C < \mu + n\sigma$, which then denotes a particular pattern A.

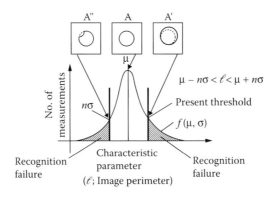

Figure 9.35 Statistical processing of image recognition data.

2. Accurate hole location by the algorithm described in (1) may be adversely effected by minute impurities such as changes in the surface gloss or the existence of flaws or deformations around the hole. To compensate for such effects, we developed the following detection algorithm.

Centered G of the image is first determined as shown in Figure 9.36. Then distances $r = \overline{Gi_n}$ (from G to the point i_n), which divide the perimeter into n equal parts, are checked to determine the three points corresponding to the value of r that most frequently appear, denoted as i_a, i_b, and i_c. The center of a circle defined by these three points is determined to be true center of the hole. This algorithm is shown in Figure 9.37. Through both the statistical processing additions and the software improvements, both a 100% recognition success rate and a detection accuracy of within 0.1 mm were achieved.

9.4.6.1.4 [A-3]: Establishing the Means of Measuring the Fitting Position

A robot with coordinate conversion function capability was chosen to be utilized by the fitting system. As the robot "fit" the panels to their respective openings, the fitting position was corrected by using coordinate data obtained from the sensors mounted on the robot hand. The algorithm selected to determine **F**, the coordinate conversion technique for the luggage compartment panel, will be described below.

The gap measurement is denoted by S_i, the surface disparity by d_i ($i = 1 \sim 4$), and the position vector matrix of the hand coordinate viewed from the base coordinate of the robot by $\tilde{M}$. The target values of S_i and d_i are defined as S_{i0} and d_{i0}, respectively, while $g_i(S_{i0}, d_{i0})$ are the ideal functions. Let the positions on the side member be $\vec{p}_{bi}$ ($i = 1-4$) and the positions on luggage side be $\vec{p}_{li}$ ($i = 1-4$). Allowing

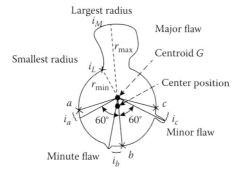

Figure 9.36 Hole center detection.

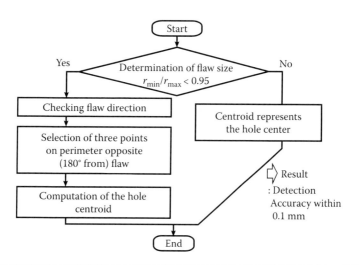

Figure 9.37 Algorithm.

$\Delta \tilde{M}$ to represent the displacement of the robot, the least squares method can be applied to approach S_{i0} and d_{i0}. $\Delta \tilde{M}$ can then be obtained by solving the following equation:

$$\frac{\partial \left\| \tilde{M} \cdot \Delta \tilde{M} \cdot \vec{p}_{li} - \vec{p}_{bi} - g_i \left(S_{i0}, d_{i0} \right) \right\|}{\partial_{aj}} = 0 \qquad (9.1)$$

where a_j ($j = 1,2,3,\ldots, 6$) are elements of $\Delta \tilde{M}$ and $\| \ \|$ indicates the absolute function.

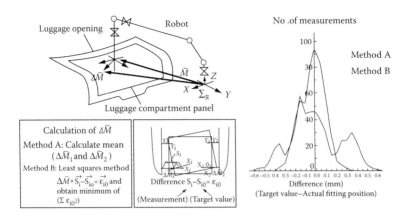

Figure 9.38 Coordinate conversion algorithm.

Defining $\tilde{M}_0 * \Delta\tilde{M} \equiv \mathbf{F}$, determines the algorithm, which determines the optimum gaps and surface disparities at each location, as shown in Figure 9.38.

9.4.6.2 B Assembly Reliability: Equipment Requirements to Overcome Technical Hurdles

9.4.6.2.1 [B-1]: Study of Bolt and Nut Designs (Bolt Type Review)

Especially for a bolt-type review, bolt designs were studied because of their significant influence on the reliability of tightening. Flat tip, round tip, and self-locating tip bolt types were studied with regard to the allowance for approach misalignment of bolt β and the bolt inclination allowance α, as shown in Figure 9.39. This study revealed that the round-tipped bolt had the best tightening reliability.

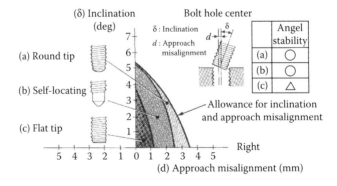

Figure 9.39 Bolt design influences.

9.4.6.2.2 [B-2]: Bolt Position Measurement Study (Rear Door Application)

When viewing a bolt, an incorrect object may be mistaken as the intended object due to the presence of multiple candidates. To eliminate such mistakes, a program was developed that chooses among multiple candidates an object that has the characteristic properties closest to those of the intended object, as shown in Figure 9.40. With this program, we achieved a 100% success rate in bolt recognition, compared with 70% in the trial operation. Images from the bolt recognizing sequence with and without the program are shown in Photo 3.

9.4.6.2.3 [B-3]: Defining the Equipment Specifications

The characteristic factors that influence the reliability of bolt tightening are shown in Figure 9.41. Those factors having significant effects are enclosed in factor boxes

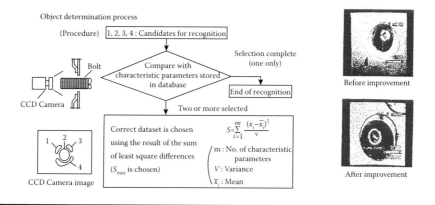

Figure 9.40 Bolt recognition program.

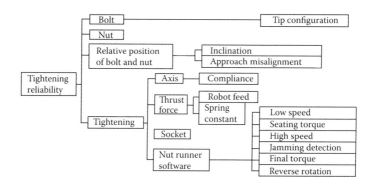

Figure 9.41 Tightening reliability cause-and-effect diagram.

in Figure 9.41. The major factor related to tightening was determined to be thrust force, thus experiments were concentrated in this area.

Initially, the thrust force was provided by a spring, but it provided excessive force in the early stage of tightening. Ultimately, the thrust force was provided by a robot, which was synchronized with the progress of tightening, as shown in Figures 9.42 and 9.43.

9.4.6.3 Engineering Aspects of System Requirements: Development of Floor Space–Efficient System Based on Review of the Transfer Equipment and Installation Equipment (Robots)

The explanation for this system will be limited to the subject of floor space efficiency. With the conventional robots on the simple mechanism of movable parallel links, the work envelope is confined to a region away from the robot body by predetermined offset. This forces the assembly line to require an undesirably large floor space.

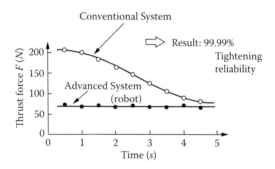

Figure 9.42 Thrust force control.

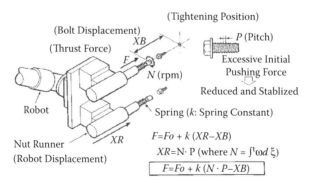

Figure 9.43 Thrust force control method.

The author reviewed the arm mechanism of the robot and employed a fixed parallel link mechanism, which enabled the robot envelope to include an area very near the robot body, as shown in Figure 9.44. The effect of floor space reduction achieved with this type of robot is shown in Figure 9.45.

9.4.6.4 Results

This system is currently integrated in the body assembly line located in the domestic plant. The conventional robot control method is the fixed point control, such as handling, spot welding, and arc brazing. On the other hand, this new method is the actual manufacturing point control, where the robot is synchronized with the progress of tightening. Remarkable improvements were achieved in this line, such as the 95% line operation availability and the replacement of approximately 10.6 workers, as shown in Table 9.3. These results were made possible by achieving

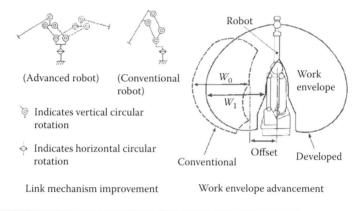

Figure 9.44 Advanced robot illustration.

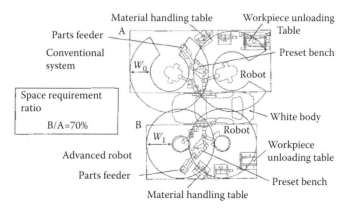

Figure 9.45 Effect of floor space reduction.

Table 9.3 System Evaluation

	Fitting Quality 3σ (mm)	Tightening Reliability (%)	Labor Savings (persons/day)
Conventional system	±1.5	99.00	–
Advanced system	±1.0	99.99	10.6

the target reliability of bolt tightening (99.99%) and the target fitting accuracy (3ς: ±1.0 mm).

9.4.7 Conclusion

Automation in its true sense was successfully realized without the need to correct the fitting in the downstream processes, satisfying the major design stipulation. Success may be attributed to mastering basic technologies and combining them into a complete system integrated into a flexible production line. In this study, the author has established the BAFM using the NJ-GPM, and its effectiveness has been verified at advanced car manufacturer A.

9.5 Total Direct Mail Model to Attract Customers

9.5.1 Introduction

Recent changes in the marketing environment have made the personal relationship between businesses and their customers even more critical. Businesses are faced with the task of constructing sales schemes that are able to flexibly and accurately grasp peculiarities and trends in customer preferences. This study looks at the effectiveness of the direct mail (DM) advertising method in bringing customers to auto dealers, based on the idea that forming personal bonds with customers is a core component of successful sales. One of the unique features of DM is the advertiser's ability to select which customers will receive the mailings. This point is the focus of this study, which attempts to identify an optimum decision-making process for selecting target customers.

The way that dealers currently select which customers will receive DM is by having individual salespeople choose them on the basis of personal experience and knowledge. Because there is no clear decision-making process, the response rate is lower than expected and dealers do not achieve the targets set forth in their sales strategies.

To address these issues, this study presents a methodology for selecting DM recipients from the dealer pool that will result in increased response (dealer visit) rates. The proposed mechanism for selecting these optimum recipients is a strategic Total Direct Mail Model (TDMM) with the aim of bringing more customers into auto dealerships (Ishiguro and Amasaka, 2012a,b). This model uses mathematical programming and statistics to aid the decision-making process and boost the effectiveness of DM advertising employing Advanced TMS, Strategic Marketing Development Model through New JIT (Amasaka, 2007b,f).

The core systems used in the model are (1) a customer purchase data analysis system to increase dealer visits, (2) a DM content optimization system to target customer preferences, (3) a system for strategically determining who should be sent DM based on who actually visits the dealership, and (4) a DM promotion system for sales staff to increase market share. These four systems are integrated and strategically applied to promote dealer visits. The author then applies this model at an actual auto dealership and is able to successfully increase dealer visits.

9.5.2 Necessity of Direct Mail

This section focuses on customer retention, a critical issue at auto dealerships, and discusses a methodology for preventing customer loss. The necessity of DM as a tool for retaining premier customers is discussed below.

9.5.2.1 Direct Mail and Customer Retention

Auto dealerships typically store customer attributes, purchase history, and other information on their clients in a database or similar storage system (Amasaka, 2004). The database is used to identify premier customers, and pinpointing their preferences is critical if the dealership is to prevent them from taking their business elsewhere. The author considers using a database to help find premier customers and then sending them DM to be an effective method of sales promotion.

Kotler and Keller (2006) identified the ability to select recipients and to offer personalized content as the advantages of DM. Piersma and Jonker (2004) studied how often to send DM to individual customers in order to establish long-term relationships between the direct mailer and the customers. Jonker et al. (2006) provided a decision support system to determine mailing frequency for active customers based on their behavioral data: their recency, frequency, and monetary (RFM) values. Bell et al. (2006) applied experimental designs to increase DM sales. Beko and Jagric (2011) presented demand models for DM and periodicals market using time series analysis.

DM is notable as an effective way of targeting specific premier customers for sales promotions (Bult and Wansbeek, 2005).

9.5.2.2 Direct Mail and the Attention-Interest-Desire-Action (AIDA) Model

The Attention-Interest-Desire-Action (AIDA) Model is a well-known advertising information management model developed by St. Elmo Lewis. The model presents the psychology of consumer purchase behavior as a series of four steps: *attention*, *interest*, *desire*, and *action* (Ferrell and Hartline, 2005; Shimizu, 2004).

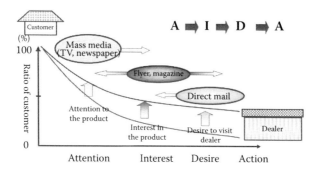

Figure 9.46 AIDA model.

The author has created a model of customer response for use in auto sales (Amasaka, 2007f, 2009). This model is based on the AIDA Model and is presented in Figure 9.46. The model allows dealerships to boost the percentage of customers that visit their showrooms by using mass media advertising, flyers, magazine ads, and DM.

9.5.2.3 Current Research on Direct Mail Activities

In studying the current status of DM activities at dealerships, the author identified the following issues:

1. *Strategy formulation*: target customers and sales concepts are unclear
2. *Direct mail content*: customer preferences are not accurately identified
3. *Recipients*: recipients are determined based on the experiential knowledge of sales staff, which leads to a high degree of variation in the quality of work performed
4. *Approach to customers who visit the dealership*: purchase behavior characteristics of individual customer are not well understood

Given these problems, DM cannot currently be thought of as effective in all cases. The author has demonstrated that the percentage of DM recipients that visit dealers is as low as 1%–5%, and that the profitability of those customers who do come in is not very high (Ishiguro and Amasaka, 2012; Ishiguro et al., 2010).

It is critical that these issues be resolved so that DM activities are more efficient (sophisticated). This will help dealerships retain their premier customers and boost their sales. A survey of the literature did not reveal any prior research that successfully addresses these problems.

9.5.3 Strategically Applying a Total Direct Mail Model

Based on previous DM research and studies done on the actual state of DM activities, the author developed TDMM, the strategic application of a TDMM. This concept model is shown in Figure 9.47.

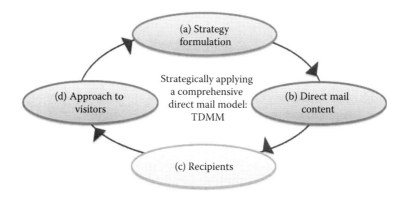

Figure 9.47 Strategically applying a total direct mail AIDA model: TDMM.

As the figure indicates, the model consists of the following four core functions:

1. *DM strategy formulation*: scientific identification of premier customers (Amasaka, 2011)
2. *DM content*: development of content based on customer preferences (Kimura et al., 2007)
3. *Mailing process*: establishing a process for determining recipients based on customer information and other data (Kojima et al., 2010; Ishiguro et al., 2010; Ishiguro and Amasaka, 2012a,b)
4. *Customer handling*: clearly identifying customer purchase behavior characteristics (Kimura et al., 2007)

Figure 9.48 shows the author's approach model to practically implementing the aforementioned strategies. The related systems are described below.

9.5.3.1 System for Analyzing Customer Purchase Data

As Table 9.4 describes, the system for analyzing customer purchase data supports dealers in formulating a DM strategy by identifying different premier customer segments.

9.5.3.1.1 Identifying Premier Customer Segments

Premier customers are first defined according to two different criteria: the likelihood that they will visit the dealership, and the amount of money that they are expected to spend on purchases. This information is based on the purchase process (whether they buy vehicles after coming to the dealer). These two indicators are then combined and are used to define the level of the premier customer.

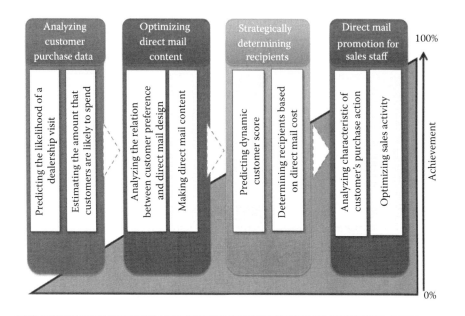

Figure 9.48 Strategically applying a total direct mail model "TDMM": approach model.

Table 9.4 System for Analyzing Customer Purchase Data

		Likelihood of Spend	
		High	*Low*
Likelihood of visit	High	Preferred customer Luxury car intention Own customer	Basic car intention Own customer
	Middle	Semipreferred customer Luxury car intention Own/other customer	Basic car intention Own/other customer
	Low	Luxury car intention Other customer	Not preferred customer Basic car intention Other customer

9.5.3.1.2 Predicting the Likelihood of a Dealership Visit

Discriminant analysis is used to score each customer in terms of how likely he or she is to visit the dealership. Past visit data are used to arrive at a discriminant score, using whether the DM will lead to a visit as the external criterion and customer attributes as the items. The score is then used to carry out a logistic regression analysis (Tsujitani and Amasaka, 1993) in order to get a regression equation (Equation 9.2) that will calculate the likelihood of a dealership visit.

$$P_i = \frac{1}{1 + \exp\left\{-\left(\alpha_0 + \sum_j \alpha_j \delta_{ij}\right)\right\}}, \quad i = 1, 2, \ldots, n \qquad (9.2)$$

where:
- i = customer ID number (n=number of the customers)
- j = customer attribute (j=number of attributes)
- P_i = likelihood that customer i will visit the dealer
- α_j = discriminant coefficient
- δ_{ij} indicates whether or not customer i has attribute j (0 or 1)

9.5.3.1.3 Estimating the Amount That Customers Are Likely to Spend

Regression analysis is used to obtain a regression equation (Equation 9.3) for this calculation, with customer purchase amount as the external criterion and customer attributes as the items. This equation is then used to estimate the amount that customers are likely to spend based on their specific attributes.

$$V_i = \beta_0 + \sum_j \beta_j \delta_{ij}, \quad i = 1, 2, \ldots, n \qquad (9.3)$$

where:
- V_i = amount that customer i is likely to spend
- β_j = partial regression coefficient

9.5.3.2 System for Optimizing Direct Mail Content

The system for optimizing DM content is used to analyze how customer preferences change in response to changes in DM design. To get this information, a key graph is used to collect free-response customer feedback on different DM designs.

Figure 9.49 shows an example of a key graph. In this example, the author was able to determine that using a black background gives the design a high-end feel and causes it to stand out. By referring to concepts arrived at using these new insights

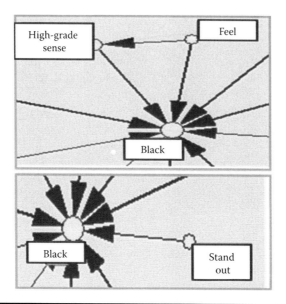

Figure 9.49 Key graph.

as well as the system for analyzing customer purchase data, DM designers can now create designs that accurately reflect sales strategies and customer preferences.

9.5.3.3 System for Strategically Determining Direct Mail Recipients

The system for strategically determining DM recipients identifies who should be sent DM among the premier customer categories identified using the system for analyzing customer purchase data. Equations 9.4 through 9.6 are used to determine the recipients.

$$\underset{x_1, x_2, \ldots, x_n \in \{0,1\}}{\text{Maximize}} \sum_i P_i V_i x_i \qquad (9.4)$$

$$\text{s.t.} \sum_i x_i = N \qquad (9.5)$$

$$L_j N \leq \sum_i \delta_{ij} x_i \leq H_j N, \quad j = 1, 2, \ldots, J \qquad (9.6)$$

where:
  x_i = whether customer i is flagged as a DM recipient (0 or 1)
  N = total number of direct mailings sent
  L_j/H_j = lower/upper limit for the percentage of direct mailings sent to customers with attribute j

Mathematical programming is used to solve Equations 9.4 through 9.6 and determine who should be sent DM. The variable x_i is the decision variable indicating whether or not customer i should be targeted as a recipient. The values of the other parameters are determined in advance.

P_i (the likelihood of a dealership visit) and V_i (the amount of money likely to be spent) are determined using the same procedure outlined for the customer purchase data analysis system (Section 9.5.2.1), once behavioral data (RFM) are added to customer attribute data.

Table 9.5 shows the difference between a static customer and a dynamic customer. It is used to determine who should be sent DM based on the current level of the premier customer (this takes into consideration different customer behaviors among different premier customer categories).

9.5.3.4 Direct Mail Promotion System for Sales Staff

The DM promotion system for sales staff reveals important insights that sales staff can use in order to handle customers when they come to the dealership. Specifically, the system surveys dealerships to determine what elements are prioritized when someone from the premier customer category purchases a vehicle.

As Figure 9.50 describes, survey data are then subjected to a cluster analysis and a principal component analysis to identify characteristics of customer purchase behavior. These results are then communicated to the sales staff.

9.5.3.5 Effectiveness of the Total Direct Mail Model

As Figure 9.51 shows, the author is currently using the TDMM presented here at Company A in an attempt to increase the percentage of customers that visit the dealer as well as the amount of money they spend when they come in. Specifically, the author is using the four systems described in Section 9.5.1 based on Company A's customer information and sales data as well as on customer preferences collected via surveys.

Table 9.5 Difference between Static Customer and Dynamic Customer

	Static (Analyzing Customer Purchase Data)	Dynamic (Strategically Determining Direct Mail Recipients)
Definition	Customer segment who like target car during the promotion term	Customer segment who like target car most now
Data	Life stage Life stage	Lifestyle Lifestyle RFM

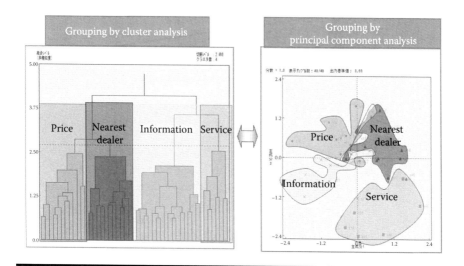

Figure 9.50 Characteristics of customer's purchase action.

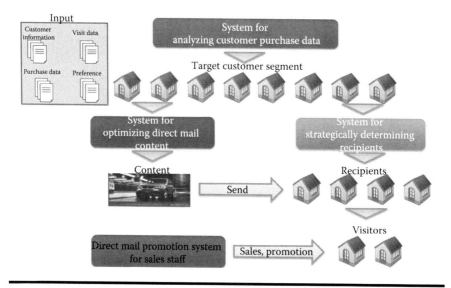

Figure 9.51 Applying TDMM.

9.5.4 Conclusion

This research presents the strategic application of a strategic TDMM for more sophisticated DM activities at auto dealerships. The model makes use of customer information and purchase data that dealerships retain but are not using effectively. The author believes that there are still more valuable ways to use the data that companies possess, and in the future, the author hopes to develop more scientific ways of using it in corporate marketing activities.

Acknowledgments

The author would like to acknowledge the generous support received from the following researchers. All those at WG'4 (Working Group No. 4) studies in the Japanese Society for Quality Control (JSQC), study group on simulation and SQC—a study of the high quality assurance CAE model for car development design, Toyota Motor Corporation, and NOK Corporation (Amasaka, 2005b; Amasaka, 2007g), and those connected with the Amasaka laboratory at Aoyama Gakuin University.

References

Amasaka, K. 2002. New JIT: A new management technology principle at Toyota, *International Journal of Production Economics*, 80: 135–144.

Amasaka, K. (ed.). 2003a. *Manufacturing Fundamentals: The Application of Intelligence Control Charts—Digital Engineering for Superior Quality Assurance*, Japan Standards Association [in Japanese].

Amasaka, K. 2003b. Proposal and implementation of the "Science SQC" quality control principle, *International Journal of Mathematical and Computer Modelling*, 38(11–13): 1125–1136.

Amasaka, K. 2004. *Science SQC, New Quality Control Principle: The Quality Control of Toyota*, Springer-Verlag, Tokyo.

Amasaka, K. 2005a. Constructing a customer science application system "CS-CIANS"—Development of a global strategic vehicle "Lexus" utilizing New JIT, *WSEAS Transactions on Business and Economics*, 2(3): 135–142.

Amasaka, K. 2005b. Interim report of WG4's studies in JSQC research on simulation and SQC(1)—A study of the high quality assurance CAE model for car development design, in *Transdisciplinary Federation of Science and Technology, 1st Technical Conference*, pp. 93–98, November 25, 2005, Nagano, Japan.

Amasaka, K. 2007a. Applying New JIT—Toyota's global production strategy: Epoch-making innovation in the work environment, *Robotics and Computer-Integrated Manufacturing*, 23(3): 285–293.

Amasaka, K. 2007b. High linkage model "Advanced TDS, TPS & TMS": Strategic development of "New JIT" at Toyota, *International Journal of Operations and Quantitative Management*, 13(3): 101–121.

Amasaka, K. 2007c. Highly reliable CAE model, the key to strategic development of Advanced TDS, *Journal of Advanced Manufacturing Systems*, 6(2): 159–176.

Amasaka, K. 2007d. New Japan Production Model, an advanced production management principle: Key to strategic implementation of New JIT, *The International Business and Economics Research Journal*, 6(7): 67–79.

Amasaka, K. 2007e. The validity of "TDS-DTM": A strategic methodology of merchandise development of New JIT—Key to the excellence design "LEXUS", *International Business and Economics Research Journal*, 6(11): 105–116.

Amasaka, K. 2007f. The validity of advanced TMS, a strategic development marketing system—Toyota's Scientific Customer Creative Model utilizing New JIT, *International Business and Economics Research Journal*, 6(8): 35–42.

Amasaka, K. (ed.). 2007g. Establishment of a needed design quality assurance framework for numerical simulation in automobile production, Working Group No. 4 studies in JSQC, study group on simulation and SQC.

Amasaka, K. 2008. An integrated intelligence development design CAE model utilizing New JIT: Application to automotive high reliability assurance, *Journal of Advanced Manufacturing Systems*, 7(2): 221–241.

Amasaka, K. 2009a. The effectiveness of flyer advertising employing TMS: Key to scientific automobile sales innovation at Toyota, *China–USA Business Review*, 8(3): 1–12.

Amasaka, K. 2009b. Chapter 4 in: *Product Design, New Edition, Quality Assurance Guidebook*, Japanese Society for Quality Control, pp. 87–101. JUSE Press, Ltd., Tokyo, Japan [in Japanese].

Amasaka, K. 2010. Proposal and effectiveness of a high quality assurance CAE analysis model: Innovation of design and development in automotive industry, *Current Development in Theory and Applications of Computer Science, Engineering and Technology*, 2(1/2): 23–48.

Amasaka, K. 2011. Changes in marketing process management employing TMS: Establishment of Toyota sales marketing system, *China–USA Business Review*, 10(6): 539–550.

Amasaka, K. 2012. Constructing optimal design approach model: Application on the advanced TDS, *Journal of Communication and Computer*, 9(7): 774–786.

Amasaka, K., Ito, T., and Nozawa, Y. 2012. A new development design CAE employment model, *The Journal of Japanese Operations Management and Strategy*, 3(1): 18–37.

Amasaka, K., Kurosu, S., and Morita, M. 2008. *New Manufacturing Theory: Surpassing JIT—Evolution of JIT*, Morikita Shuppan, Tokyo [in Japanese].

Amasaka, K., Nagaya, A., and Shibata, W. 1999. Studies on design SQC with the application of Science SQC—Improving of business process method for automotive profile design, *Japanese Journal of Sensory Evaluations*, 3(1): 21–29.

Amasaka, K. and Onodera, T. 2014. Developing a Higher-Cycled Product Design CAE Model: The evolution of automotive product design and CAE, *International Journal of Technical Research and Applications*, 2(1), 17–28.

Amasaka, K. and Sakai, H. 2008. Evolution of TPS fundamentals utilizing New JIT strategy proposal and validity of Advanced TPS at Toyota, *Journal of Advanced Manufacturing Systems*, 9(2): 85–99.

Amasaka, K. and Sakai, H. 2009a. TPS-QAS, new production quality management model: Key to New JIT—Toyota's global production strategy, *International Journal of Manufacturing Technology and Management*, 18(4): 409–426.

Amasaka, K. and Sakai, H. 2009b. Proposal and validity of New Japan global production model "G-NJPM" utilizing New JIT: Strategic development of advanced TPS surpassing TPS, in *The Japanese Operations Management and Association, the 1st Annual Conference*, pp. 15–30, June 20, 2009, Aoyama Gakuin University, Japan [in Japanese].

Amasaka, K. and Sakai, H. 2011. The new Japan global production model "NJ-GPM": Strategic development of *Advanced TPS*, *The Journal of Japanese Operations Management and Strategy*, 2(1): 1–15.

Asahi Shinbun. 2005. The manufacturing industry: Skill tradition feels uneasy, May 2, 2005, Tokyo, Japan [in Japanese].

Asami, H., Ando, T., Yamaji, M., and Amasaka, K. 2010. A study on automobile form design support method "AFD-SM", *Journal of Business and Economics Research*, 8(11): 13–19.

Asami, H., Owada, H., Murata, Y., Takebuchi, S., and Amasaka, K. 2011. The A-VEDAM model for approaching vehicle exterior design, *Journal of Business Case Studies*, 7(5): 1–8.

Beko, J. and Jagric, T. 2011. Demand models for direct mail and periodicals delivery services: Results for a transition economy, *Applied Economics*, 43(9): 1125–1138.

Bell, G. H., Ledolter, J., and Swersey, A. J. 2006. Experimental design on the front lines of marketing: Testing new ideas to increase direct mail sales, *International Journal of Research in Marketing*, 23(3): 309–319.

Bult, J. R. and Wansbeek, T. 2005. Optimal selection for direct mail, *Marketing Science*, 14(4): 378–394.

Dijkastra, E. W. 1959. *Communication with an Automatic Computer*, Excelsior, Amsterdam.

Dijkastra, E. W. and Feijen, W. H. J. 1991. *A Method of Programming*, Addison-Wesley Longman, Boston.

Ferrell, O. C. and Hartline, M. 2005. *Marketing Strategy*, Thomson Learning/South-Western College, Mason, OH.

Fujieda, S., Matsuda, Y., and Nakahata, A. 2007. Development of automotive color designing process, *Journal of Society of Automotive Engineers of Japan*, 61(6): 79–84.

Ishiguro, H. and Amasaka, K. 2012a. Proposal and effectiveness of a highly compelling direct mail method: Establishment and deployment of PMOS-DM, *International Journal of Management and Information Systems*, 16(1): 1–10.

Ishiguro, H. and Amasaka, K. 2012b. Establishment of a strategic Total Direct Mail Model to bring customers into auto dealerships, *Journal of Business and Economics Research*, 10(8): 493–500.

Ishiguro, H., Kojima, T., and Matsuo, I. 2010. A highly compelling direct mail method "PMOS-DM": Strategic applying of statistics and mathematical programming, in The 4th Spring Meeting of Japan Statistical Society Poster Session, (Student Poster Award) March 7, 2010, Aoyama Gakuin University, Tokyo, Japan [in Japanese].

Ito, T., Yamaji, M., and Amasaka, K. 2010. Optimized design using high quality assurance CAE—Example of the automotive transaxle oil seal leakage, in *Proceedings of the 9th WSEAS International Conference on System Science and Simulation in Engineering*, pp. 227–232. October 1, 2010, Iwate Prefectural University, Iwate, Japan.

Jonker, J. J., Piersma, N., and Potharst, R. 2006. A decision support system for direct mailing decision, *Decision Support Systems*, 42: 915–925.

Kameike, M., Ono, S., and Nakamura, K. 2000. The helical seal: Sealing concept and rib design, *Sealing Technology*, 2000(77): 7–9.

Kawano, K., Konjiki, F., and Ago, Y. 2004. Psychological effects and sexual differences in EEG values during collage making, *Journal of International Society of Life Information Science*, 22(1): 60–64.

Kimura, T., Uesugi, Y., Yamaji, M., and Amasaka, K. 2007. A study of scientific approach method for direct mail, SAM-DM: Effectiveness of attracting customer utilizing Advanced TMS, in *Proceedings of the 5th Asian Network for Quality Congress*, pp. 938–945, October 16, 2007, Incheon, Korea.

Koizumi, K., Kanke, R., and Amasaka, K. 2013. forthcoming. Research on automobile exterior color and interior color matching, In *Proceedings of the International Conference on Management and information Systems*, PP. 201–211, September 23, 2013, Bangkok, Thailand.

Kojima, T., Kimura, T., Yamaji, M., and Amasaka, K. 2010. Proposal and development of the direct mail method "PMCI-DM" for effectively attracting customers, *International Journal of Management and Information Systems*, 14(5): 15–22.

Kotler, P. and Keller, K. L. 2006. *Marketing Management*, 12th edn., Prentice Hall, Upper Saddle River, NJ.

Kume, H. 1990. *Quality Management in Design Development*, Union of Japanese Scientists and Engineers, Tokyo, Japan [in Japanese].

Magoshi, Y., Fujisawa, H., and Sugiura, T. 2003. Simulation technology applied to vehicle development, *Journal of Society of Automotive Engineers of Japan*, 57(3): 95–100.

Muto, M., Miyake, R., and Amasaka, K. 2011. Constructing an Automobile Body Color Development Approach Model, *Journal of Management Science*, 2: 175–183.

Muto, M., Takebuchi, S., and Amasaka, K. 2013. Creating a new automotive exterior design approach model—The relationship between form and body color qualities, *Journal of Business Case Studies*, 9(5): 367–374.

Nihon Keizai Shimbun. 2001. IT innovation of manufacturing, January 1, 2001, Tokyo, Japan [in Japanese].

Nozawa, Y., Ito, T., and Amasaka, K. 2013. High precision CAE analysis of automotive transaxle oil seal leakage, *Chinese Business Review*, 12(5): 363–374.

Nozawa, Y., Yamashita, R., and Amasaka, K. 2012. Analyzing cavitation caused by metal particles in the transaxle: Application of high quality assurance CAE analysis model, *Journal of Energy and Power Engineering*, 6(12): 2054–2062.

Ohno, T. (1977). *Toyota Production System*, Diamond-Sha, Tokyo, Japan [in Japanese].

Okabe, Y., Hitumi, S., Yamaji, M., and Amasaka, K. 2006. Research on the automobile package design concept support methods "CS-APDM", in *Proceedings of the 11th Annual International Conference on Industrial Engineering Theory, Application and Practice*, pp. 268–273, September 9, 2006. Tokai University, Shizuoka, Japan.

Onodera, T. and Amasaka, K. 2012. Automotive bolts tightening analysis using contact stress simulation: Developing an optimal CAE design approach model, *Journal of Business and Economics Research*, 10(7): 435–442.

Onodera, T., Kozaki, T., and Amasaka, K. 2013. Applying a highly precise CAE technology component model: Automotive bolt-loosening mechanism, *China–USA Business Review*, 12(6): 597–607.

Piersma, N. and Jonker, J. J. 2004. Determining the optimal direct mailing frequency, *European Journal of Operational Research*, 158(1): 173–182.

Sakai, H. and Amasaka, K. 2006a. TPS-LAS model using process layout CAE system at Toyota, advanced TPS: Key to global production strategy New JIT, *Journal of Advanced Manufacturing Systems*, 5(2): 1–14.

Sakai, H. and Amasaka, K. 2006b. Strategic HI-POS, intelligence production operating system—Applying advanced TPS to Toyota's global production strategy, *WSEAS Transactions on Advances in Engineering Education*, 3(3): 223–230.

Sakai, H. and Amasaka, K. 2007a. Human Digital Pipeline method using total linkage through design to manufacturing, *Journal of Advanced Manufacturing Systems*, 6(2): 101–113.

Sakai, H. and Amasaka, K. 2007b. Development of a robot control method for curved seal extrusion for high productivity for an advanced Toyota Production System, *International Journal of Computer Integrated Manufacturing*, 20(5): 486–496.

Sakai, H. and Amasaka, K. 2007c. The Robot Reliability Design and Improvement Method and the advanced Toyota Production System, *Industrial Robot: International Journal of Industrial and Service Robotics*, 34(4): 310–316.

Satake, I., Ando, K., Kuwano, K., Sato, T., Hattori, H., and Kajiwara, K. 2004. Study on relationship between automotive exterior color and automotive shape category, *Journal of the Color Science Association of Japan*, 28(2): 102–110.

Sato, Y., Toda, A., Ono, S., and Nakamura, K. 1999. A study of the sealing mechanism of radial lip seal with helical ribs—Measurement of the lubricant fluid behavior under sealing contact, SAE Technical Paper Series, 1999-01-0878.

Shimizu, K. 2004. *Theory and Strategy of Advertisement*, Sousei-sha Publishers, Tokyo, Japan [in Japanese].

Takebuchi, S., Asami, H., and Amasaka, K. 2012a. An automobile exterior design approach model linking form and color, *China–USA Business Review*, 11(8): 1113–1123.

Takebuchi, S., Asami, H., Nakamura, T., and Amasaka, K. 2010. Creation of Automobile Exterior Color Design Approach Model "A-ACAM", in *Proceedings of the 40th International Conference on Computers and Industrial Engineering*, pp. 1–5. Awaji Island, Japan.

Takebuchi, S., Nakamura, T., Asami, H., and Amasaka, K. 2012b. The Automobile Exterior Color Design Approach Model, *Journal of Japan Industrial Management Association*, 62(6E): 303–310.

Takimoto, H., Ando, T., Yamaji, M., and Amasaka, K. 2010. The proposal and validity of the customer science dual system—The key to corporate management innovation, *China–USA Business Review*, 9(3): 29–38.

Tsujitani, M. and Amasaka, K. 1993. Analysis of enumerated data (1): Logit transformation and graph analysis, *Quality Control*, 44(4): 61–66 [in Japanese].

Yamada, H. and Amasaka, K. 2011. Highly reliable CAE analysis approach: Application in automotive bolt analysis, *China–USA Business Review and Chinese Business Review*, 10(3): 199–205.

Yamaji, M., Hifumi, S., Sakalsiz, M., and Amasaka, K. 2010. Developing a strategic advertisement method "VUCMIN" to enhance the desire of customers for visiting dealers. *Journal of Business Case Studies*, 6(3): 1–11.

Yoo, S. L. C., Park, J. C., and Kang, M. 2010. Design of the autonomous intelligent simulation model (AISM) using computer generated force, in *Proceedings of the 9th WSEAS International Conference on System Science and Simulation in Engineering (ICOSSSE '10)*, pp. 302–307. Iwate Prefectural University, Japan.

Chapter 10

New Manufacturing Management Employing New JIT

10.1 New Software Development Model for Information Sharing

10.1.1 Introduction

Today, ensuring the success of software development can be considered the crux of the information service industry, and this success is dependent on how well information is shared between customers and vendors. A close look at the way recent development projects work reveals that information sharing is taking place in various forms—most commonly during regular monthly meetings. However, there exists a pressing issue here in that the success of these development projects is often evaluated in subjective and unspoken ways. An objective method of evaluation is needed in order to ensure that information is shared intelligently in future development projects.

In response to this need, the authors have created a New Software Development Model (NSDM) (Nakamura et al., 2011) to assess the success of information sharing that is based on quantitative evaluation of how well information is exchanged between customers and vendors during individual development projects utilizing Science Statistical Quality Control (SQC) (Amasaka, 2004b, refer to Section 5.4).

NSDM makes it possible to specify measures to improve those aspects of information sharing that are less successful. Specifically, Science SQC was applied to derive a mathematical model for calculating the impact of various factors essential for successful information sharing on the overall success of a given project employing New Just in Time (JIT) (Amasaka, 2002, refer to Chapters 7 through 9). NSDM was then converted into a piece of software. The authors then applied NSDM to three Japanese vendors to verify its effectiveness.

10.1.2 Current Status and Problems in Software Development

10.1.2.1 Current Status of Software Development

In recent years, sales in the information service industry have exceeded 10 trillion yen, with a rapid growth rate averaging 8% over the last 10 years. About 67% of those sales are related to software development (Ministry of Economy, Trade and Industry, 2010). Vendors that work in software development are responsible for developing software that uses information technology in a way that improves the quality of their customers' work tasks.

In order to fulfill this responsibility, there must be a shared awareness between vendors and the customers that they develop software for (Faraj and Spoull, 2000; Lu et al., 2011; Mihara et al., 2010; Sakai et al., 2011; Sakai and Amasaka, 2012). However, there are problems with this in that there is often a gap between customers' and vendors' awareness of customer work tasks and systems (Nakamura et al., 2011). Closing this gap in awareness with intense information sharing between the two parties is essential for successful development projects (Amasaka, 2000, 2004b; Kirsch et al., 2002).

10.1.2.2 Information Sharing in Software Development

Information sharing in software development currently takes place during the course of reviews held at the end of each development process or during regular monthly meetings (Kawai, 2005). However, what is important is not whether these methods of information sharing are taking place, but whether they are being utilized effectively. The problem is that there is currently no way to objectively evaluate how successful information sharing is on any given development project.

Two studies on information sharing in software development within Japan and abroad have been conducted: Development and Evaluation of Information Sharing Infrastructure for Software Development Projects (Murata et al., 2001) and Customer–Developer Links in Software Development (Keil and Carmel, 1995; Gallivan and Keil, 2003). However, both of these studies targeted methodologies for information sharing, and to the author's knowledge, neither offered examples of ways to quantitatively and objectively assess the degree of success of information sharing between customers and vendors during actual development projects.

10.1.3 Constructing an NSDM to Assess the Success of Information Sharing between Customers and Vendors

10.1.3.1 Identifying Key Problem and Solution Factors in Software Development Information Sharing

Problems with information sharing were identified through interviews with Japanese vendors, and the information was organized using a system diagram. Specifically, gaps in awareness between customers and vendors were labeled "results," and arrows were drawn from the problems affecting those results. Certain problems were labeled "root problems," and those problems that were caused by the root problems were labeled "triggered problems."

This process allowed the authors to identify not only problems with a direct impact on information sharing, but also root problems that at first glance seemed to have nothing to do with information sharing (such as the knowledge or experience of members involved in the project, project management, and other factors) but in actual fact were triggering the gaps in awareness between customers and vendors.

Using the system diagram organizing problems in information sharing, 30 key factors essential to information sharing (information-sharing factors) were identified from the root problems. For example, the key factor "advantage over competitors" was taken from the problem "each vendor has different strengths and weaknesses." The identified information-sharing factors are shown in Table 10.1.

10.1.3.2 Collecting Data to Assess the Impact of Information-Sharing Factors on Overall Success

In order to understand the impact of the 30 information-sharing factors on the overall level of information-sharing success, data were collected from 11 vendor companies in Japan. These companies are customers, manufacturing vendors, or independent software vendors. The companies were first asked to rank the 30 factors on a seven-point scale in terms of their level of success in development projects (1 being extremely inadequate and 7 being fully adequate).

The results were subjected to a cluster analysis and covariance structure analysis to determine the relative impact of each information-sharing factor on overall success. The companies were also invited to record (in free-response format) what they were doing during projects in order to boost the success of each information-sharing factor. From this data, the authors identified success-boosting factors: those factors that were needed in order to boost the success of the other factors, as shown in Table 10.2.

Table 10.1 Information-Sharing Factors

O_1. Successful reports	O_{16}. Customer understanding of systems
O_2. Shared customer–vendor awareness	O_{17}. Negotiating ability
O_3. Clear documents	O_{18}. Appropriate personnel/task assignments
O_4. Strong training programs	O_{19}. Advantage over competitors
O_5. Self-improvement by development of team members	O_{20}. Clear customer needs
O_6. Successful databases	O_{21}. Customer needs successfully met
O_7. Balanced costs and budget	O_{22}. Growing customer needs considered
O_8. Awareness of customer work tasks	O_{23}. Clear scope of systematization
O_9. Influence of feedback from customer workers	O_{24}. Appropriate project manager assignments
O_{10}. Influence of customer managers on customer workers	O_{25}. Unique project features considered
O_{11}. Decision-making ability of customer managers	O_{26}. Clear system functions
O_{12}. Clear organizational structure	O_{27}. Clear development tasks
O_{13}. Ability to propose technologies	O_{28}. Databases put to good use
O_{14}. Good progress management by the project manager	O_{29}. Development environment risks considered
O_{15}. Customer understanding of proposals	O_{30}. Risks associated with new technologies considered

10.1.3.3 Classifying Information-Sharing Factors

Using the information on the success of each information-sharing factor (seven-point evaluation), the author used a cluster analysis to group the 30 factors into five latent factors, as shown in Figure 10.1.

For example, all of the factors in the group that includes O_7 (balanced costs and budget) and O_{19} (advantage over competitors) can be considered critical during the estimation process; this group was thus labeled latent factor L_1 (success of

Table 10.2 Success-Boosting Factors

E_1. Holding video conferences	E_{37}. Achievement of quality, cost, and delivery targets
E_2. Holding meetings other than those regularly scheduled	E_{38}. Advanced technical skills
E_3. Contact via email and telephone	E_{39}. Hardware-software compatibility
E_4. Contact via online message boards	E_{40}. Use of offshore resources
E_5. Use of past case examples	E_{41}. Acceptance of documented requests
E_6. Detailed work breakdown structure (WBS)	E_{42}. Clear systemization tasks
E_7. Creating meeting minutes	E_{43}. Package compatibility analyzed
E_8. Making a database of company expertise	E_{44}. Required specifications checked regularly
E_9. Use of personnel familiar with work tasks	E_{45}. Customer test support carried out
E_{10}. Using tables to share problems and management tasks	E_{46}. Requests reviewed
E_{11}. Standards written on specification sheets	E_{47}. Attainable cost and delivery targets
E_{12}. Simplified specification sheets	E_{48}. Information obtained on system operation
E_{13}. Specification sheets checked during reviews	E_{49}. Budget includes room for specification changes
E_{14}. Training sessions held	E_{50}. Demands are prevented from ballooning
E_{15}. Enrollment system for distance-learning programs	E_{51}. Scope covered by package shown
E_{16}. On-the-job training	E_{52}. Careful risk management
E_{17}. Individual study	E_{53}. Proposals coordinated during internal meetings
E_{18}. Standards for price proposals	E_{54}. System functions presented by team leaders

(Continued)

Table 10.2 (Continued) Success-Boosting Factors

E_{19}. Good communication when determining costs	E_{55}. Defining requirements given top priority
E_{20}. Costs determined based on past project results	E_{56}. Reviews carried out quickly
E_{21}. Costs determined based on project manager experience	E_{57}. Guidance offered using database information
E_{22}. Open communication in the workplace	E_{58}. System in place for easy database management
E_{23}. Interviews with customer workers	E_{59}. Past project information documented
E_{24}. Feedback from customer workers used to make decisions	E_{60}. Database information is well understood
E_{25}. Customer workers invited to review system proposals	E_{61}. Work manuals provided
E_{26}. Study sessions held	E_{62}. Development experience in similar work tasks
E_{27}. Participation in lectures	E_{63}. Information on appropriate work tasks clearly understood
E_{28}. Proposal sheets and technological materials shared	E_{64}. Customer understanding of proposals
E_{29}. Progress management through one-on-one communication	E_{65}. Influence of customer managers on customer workers
E_{30}. Progress management through meetings	E_{66}. Decision-making ability of customer managers
E_{31}. Project leader manages progress and reports to project manager	E_{67}. Clear organizational structure
E_{32}. Real-time progress management (use of shared systems)	E_{68}. Appropriate personnel/task assignments
E_{33}. Daily progress management	E_{69}. Customer understanding of systems
E_{34}. Use of experienced and knowledgeable personnel	E_{70}. Appropriate project manager assignments

Table 10.2 (Continued) Success-Boosting Factors

E_{35}. Negotiations held on a level playing field	E_{71}. Development environment risks considered
E_{36}. Different negotiators used for different hierarchical levels	E_{72}. Risks associated with new technologies considered

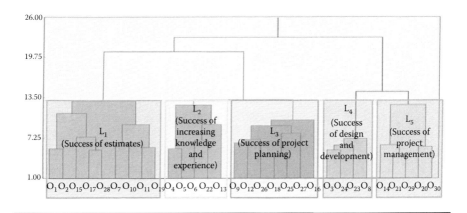

Figure 10.1 The result of cluster analysis.

estimates). Interpretations were assigned to the other groups of information-sharing factors in the following way: L_2 (success of increasing knowledge and experience), L_3 (success of project planning), L_4 (success of design and development), and L_5 (success of project management).

10.1.3.4 Constructing an NSDM of Successful Information Sharing in Software Development

The next step was constructing an NSDM of the requirements for successful information sharing using Science SQC, as shown in Figure 10.2. This was done using the information-sharing factors identified in the system diagram, the latent factors derived through the cluster analysis, and the success-boosting factors identified through data collection. The model consists of four layers: (1) overall success, (2) latent factors, (3) information-sharing factors, and (4) success-boosting factors. Each of the layers presents a more detailed picture of the factors required for the success of the layer above.

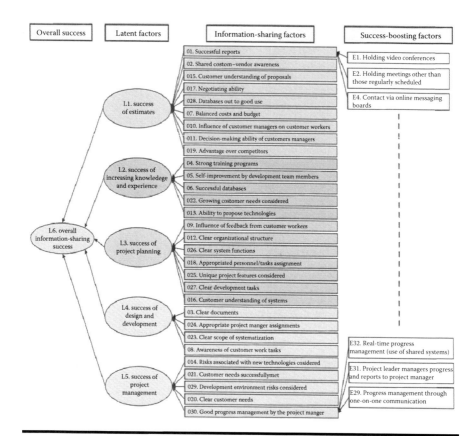

Figure 10.2 The hypothetical NSDM of successful information sharing in software development.

10.1.3.5 Collecting Data to Assess the Impact of Success-Boosting Factors on Information-Sharing Factors

Data were again collected from 11 vendor companies in Japan in an effort to use Quantification Theory Type I to assess the relative impact of the 72 success-boosting factors on the 30 information-sharing factors. The companies were first asked to evaluate the success of the 30 factors in development projects using a seven-point scale. The results of the survey were used as objective variables during the Quantification Theory Type I analysis.

The authors then had the companies select the success-boosting factors that were being implemented in order to improve the success of each of the 30 factors. In the Quantification Theory Type I analysis, factors that were being implemented were converted to a 1, and factors that were not being implemented were converted to a 0. The data were then used as explanatory variables for qualitative data.

10.1.3.6 Assessing the Impact of Success-Boosting Factors on Overall Information-Sharing Success

Based on the evaluations of the various collected data sets, a covariance structure analysis was conducted to assess the relationship between the first and second layers of the model and between the second and third layers of the model. Quantification Theory Type I was used to assess the relationship between the third and fourth layers.

The authors then used the results to construct a mathematical model that would calculate (1) the impact of the information-sharing factors on overall success and (2) the impact of success-boosting factors on information-sharing factors. However, the Quantification Theory Type I analysis was abbreviated because for the nine information-sharing factors where there was a one-to-one relationship between the target variables and dependent explanatory variables, the success of the success-boosting factors was the same as the success of the information-sharing factors.

This model formula derived from this analysis is shown below.

(Mathematical modeling formula to calculate the degree to which Layer 2 impacts Layer 1)

$$L_6 = 0.541L_1 + 0.421L_2 + 0.506L_3 + 0.323L_4 + 0.410L_5 \ldots \quad (10.1)$$

(Mathematical modeling formula to calculate the degree to which Layer 3 impacts Layer 2)

$$L_1 = 0.242O_1 + 0.317O_2 + 0.330O_{15} + 0.185O_{17} + 0.239O_{28}$$
$$+ 0.438O_7 + 0.298O_{10} + 0.427O_{11} + 0.421O_{19} \ldots \quad (10.2)$$

$$L_2 = 0.368O_4 + 0.238O_5 + 0.490O_6 + 0.513O_{22} + 0.553O_{13} \ldots \quad (10.3)$$

$$L_3 = 0.318O_9 + 0.334O_{12} + 0.436O_{26} + 0.362O_{18} + 0.311O_{25}$$
$$+ 0.392O_{27} + 0.465O_{16} \ldots \quad (10.4)$$

$$L_4 = 0.488O_3 + 0.439O_{24} + 0.667O_{23} + 0.353O_8 \ldots \quad (10.5)$$

$$L_5 = 0.586O_{14} + 0.384O_{21} + 0.523O_{29} + 0.387O_{20} + 0.291O_{30} \ldots \quad (10.6)$$

(Mathematical modeling formula to calculate the degree to which Layer 4 impacts Layer 3)

$$O_1 = 0.633E_1 + 0.967E_2 + 0.350E_3 + 0.300E_4 + 3.100\ldots \quad (10.7)$$

$$O_2 = 0.619E_5 + 0.905E_6 + 0.238E_7 + E_8 + 0.667E_9 + 1.429E_{10} + 2.143\ldots \quad (10.8)$$

$$O_3 = 1.909E_{11} + 1.182E_{12} + 0.182E_{13} + 2.818\ldots \quad (10.9)$$

$$O_4 = 1.808E_{14} + 0.185E_{15} + 1.705E_{16} + 2.274\ldots \quad (10.10)$$

$$O_5 = 0.808E_{14} + 0.202E_{15} + 0.952E_{17} + 3.077\ldots \quad (10.11)$$

$$O_6 = 1.389E_{57} + 1.611E_{58} + 1.833E_{59} + 1.056\ldots \quad (10.12)$$

$$O_7 = 0.415E_{18} + 1.551E_{19} + 0.095E_{20} + 1.966E_{21} + 2.014\ldots \quad (10.13)$$

$$O_8 = E_{61} + 1.500E_{62} + 0.500E_{63} + 3.000\ldots \quad (10.14)$$

$$O_9 = 1.825E_{22} + 0.643E_{23} + 0.544E_{24} + 1.490E_{25} + 1.722\ldots \quad (10.15)$$

$$O_{13} = 0.833E_{26} + 0.333E_{27} + 0.611E_8 + 0.667E_{28} + 3.556\ldots \quad (10.16)$$

$$O_{14} = 0.147E_{29} + 0.500E_{30} + 0.471E_{31} + 1.176E_{32} + 0.324E_{33} + 4.382\ldots \quad (10.17)$$

$$O_{17} = 1.778E_{34} + 2.111E_{35} + 0.556E_{36} + 1.667\ldots \quad (10.18)$$

$$O_{19} = 0.545E_{37} + 1.136E_{38} + 2.614E_{39} + 0.182E_{40} + 1.977\ldots \quad (10.19)$$

$$O_{20} = 2.824E_9 + 0.510E_{41} + 1.765E_{42} + 0.765E_{43} - 0.765\ldots \quad (10.20)$$

$$O_{21} = 0.914E_{37} + 0.879E_{38} + 0.879E_{44} + 0.793E_{45} + 1.103E_{46} + 2.121\ldots \quad (10.21)$$

$$O_{22} = 0.822E_{47} + 1.137E_{48} + 1.449E_{44} + 0.332E_{49} + 1.586E_{50} + 2.171\ldots \quad (10.22)$$

$$O_{23} = 1.759E_{42} + 0.379E_5 + 0.897E_{51} + 2.655\ldots \quad (10.23)$$

$$O_{25} = 0.167E_5 + 0.750E_{47} + 1.500E_{44} + E_{52} + 1.167E_{53} + 3.833\ldots \quad (10.24)$$

$$O_{26} = 0.678E_{54} + 0.524E_{55} + 0.853E_{10} + 0.245E_{56} + 4.161\ldots \quad (10.25)$$

$$O_{27} = 2.556E_6 + 1.963E_{10} + 1.593E_{54} + 0.481\ldots \quad (10.26)$$

$$O_{28} = 1.158E_{60} + 2.105E_{58} + 1.684E_{59} + 2.316\ldots \quad (10.27)$$

The derived mathematical modeling formula revealed that the latent factor with the greatest impact on the overall success of information sharing was L_1 (success of estimates). In looking at the information-sharing factors that make up latent factor L_1, it was found that O_7 (balanced costs and budget) had the most impact.

These results shed light on the feedback received during domestic vendor interviews that "oftentimes, contract performance is emphasized more than profit." Also, because none of the factors demonstrated extremely weak impact in terms of overall success, it was concluded that the latent factors, information-sharing factors, and success-boosting factors on which the hypothetical model was based were accurately identified.

10.1.3.7 Developing Software to Implement a Diagnostic System for Information-Sharing Success

NSDM was developed in order to implement a diagnostic system for information-sharing success, based on the mathematical model formula obtained above. The diagnostic system has vendors indicate whether they are implementing each of the 72 success-boosting factors in their own projects. It then uses percentages for the successfulness of latent factors L_1–L_5 and information-sharing factors O_1–O_{30} in the project and calculates standard values based on averages for domestic vendors.

The results are presented as a radar chart, as shown in Figure 10.3. The system also proposes specific improvement strategies that can be used to target up to three information-sharing factors with the lowest scores calculated by the mathematical model formula, as compared to standard values. The model thus allows vendors to objectively evaluate the success of information sharing between themselves and customers while identifying specific improvement strategies that can be used to further improve their current level of success.

This system targets individual software development projects and can be used during reviews of project management plans and once development projects are complete.

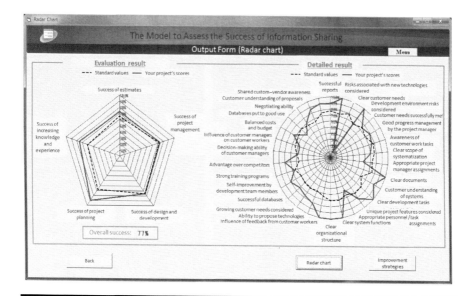

Figure 10.3 Diagnostic system for information-sharing success employing NSDM (radar chart output).

10.1.3.8 Verifying the Diagnostic System for Information-Sharing Success

This diagnostic system was actually applied to projects at domestic vendors and feedback was collected to verify its effectiveness, as shown in Figure 10.3. The authors had on-site system engineers use the model as if they were applying it to their current projects and then implement the suggested improvement strategies. The result was a 7% increase in the overall success of information sharing following the improvements. Evaluations of the diagnostic model for information-sharing success (based on these results) were as follows.

Favorable evaluations included (1) "radar charts allow easy visualization of the success of information sharing" and (2) "the prioritized list of suggested improvement strategies made it easy to begin carrying out improvements." Areas for improvement included (1) "it would be nice to see more specific factors that reflected specific work tasks" and (2) "I'd like to see an analysis that takes customer feedback into account as well."

The results of the investigation affirmed that the diagnostic system was effective in objectively evaluating the success of information-sharing employing NSDM.

10.1.4 Conclusion

NSDM using a diagnostic system was constructed to objectively evaluate the success of information sharing and offer specific strategies for improvement.

Firstly, the authors identified problems in information sharing and factors essential for successful information sharing to establish this model. Secondly, we derived a mathematical model for calculating the impact of the factors on the overall success of a given project. Finally, the model was then converted into a piece of software.

The vendor can improve the quality of the information sharing between the customer and the vendor by using NSDM. As a result, they are able to identify gaps in the awareness of customers and vendors during the software development process.

10.2 Automotive Product Design and CAE for Bottleneck Solution

10.2.1 Introduction

The technological challenge currently facing Japanese companies is to enable the simultaneous achievement of quality, cost, and delivery (QCD) by innovating product design processes, in order for them to prevail in the worldwide quality competition. It is therefore necessary to undertake principle-based research for quality assurance (QA) in product design and development by utilizing the latest numerical simulation technology (computer-Aided engineering, CAE) employing TDS through New JIT (refer to Section 7.2).

In order to realize this development, it is important to increase the ratio of the application of CAE using Advanced TDS (refer to 8.1, 8.3 and 9.2). CAE needs to explore the technological mechanism to be useful, and to create the generalized model. Then, the author contributes to the generalization model by finding a solution to the case with an unknown mechanism (Amasaka, 2007c, 2008a, 2010a, 2012; Amasaka et al., 2012). To achieve this aim, the advanced manufacturing industry has been faced with the urgent task of drastically reducing their product design period (the lenght of design time) in order to respond quickly to changing consumer needs.

One of the most important challenges for manufacturers is enhancing CAE analysis in order to achieve product design processes that are of high quality and very brief. To address these issues, the author conducted research on the Higher-Cycled Product Design CAE Model employing the Highly Reliable CAE Analysis Technology Component Model (Amasaka, 2007d, 2007c, 2008a).

At present, advanced companies in Japan and overseas in the automobile and other industries are endeavoring to survive in today's competitive market by expanding their global production. In this study, the author addresses the technological problem "loosening bolts" and others of product design bottlenecks at auto manufacturers (Amasaka et al., forthcoming). In this research, the aim is to grasp the dynamic behavior of the technical problem by using experiments as the empirical approach and the numerical simulation.

10.2.2 Background

10.2.2.1 Automotive Product Design and Production

Recently, Japanese automotive enterprises have been promoting global production to realize uniform quality worldwide and production at optimal locations in response to severe competition (Amasaka, 2007c). The mission of the automotive manufacturers in this rapidly changing management technology environment is to be properly prepared for the "worldwide quality competition" so as not to be pushed out of the market, and to establish a new management technology model which enables them to offer highly reliable products of the latest product design that are capable of enhancing the value for the customer.

Focusing on management technology for development and production processes, it is clear that there has been excessive repetition of prototyping, testing, and evaluation for the purpose of preventing the "scale-up effect" in the bridging stage between design and development and mass production. This has resulted in unstable built-in QA in the product design stage and an increase in the development period and cost. Therefore, it is now vital to reform conventional product design processes (Nikkei Shinbun, 2005). This is to be achieved through a changeover from product design through actual product confirmation and improvement to a process of prediction evaluation–oriented development through the effective use of the latest numerical simulation technology (CAE) (Amasaka, 2007c, 2008a).

Figure 10.4 shows transitions in the automotive product design process in Japan. For model changes in the past (development of product design to production: ~4 years), after completing the designing process, problem detection and improvement were repeated mainly through the process of prototyping, testing, and evaluation (refer to Chapter 2). In some current automotive development,

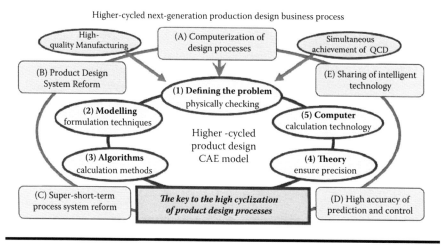

Figure 10.4 Higher-cycled product design CAE model.

vehicle prototypes are not manufactured in the early stage of product design due to the utilization of CAE and simultaneous engineering (SE), resulting in a substantially shorter development period (Amasaka, 2007a, 2010a).

It is now possible to utilize CAE for comparative evaluation, rather than the conventional supplementary "observation" role during the testing of prototypes. This improvement means that CAE is utilized to the same extent as prototype testing. The vehicle product design and production process has been shortened to 1 year and there has been a transition to a super-short-term concurrent product design process based on the utilization of CAE and solid computer-aided design (CAD), allowing individual processes to progress simultaneously. This would be virtually impossible using the conventional repetitive testing of prototypes.

10.2.2.1.1 Problems When Applying CAE to Product Design Reform

Amasaka (2007c) has identified problems that occur when applying CAE to product design reform and the main points are summarized in Figure 8.14 from the standpoint of the simultaneous achievement of QCD (refer to Section 8.3). It is clear that it is necessary to eradicate the tendency toward prototypes and experimental evaluation in favor of predictive evaluation based on CAE analysis.

However, in order to achieve this, the discrepancy (gap) between CAE analysis data and experimental evaluation data (analysis error) must be reduced from over 10% to 1%–2%. Therefore, the author has conducted factorial analysis, as shown in Figure 8.14, from the perspective of CAE technology for application, analysis, evaluation, and management.

In order to realize the simultaneous achievement of QCD, it is necessary to enhance CAE analysis technology (Amasaka, 2007a,c; Whaley et al., 2000). It is therefore important to determine the mechanisms behind the main technological problems and accurately represent them with numerical calculations (simulation), as shown in Figure 8.14. Specifically, this means that it is vital to improve CAE analysis in order to more accurately and precisely reproduce the results of prototype testing.

10.2.2.1.2 Product Design Process Utilizing CAE

10.2.2.1.2.1 Product Design Process Utilizing CAE — The time between product design and production has been drastically shortened in recent years with the rapid spread of global production. QA has become increasingly critical, making it essential that the product design process—a critical component of QA—be reformed to ensure quality (Amasaka, 2010b).

Figure 9.7 shows the typical product design process currently used by many companies (refer to Section 9.2). It shows that companies first create product design instructions based on market research and planning (refer to Section 9.2). They then use these instructions to make specific product design specifications (drawings) and to promptly convert them to digital format so that they can be suitably processed and applied. The data are primarily used in numerical simulations known as CAE (Amasaka, 2007a; Amasaka et al., 2012).

CAE analysis enables engineers to determine whether the product design specifications result in sufficient product quality and if not, the specifications are promptly optimized. Manufacturers then carry out effective preproduction, testing, and evaluating prototypes (prototype testing). A final evaluation of the development design is conducted, and if there are no problems, companies move to production preparation and process design in order to get ready for full-scale production.

10.2.2.1.2.2 Current Status and Issues — CAE and other numerical simulations have been applied to a wide variety of business processes in recent years, including research and development, design, preproduction and testing/evaluations, production technology, production preparation, and manufacturing. These and other applications are expected to have effective results. The product design process, for example, is typically one guided by unspoken experiential knowledge and rules of thumb, leading to prototype testing guided by repeated trial-and-error efforts.

In this age of global quality competition, using CAE for the predictive evaluation method in design work is expected to contribute a great deal to shortening development design time and improving quality. Despite these high expectations, conventional forms of CAE analysis resulted in figures that deviated as much as 10%–20% from those obtained through prototype testing evaluations. This means that many companies are now stuck with applying CAE only to the monitoring task of comparative evaluations of old and new products—despite the enormous amount of funds they have invested in CAE development (Amasaka, 2008a).

There are two absolute requirements for precise (highly reliable) CAE analysis methods that can both prevent the critical technical problems plaguing manufacturers from recurring and contribute to new product designs. The first is reducing the deviation from prototype testing evaluation figures to 5% or less, and the second is evaluating the absolute values needed for tolerance designs.

10.2.3 Proposal of Higher-Cycled Product Design CAE Model

Currently, to continuously offer attractive, customer-oriented products, it is important to establish a new development design model for intelligent product design that predicts customer needs. In order to do so, it is crucial to reform the product design business process. Manufacturing is a battle against irregularities, and it is imperative to renovate the business process in the product design system and to create a technology so that serious market quality problems can be prevented in advance by means of accurate prediction/control.

For example, as a solution to technical problems, approaches taken by product design engineers, who tend to unreasonably rely on their own past experience, must be clearly corrected. In the business process from product design to production, the product design cost is high and time period is prolonged due to the "scale-up effect" between the stages of experiments (tests and prototypes) and mass production. In order to tackle this problem, it is urgently necessary to reform the conventional product design process.

Focusing on the successful case mentioned above, the author deems it a requisite for leading manufacturing corporations to balance high-quality development design with lower cost and shorter product design time by incorporating the latest CAE and statistical science (Amasaka, 2010a). Against this background, it is vital not to stick to the conventional product design method, but to expedite the next-generation product design business process in response to a movement toward digitizing product design methods.

In order to achieve the necessary advancements in theoretical CAE analysis, the authors propose a Higher-Cycled Product Design CAE Model required for CAE analysis software as the key to the high cyclization of product design processes as described in Figure 10.4. This model is aimed at the simultaneous achievement of QCD by high-quality manufacturing, which is essential to realize customer satisfaction (CS).

As shown in the figure, in order to eradicate QA that relies on actual product design of vehicle evaluation, it is vital to have a thorough understanding of the following: (1) problem setting (physically checking the actual item), (2) algorithms (calculation methods), (3) modeling (formulization techniques: statistical calculation and model application), (4) theory (ensure precision: establishing theories required to clarify problems), and (5) computing (calculation technology: selection of calculators).

10.2.4 Developing Highly Reliable CAE Analysis Technology Component Model

In order to achieve the necessary advancements in theoretical CAE analysis by developing the Higher-Cycled Product Design CAE Model, Amasaka (2007c) has developed the Highly Reliable CAE Analysis Technology Component Model required for CAE analysis software, as shown in Figure 8.15 (refer to Section 8.3).

What is urgently needed is innovation to promote the advance from the conventional evaluation-based development, which uses the prototyping and experiment process (a method based on the confirmation of real goods for improvement) that had long supported the highly reliable design, to a CAE prediction–based design process.

To accomplish this model, the developers engaging in CAE (car body manufacturers—Hino Motors, Ltd. and Toyota Motor Corp.; parts manufacturers—NHK Spring Co., Ltd.; software makers—Mizuho Information and Research Institute, Inc.; system developers—Mathematical Systems Inc., Tsukuba University, and Aoyama Gakuin University, etc.) have jointly sorted out the free responses to the questions given and summarized the required technical requests.

Figure 8.15 clearly shows that there is a sufficiently wide range of options available for elemental technologies in order to introduce CAE analysis on a general level. However, from the standpoint of implementing CAE as a problem-solving method on a working level, the number and wide selection of these elemental technologies are not sufficient. This may be because reliable CAE is a process consisting of multiple elemental technologies (Alba, 2005).

Then, the suitability of the algorithm itself, its range of application, and performance (expected accuracy) should be indicated theoretically. It goes without saying that the functionality of the calculator used to realize the algorithm will significantly influence the success of CAE analysis.

When implementing CAE analysis, select the problem to be solved and then start by modeling the problem as an algorithm. Next, during actual CAE analysis, use a calculator to analyze the model. However, the actual analysis method must be reproducible as software. That is to say, it must be set as an algorithm.

The technological components that constitute the CAE analysis process must be mutually compatible and must supplement each other's weak points. Even theoretically, superior algorithms will not produce the desired results if not applied effectively on a calculator. The performance of an algorithm also significantly affects its suitability for modeling.

Even if the problem is correctly set, the algorithm will not function effectively if it is modeled incorrectly. The compatibility between the algorithm and the calculator cannot be ignored. An algorithm that makes the best use of the characteristics of the calculator will produce the best results. This means that if a suitable combination of technological elements is not used, the whole CAE analysis process will fail to function correctly.

Successful CAE analysis depends on the overall capability provided by a wide range of technological components (see Figure 8.15). The requirements for the business processes involved in producing intelligent CAE analysis software and its effective deployment are also vital elements. Specifically, it is important to (i) accurately model the mechanisms of the problem, (ii) select suitable analysis methods as a basis for simulation, and (iii) select suitable tools for data analysis and apply them in conjunction with statistical analysis methods. It is also important to (iv) effectively share know-how to (v) develop a comprehensive knowledge base.

In other words, success in CAE depends on the collective strengths of the elemental technologies, and this is what the author asserts here. Skilled CAE engineers are not experts in all the fields of the elemental technologies, but they understand their characteristics and interactions as implicit knowledge and thus conduct selection and combination to obtain favorable interactions and consequently the desired results. The formulation of such implicit knowledge confined to the personal know-how of the engineers is an indispensable step to be taken for sophistication of CAE as a problem-solving method, and therefore it is positioned as a major theme in Amasaka (2007a,c).

10.2.5 Application

In this section, the authors present examples showing the results of research and effectiveness where the Higher-Cycled Product Design CAE Model employing the Highly Reliable CAE Analysis Component Model mentioned above has been applied in current technological problems, loosening bolts, and others of product design bottlenecks in automotive product design.

10.2.5.1 Automotive Bolt-Tightening Analysis Simulation Using Experiments and CAE

There are more than 3000 bolts in an automobile. Bolts are the most used parts in automobile manufacturing and are necessary for keeping parts together. For bolts, it is important that axial force and torque are in the prescribed sizes for displayed tightening force. Therefore, cracks occur for bolts and substrates as a result of irregular forces. Important quality issues arise if this looseness or cracks occur while driving. Therefore, the authors are experimenting with and simulating bolt tightening in order to avoid these quality problems.

In this study, the creation of a Bolt-Tightening Simulator, one of the main elemental technologies for automotive product design, is featured as part of the in-depth study conducted by utilizing the Higher-Cycled Product Design CAE Model (Amasaka, 2007a; Ueno et al., 2009; Takahashi et al., 2010; Kozaki et al., 2011; Onodera and Amasaka, 2012).

10.2.5.1.1 Bolt-Tightening Experiment

The first, experimental data for bolt tightening is to be collected with a view to reproducing the conditions of bolt tightening by means of CAE. In the case of defective bolt-tightening operations such as looseness, the tightening conditions need to be viewed as initial conditions (Izumi et al., 2005; Zhang and Jiang, 2004; Aragóna et al., 2005). In the above setting, a strain gauge is attached to the material to which hexagon flange bolts are tightened, and an experiment was conducted as shown in Figure 10.5 in order to confirm an important parameter of the friction coefficient based on the dynamic behavior of angle × torque × axial force of (a) the threaded portion and (b) the bearing surface.

The bolt-tightening experiment was conducted using hexagonal flange bolts and hexagonal nuts with flange, which are actually used in the automotive industry

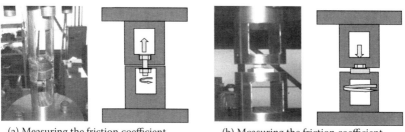

(a) Measuring the friction coefficient (b) Measuring the friction coefficient

Figure 10.5 Measuring the friction coefficient of (a) the threaded portion and (b) the bearing surface to tighten bolts

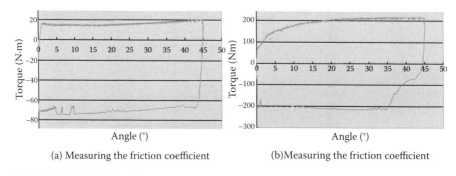

(a) Measuring the friction coefficient (b) Measuring the friction coefficient

Figure 10.6 The relationship between angle and torque: (a) measuring the friction coefficient of the threaded portion; (b) measuring the friction coefficient of the bearing surface.

for commercial use. The friction coefficient of the threaded portion (a) was measured using a two-axial fatigue testing machine, which is capable of applying axial force and twist force at the same time.

The bolt head was pulled with the force of 20, 40, and 60 kN, conducting the experiment for each tensile force five times in total, while torque and axial force were measured. The friction measurement for coefficient of the bearing surface (b) was similarly conducted using the two-axial fatigue testing machine, applying a compression load of 20, 40, and 60 kN onto the substrate while rotating the nut so as to confirm the torque and axial force.

Based on the results of the friction coefficient measurement of (a) the threaded portion and (b) the bearing surface, torque and axial force were obtained. Applying theoretical equations to these experiment results, the friction coefficients were calculated. Figure 10.6 shows the relationship between angle and torque for 60 kN of tensile force as well as compression load.

When calculating the friction coefficient of the threaded portion, μ_s, Equation 10.28 was used. The friction coefficient of the bearing surface μ_w was

Table 10.3 Bolt and Nut Measurements

	Bolt	Nut
Major diameter of external thread (mm)	12	12
Pitch diameter (mm)	11.19	11.19
Minor diameter of external thread (mm)	10.65	10.65
Pitch of threads (mm)	1.25	1.25
Width across flat (mm)	16.91	16.91
Flange diameter (mm)	26	26
Lead angle (°)	2.03	2.03
Thread angle (°)	30	30

Source: ISO 4161/15701/15702/10663, BS (British Standards)/ES (European Standards), 1991.

Note: The authors measured or calculated.

calculated using Equation 10.29 (Suzuki, 2005; Tanaka et al., 1981). The bolt and nut dimensions necessary for calculation of these coefficients are given in Table 10.3.

$$\mu_s = \frac{2T_s/d_2 - F\tan\beta}{2T_s/d_2 \tan\beta + F} \cos\alpha \qquad (10.28)$$

$$m_w = \frac{2T_w}{Fd_w} \qquad (10.29)$$

where:
μ_s = the friction coefficient of the threaded portion
T_s = the torque of the threaded portion
d_2 = pitch diameter
F = axial force
β = lead angle
α = thread angle
μ_w = friction coefficient of the bearing surface
T_w = torque of the bearing surface
d_w = equivalent diameter of torque on bearing surfaces

The dimensions of the bolt are actually measured here, and units are in millimeters. However, pitch and other dimensions that are difficult to measure are taken from the ISO standard (ISO 4161/15071/5072/10663, 1999). Using equations and dimensions such as those mentioned, the friction coefficients of the

Table 10.4 Friction Coefficients

Friction Coefficient of the Threaded Portion			Friction Coefficient of the Bearing Surface		
Angle (°)	Torque (N·m)	Friction Coefficient	Angle (°)	Torque (N·m)	Friction Coefficient
40	18.63	0.02	40	215.26	0.35
40	23.05	0.03	40	221.63	0.37
40	35.30	0.06	40	198.09	0.33
40	28.44	0.04	40	208.39	0.34
40	32.85	0.05	40	226.53	0.37

threaded portion and bearing surface were then calculated. As an example, the friction coefficients of the threaded portion and bearing surface when the tensile force and compression load in the axial direction are both 60 kN at the angle of 40° are shown in Table 10.4. Using these calculated friction coefficients, bolt-tightening simulation was conducted and the relationship between simulative axial force and torque was confirmed.

The authors conducted a bolt vibration test to define the bolt mechanism. An external force is added in the bolt vibration test to gauge changes to the axial force, which is the bolt-tightening force, and in the stress distribution on the nut bearing surface and contact surface of the base material. Specifically, oscillation amplitude was applied to the bolt–nut joint, and the reduction in axial force of the joint and stress distribution at the bearing surface were measured by using the strain gauge–equipped test bolt and the surface pressure sheets for measuring the stress distribution.

The test piece was a M12 × 1.25 9T hexagonal bolt and nut with flange, which is often used to tighten parts in automotive applications, as shown in Table 10.3. The stress distribution was measured at the nut bearing surfaces using pressure sheets with 20 and 35 kN bolt-tightening loads at 1.8 and 2.8 kN of oscillation load, respectively. Figure 10.7 shows the transition of bolt load with the initial applied bolt load of 20 kN of external stress in a normal direction. Each measurement result is plotted with the bolt load on the vertical axis and the sampling number (5 Hz) on the horizontal axis.

10.2.5.1.2 Technological Elements of Bolt-Tightening Simulator

The main issues for creation of CAE software are grasping (1) the actual state of contact between the threaded portion and the grooves and (2) the friction coefficient of the threaded portion and bearing surface. For structural analysis

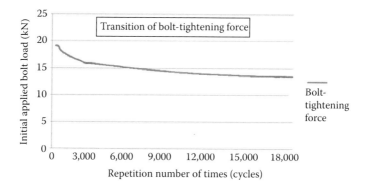

Figure 10.7 Bolt load of 20 kN of external stress in a normal direction.

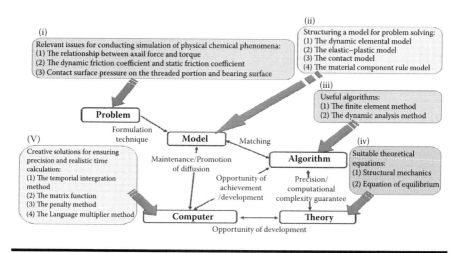

Figure 10.8 Highly Reliable CAE Analysis Technology Component Model for bolt-tightening simulator.

to function properly, there are two minimum requirements. These are necessary to simulate the dynamic behavior of angle×torque×axial force for both (a) the threaded portion and (b) the bearing surface during bolt tightening. The requirements are to provide contact between the threaded portion and groove and between the threaded portion and bearing surface. This should be done on the basis of elastic static analysis and elastic dynamic analysis results, calculated using the finite element method (FEM).

Figure 10.8 shows the Highly Reliable CAE Analysis Technology Component Model of the Bolt-Tightening Simulator when the Highly Reliable CAE Analysis Technology Component Model (mentioned in Figure 8.15) is applied. The figure shows the following. (i) Relevant issues for conducting simulation of physical chemical phenomena are (1) the relationship between axial force and torque, (2) the dynamic friction coefficient and static friction coefficient, and (3) contact surface pressure on the threaded portion and bearing surface. (ii) Models for solving these issues are (1) the dynamic elemental model, (2) the elastic-plastic model, (3) the contact model, and (4) the material component rule model. (iii) Useful algorithms are (1) the finite element method and (2) the dynamic analysis method. (iv) Suitable theoretical equations are (1) structural mechanics and (2) equation of equilibrium. (v) Finally, examples of creative solutions for ensuring precision and realistic time calculation are (1) the temporal integration method, (2) the matrix function, (3) the penalty method, and (4) the Lagrange multiplier method.

10.2.5.1.3 Development of the Bolt-Tightening Simulator

Many bolts are used in the assembly of automobiles, and they must be tightened in such a way as to ensure the strength (safety) of the areas in which they are used. With this in mind, the author applied the Bolt-Tightening Simulator and the analysis utilizes the axis symmetry two-dimensional (2-D) model, as shown in Figure 10.9.

The analysis parameters are the same as the experiment parameters. Regarding the analysis parameters, axial force is applied to part of the upper substrate and the bolt bounded from all directions in the same way as it is applied to part of the bolt head and lower substrate. The friction coefficient was calculated by conducting experiments on parts of the contact bearing surfaces and the contact surface of the threaded portions and was then inputted into CAE simulation software.

Next, the authors utilize three-dimensional (3-D) finite element models simulating the helical structure of the threaded portion, as shown in Figures 10.10 and 10.11. In the diagrams, the areas around the bolt and nut and the contact sections

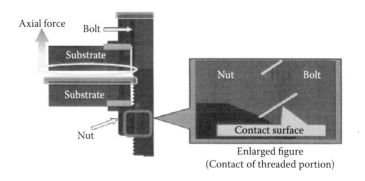

Figure 10.9 The axis symmetry 2-D model.

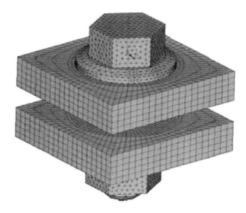

Figure 10.10 3-D finite element model.

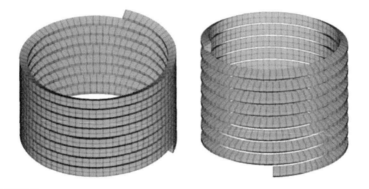

Figure 10.11 3-D finite element model of the helical structure.

are represented by a fine mesh compared to the 2-D finite element model. The analysis process involves the following steps: (1) Place parts to be clamped (two substrates) between the bolt and nut. (2) Apply axial force to the bolt and determine the distribution of stress during contact. (3) Fix the edge of the lower substrate and apply upward perpendicular force (axial force) to the upper substrate. (4) Determine the pressure on the contact surface (bolt/nut bearing surfaces and between the bolt crest and nut).

The results of 3-D analysis when axial force was applied to the upper substrate in a perpendicular direction are shown in Figures 10.12 and 10.13. The diagrams show that there is a greater degree of stress on the surface of the substrate in contact with the bolt flange than on the periphery of the substrate. This stress distribution shows that stress is concentrated in the perpendicular section between the bolt head and the flange, making it possible to examine more thoroughly the results of analysis conducted using the 2-D axial symmetry model (Figure 10.9).

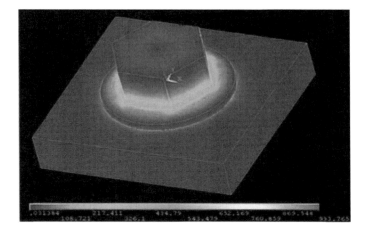

Figure 10.12 Stress of the surface of the substrate.

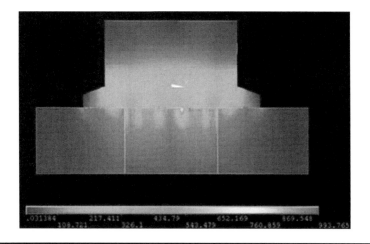

Figure 10.13 Stress of the surface of the substrate (cross-section).

Figure 10.14 shows a stress diagram (cross-section) with the thread contact section enlarged. It can be seen from the diagram that the lower section has a higher stress value than the upper section.

The results of the experiment are represented by the dotted line (experiment results) and the results of the simulation analysis are represented by the solid line (CAE results) in Figure 10.15. This chart shows that the results of the simulation analysis closely match the results of the experiment. Therefore, the author immediately applied the knowledge gained from this analysis and conducted bolt-tightening simulation analysis on 3-D models to enable analysis of looseness.

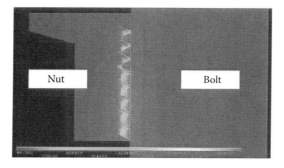

Figure 10.14 Stress of the thread contact section.

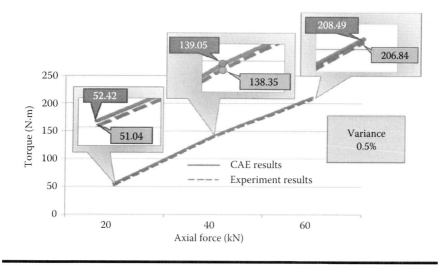

Figure 10.15 Comparison result.

10.2.5.1.4 Highly Precise CAE Analysis of the Bolt-Loosening Mechanism

Figure 10.16 show the results of the 2-D analyses conducted as a stepping stone to conducting a 3-D analysis. The following stress distribution diagrams indicate the presence of higher stress in the inner portion of the nut as well as in the base material. Because the purpose of the axially symmetrical 2-D model analysis was to understand the boundary conditions, model integrity, and contact relationships, detailed analysis results such as stress distribution at the nut bearing surfaces will be discussed in the 3-D model analysis.

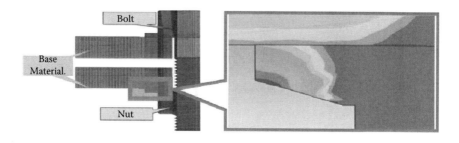

Figure 10.16 The results of axis symmetry 2-D model.

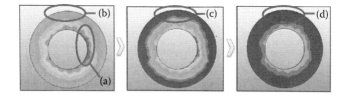

Figure 10.17 Transition of stress distribution at the nut bearing surfaces from CAE analyses.

Furthermore, the authors will discuss the 3-D analysis results at the bolt bearing surface when an axial force of 20 kN was applied to the bolted joint along with ±1.8 kN of oscillation load applied to the base material on the oscillation side. Figure 10.17 shows the change in stress distribution at the bolt bearing surface due to oscillation load.

High stress is observed at the nut bearing surface, particularly at the starting point and the end point of the helix structure of the bolt and nut (Figure 10.17a). This is similar to the observation made during the tightening analysis discussed earlier. Also, it is evident that the outer part of the nut bearing surfaces experience less stress (Figure 10.17b,c,d).

This indicates that with applied vibration, stress reduces more significantly at the outer edges with less stress compared to the inner portion with higher stress. This result indicates that a reduction in contact force at the bearing surface of the area closer to the outer edge of a nut leads to slippage at the bearing surface when a bolt's axial force reduces with externally applied force.

Here, we will compare stress distribution of the nut bearing surfaces during the vibration test and the stress distribution of nut bearing surfaces from the 2-D and 3-D analyses. The stress at the nut bearing surface is higher in the area closer to the helix of the nut and shows a stress distribution similar to the CAE simulation result. Although there are differences in the values that can be attributed to the distortion in the base material, wear, and roughness of the contact surfaces, the simulation was successful in reproducing similar stress values and stress distribution changes.

Therefore, by conducting CAE analysis according to the highly precise CAE analysis approach proposed in this paper, the authors were able to show the effectiveness of the approach by reproducing an actual phenomenon. It is also plausible that the approach can contribute to bolt-loosening and bolt-fracture analysis by understanding the changes in stress distribution at the nut bearing surfaces. The 3-D analysis procedure is conducted in the next step. First, the two bolted work pieces are placed between the bolt and the nut using pitch 0.50 and 1.75 mm. The author discusses the 3-D analysis results at the bolt bearing surface when an axial force of 35 kN is applied to the bolted joint along with ±1.8 kN of oscillation load applied to the base material on the oscillation side.

Axial force is applied to the bolt (securing the edge of the lower piece) and perpendicular to the upper piece. Finally, the analysis looks at the pressure on the contact surface (between the bolt/nut bearing surface and the work pieces and between the bolt threads and the nut) and at the behavior of the pieces in terms of reduction in axial force when pressure changes.

Figure 10.18 shows a lack of uniformity in terms of nut bearing surface response and a localized strong response where the nut thread structure begins. In looking at the differences between the two pitch values, the response distribution is more concentrated around the initial thread structure on the bolt with the longer pitch (1.75 mm).

These results suggest that differences in pitch also lead to differences in non-uniformity on nut bearing surfaces and a more pronounced gap between areas of strong and weak response. In terms of new information, the 3-D analysis revealed that the length of bolt and nut pitch impacts the reduction in contact force on contact surfaces.

Comparing the axial force reduction behavior measurements obtained through prototype testing and those obtained through CAE analysis verified the precision

(a) Pitch 0.5 mm   (b) Pitch 1.75 mm

Figure 10.18 3-D analysis of pitch differences of two kinds: (a) pitch difference 0.5 mm; (b) pitch difference 1.75 mm.

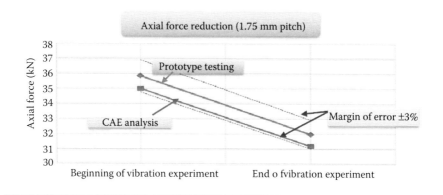

Figure 10.19 Comparing the results.

of the CAE analysis results. Figure 10.19 compares the results of prototype testing and CAE analysis for the bolt and nut with 1.75 mm pitch under a fastener load of 35 kN. The dashed line shows a margin of error of 3%, indicating that a high-quality CAE analysis was achieved. Similar results were achieved for the 0.50 mm-pitch bolt and nut.

10.2.5.2 Application to Similar Problems Using Higher-Cycled Product Design CAE Model

With its effectiveness verified, the author was able to apply the Higher-Cycled Product Design CAE Model to critical development design technologies for automotive production, including predicting and controlling the special characteristics of urethane foam molding of seat pads, antivibration design of door mirrors, aerodynamics of body lifting power, and transaxle oil seal leakage (Amasaka, 2010a, 2012; Amasaka et al., 2012; Nozawa et al., 2012).

In each of these cases, discrepancy was 3%–5% versus prototype testing. Based on the achieved results, the model is now being used as an intelligent support model for optimizing product design processes.

10.2.6 Conclusion

In this paper, the authors proposed a Higher-Cycled Product Design CAE Model employing Highly Reliable CAE Analysis Technology Component Model as part of principle-based research aimed at the evolution of product design processes to ensure high quality assurance. The model's effectiveness has been demonstrated with a precision CAE analysis of automotive loosening bolts and others with no discrepancy between the experiment and CAE.

10.3 Human-Integrated Assistance System for Intelligence Operators

10.3.1 Introduction

The Japanese manufacturing industry is now developing global production by establishing production sites in various countries. High quality assurance is regarded as the strong point of manufacturing in Japan (Abegglen, 1958). However, this situation is under threat (Hunt, 1992).

The author considers it vital to make production operators more independent and creative in addition to their engineering capabilities and skills so as to become an intelligence operator employing New JIT. In view of the need to develop a new, creative, human-oriented production system for meaningful work, the authors propose this Human-Integrated Assistance System for creative, meaningful work leading to improved productivity. It supports autonomous development of *kaizen* as the core of this system for the global production strategy.

In the wake of the recent rapid expansion of globalization, short-term training of production operators is an especially critical issue, particularly for ensuring productivity at the start-up of local production. To deal with this issue, it is urgent to apply this system to analyze the factors that contribute to the variations in the skill acquisition level of local operators. This is done with a view to establishing a training method that can support them to stably perform work despite the differences in their ability.

In view of the need to develop a new creative human-oriented production system for meaningful work, the authors propose the Human-Integrated Assistance (HIA) System for creative, meaningful work leading to intelligence productivity improvement. It supports autonomous development of *kaizen* as the core of this system for the global production strategy (Sakai and Amasaka, 2008; Amasaka et al., 2008).

In definite terms, a brand new tool, the *visual manual*, is characterized by (1) convenience, (2) accumulation of know-how, and (3) utilization of CAD and CAE data for further development of advanced skills for intelligence operators. Given these circumstances, the authors have created a new Intelligent IT System that incorporates a training curriculum adjusted to the skill acquisition level of each operator, thereby bringing the training program to a higher level. During the course of implementation, the authors also adopted an aptitude test for assessing the aptitude and inaptitude of operators.

This was designed for the establishment of an efficient training system. Its effectiveness has been tested at advanced car manufacturer A, a leading automotive manufacturer, as a system that brings about autonomous, voluntary skill improvement in intelligence operators.

10.3.2 Need to Develop Intelligence Operators for a Highly Reliable Production System and Accelerated Training of Production Operators

At present, the Japanese manufacturing industry is rapidly deploying global customer-oriented production. This rapid spreading of production sites abroad, however, has brought about many new problems for Japanese production, which has developed a reputation in the past for assuring high quality. As seen in the case of automobiles, the highly reliable production system with target productivity equal to that in Japan is spreading worldwide, but the actual productivity (availability) is still often lower (Gabor, 1990).

The major cause of this situation, it has been revealed, is the dependence of the conventional production system on operators' *kaizen* awareness or individual capabilities. This has limited its applicability to overseas operators familiar with different systems and from different cultural backgrounds. In Japan, for example, the reliability of a production line is gradually improved through repeated *kaizen* actions after a new start-up, resulting in possibilities for high product QA.

To solve such problems, early fostering of intelligence operators is the key to successfully achieving global production with high quality assurance. In order to attain high quality assurance worldwide, it is necessary to ensure all intelligence operators acquire a consistent level of manufacturing skill. In other words, correct *kaizen* and maintenance of the production equipment through fostering of intelligence operators is indispensable to worldwide application of the highly reliable production system. Furthermore, for the improvement of production quality, accelerated skill training, among other things, will have greater importance and be a key to successful production activities.

In connection with the recent increase of domestic and overseas production, a large number of new recruits and limited-term workers are being employed. Generally, these employees learn from their seniors by watching their work on an on-the-job training (OJT) or off-the-job training (OFF-JT) basis. Some acquire the required skills quickly (within a week) and start to work on the production line immediately, whereas others cannot reach a required skill level even after 3 weeks or perform unstable operations in building the product, resulting in being removed from the line until they reach the required skill level while working as supporting operators.

Such a situation where quality build up, which plays a central role in QCD activities, cannot be sufficiently achieved has been observed from time to time. In order to realize the production strategy of uniform quality worldwide and simultaneous start-up, it is urgent to carry out short-term training for production operators so as to level out the variations in their skill acquisition.

10.3.3 Strategic Implementation of Advanced TPS for Intelligence Operators

While manufacturing in workshops is being transformed thanks to digital engineering, the engineering capability in manufacturing workshops often drops, thereby weakening scientific production control for quality incorporation in processes (Amasaka, 2002, 2003, 2008b). It is an urgent task to further advance the Toyota Production System strategically for higher-cycled next-generation production business processes, apart from the conventional experience in success from the viewpoint of global production.

The author, therefore, considered the necessity of including and organically integrating the four elements (Amasaka, 2005) (production based on information, information technology; production based on management, process management; production based on technology, production technology; and production based on partnership, human management) with strategic application of Total Production System (TPS) through New JIT (Amasaka, 2002) and clarified Advanced TPS (Amasaka, 2005) for the global production strategy (refer to Sections 5.1 and 8.1).

With respect to the requirements shown in Figure 8.13 (refer to Section 8.3), the mission of Advanced TPS in global deployment is to realize customer satisfaction (CS), employee satisfaction (ES), and social satisfaction (SS) through production that achieves high quality assurance (Amasaka, 2007b). In implementing Advanced TPS for uniform quality worldwide and production at optimal locations (concurrent production), renewal of production management systems appropriate for digitized production and creating attractive workshop environments tailored to increasing the number of older and female workers are fundamental requirements.

In more definite terms, the following are required; (a) strengthening of the process capability maintenance and improvement by establishing an intelligent quality control system, (b) establishment of a highly reliable production system that achieves high quality assurance, (c) reformation of the work environment for enhancement of intelligent productivity, and (d) development of intelligent operators (skill level improvement) and establishment of an intelligent production operating system. These factors combined will realize higher-cycled next-generation business processes for early implementation of uniform quality worldwide and production at optimum locations.

10.3.4 Proposal of Human-Integrated Assistance (HIA) System: Developing Intelligence Operators for Production

Recently, to realize the higher productivity from production operators required for global production, the authors have paid attention to the need for a new human-oriented production system (Iiyoshi and Hannafin, 2002). This provides for creative, meaningful work, based on a production philosophy as the engineering nucleus of the Advanced TPS (Aspy et al., 2000; Aspy and Wai, 2001; Sakai and Amasaka, 2005, 2006). Three basic requirements of production are as follows:

1. Establishing a production system that ensures overall line reliability and maintainability and improves advanced production equipment operation technologies (fault diagnosis, maintenance, and preventive maintenance of production equipment) to prevent availability degradation and quality defects.
2. Upgrading engineering and other skills to levels comparable with Japanese staff in a wide range of areas. This includes manual lines for small-lot production in automotive and other general assembly industries involving many hard-to-automate jobs, especially in developing countries.
3. Attaining mastered skills equivalent to those of production operators in advanced countries equipped with sophisticated mass-production-type equipment that allows intelligent diagnosis by individuals in the quality incorporating stage.

For systematization to satisfy these three requirements, it is necessary to improve the sophisticated production equipment operating skills and mastered skills of production operators themselves. Developing the intelligence operators for upgrading and unifying the production capabilities of worldwide operators is required before the advanced production process, which is continuously driven by IT and digitization, creates a black box that depends on personal discretion (Chawla, 2002; Sakai and Amasaka, 2007).

Therefore the author proposes the HIA system, which realistically copes with these requirements (Sakai and Amasaka, 2008). Figure 10.20 shows the construction

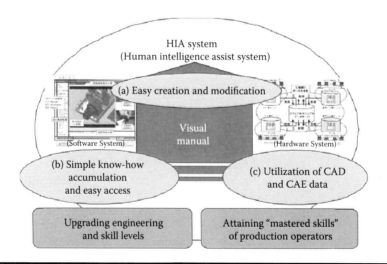

Figure 10.20   HIA system construction.

of the HIA System. As will be explained in Section 10.3.5, the HIA System covers system (tools, execution plan, skill requirements, and evaluation necessary) and methodology (formulating and establishing shortened training methods) for early fostering of operators with mastered skills as one of the two requirements explained earlier for a global production strategy.

10.3.5 HIA System Construction: The Key to Advanced TPS

Sharing of the same level of specialized job knowledge by worldwide intelligence operators with different cultural and language backgrounds is becoming increasingly important for global production (Hounshell, 1984; Roos et al., 1990; Leonard-Barton, 2003). This makes it necessary to develop a user-friendly method with worldwide applicability. The authors, therefore, have devised the visual manual as a new communication tool. This is a key technology in the aforementioned HIA System, of which the three basic requirements are summarized below.

10.3.5.1 Easy Creation and Modification

Conventional, hard-to-understand operation manuals written only in character text should be replaced with user-friendly intelligent manuals allowing easy understanding of the content by intelligence operators. The contents of such visual manuals should be revised as needed to present new ideas clearly.

Figure 10.21 shows the visual manual creation process. The entry sheet is created by modifying a world-standard Microsoft Excel sheet. This sheet is for the input of

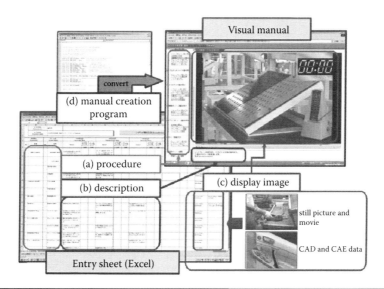

Figure 10.21 Visual manual creation process.

(a) procedure, (b) description, and (c) display image, and the input information is then converted to a visual manual using (d) the manual creation program. Anybody can easily revise the visual manual by using this sheet to modify the display image and explanatory text.

Through this method, manuals that have conventionally been used only for basic knowledge learning can be improved to facilitate accumulation of specialized knowledge.

10.3.5.2 Simple Know-How Accumulation and Easy Access

It is necessary for intelligence operators at worldwide production sites to be able to have access to the necessary visual manual content for immediate use. Furthermore, revision of and additions to the visual manual require simultaneous data storage and distribution.

Figure 10.22 shows the system for visual manual selection. Depending on the selection method, a series of screens matching each particular situation can be selected. Method (a) clearly indicates the contents (title of each item is in a list) for the search of necessary content. On the other hand, Method (b) employs a search program using a key word for selection of content from the search result.

The visual manual data is handled in the general HTML format, enabling it to be used anywhere in the world, and the data can be delivered locally and globally through the internet. With regard to the hardware system, a plant server is installed at each plant to establish a server and client system. It is used to view the data on each client system (PC) at each office or work site through the internet, and a note

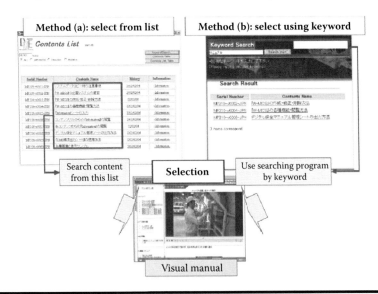

Figure 10.22 System for visual manual selection.

can be written as required on the spot. Then, each plant server is controlled in synch with the central server so as to make the system simultaneously update and distribute any modifications.

Thus, the intelligence operators at local and overseas plants can virtually experience the production method related to each production process and also obtain the related know-how, knowledge, and information concerning the same production process.

10.3.5.3 Utilization of CAD and CAE Data

Explanation of work instructions by character text only creates problem such as lack of clarity and difficulty accessing the required item. On the other hand, visually appealing work instructions provide a perfect description of the scene so as to enable members that use different languages to obtain unified understanding of the same material. As the skill level and training method may vary from trainer to trainer, use of the still picture and movie images of the CAD and CAE data in the visual manual will convey consistent information at a higher level.

Figure 10.23 shows the flows of CAD and CAE data. The product data are mainly provided in still images for use by (a) the design division. Movie images, on the other hand, are used as visual manual data in the (b) R&D and (c) production engineering divisions and in the manufacturing shop.

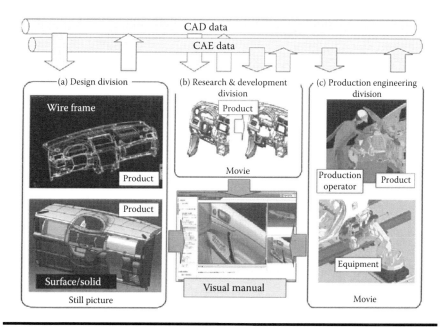

Figure 10.23 Flows of CAD and CAE data.

Thus, CAD and CAE data information has been made available on a network to production engineering and up to the manufacturing process, and it has become possible to instantaneously create procedures for the manufacturing process based on the design data, which in turn has made it possible to shrink the lead times from design to production.

10.3.6 Formulating and Establishing Shortened Training Methods

10.3.6.1 Current Issues Regarding Training Methods

10.3.6.1.1 Advancement of HIA System for Overcoming Variations in Training Methods

Up until now, highly skilled trainers with extensive experience have conducted technical training for production operators on a one-to-one basis. Consequently, the dissemination and deployment of operator training, particularly in Japan, has largely depended on the personal ability and qualifications of the trainers. However, replacement of these highly skilled trainers can easily cause confusion due to the different training methods provided.

This factor accounts for the variation in the level of technique/skill acquisition of production operators. Therefore, the advancement of the HIA system is vital for training production operators around the world who speak different languages.

10.3.6.1.2 Measures for Countries with Low Retention Rates

Conventionally, the unique Japanese-style training (experience-based learning) method on a one-to-one (face-to-face) basis where a sufficient amount of time is available has been successful. However, due to the low retention rate of employees in some countries, such methods do not seem effective for labor-intensive transmission of know-how (implicit knowledge).

Against this background, a new tactic is needed to drastically shorten the training period compared to the conventional method.

10.3.6.1.3 Overcoming the Variation in Skill Levels of Overseas Production Operators

Newly recruited overseas production operators learn from their seniors by watching their work on an OJT or OFF-JT basis, but the time spent acquiring the needed skills varies a great deal from operator to operator. Analysis of the factors involved has revealed that, at the first training experience in operations such as high-precision fitting or parts assembly, whether or not operators can understand the fundamental skill required of such operations is largely dependent on their aptitude and inaptitude. Hence, it is vital to correctly assess the aptitude and inaptitude of individuals' ability in order to conduct efficient training.

10.3.6.2 Measures for Shortened Training

10.3.6.2.1 Repetitive Training Utilizing the Visual Manual

Our proposed visual manual can be utilized as a training material that supports the off-the-line training of production operators and enhances their understanding of manufacturing-related skills. It is a new communication tool for production operators with different cultural and linguistic backgrounds. It enables them to share the same level of specialized operation knowledge, virtually experience the manufacturing methods specific to each process as well as the related know-how, and share knowledge and information regarding the production process with those working with the same process.

10.3.6.2.2 Intelligent IT System

A video-based Intelligent IT System was created for the self-study of operators so that they can understand and recognize the work postures and other technical know-how (Nomiya and Uehara, 2004), which have been taught orally up until now. This system is designed for the accurate transmission of operation skills and repetitive training until the trainee fully understands the work procedure. Specifically, this system provides the footage of work operations performed after the repetitive image training via the visual manual mentioned in Section 10.3.6.2.1.

As shown in Figure 10.24, a series of work operations demonstrated by a highly skilled trainer are video-recorded, and the same operations performed by a newly recruited production operator are also recorded and uploaded as video data. These two files are processed to be played side by side on a PC display.

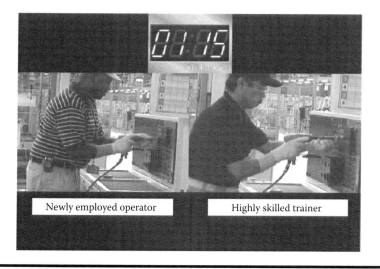

Figure 10.24   Image comparisons of a highly skilled trainer and a new operator.

398 ■ *New JIT, New Management Technology Principle*

In this way, production operators can view their weak points objectively and work on correcting them repeatedly so as to improve their skills to the desired level in a short period of time.

10.3.6.2.3 Assessment of Aptitude/Inaptitude by Aptitude Test

It has been determined that, of the production operators assigned to assembling processes, those who are assessed in advance as apt have a tendency to improve their skill in a short time after the assignment. The correlation between fundamental skill training items in the assembling process and skill acquisition on the site has also been identified. Based on these notions, the aptitude test sheet shown in Figure 10.25 was devised for the advanced assessment of aptitude and inaptitude.

This illustrates the example of an assembling process in which a test is carried out for eight fundamental skill items, and the total points for each item are used for the assessment. Based on the results, those who are assessed as apt immediately start receiving training at the actual production line, whereas those who proved to be inapt go through repeated dynamic training, which simulates a similar movement to the actual operations at the production line, to accelerate their skill development.

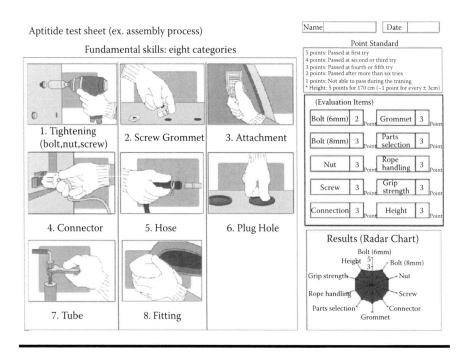

Figure 10.25 Aptitude test sheets (ex. assembly process).

10.3.6.3 Optimization of Training Steps

Figure 10.26 presents the flowchart outlining the training steps. Repetitive training using the visual manual mentioned in Section 10.3.6.2.1 will be repeated along with the self-study video described in Section 10.3.6.2.2 to make the trainee understand and recognize their weak points until they reach the skill level designated for each process on the specified evaluation sheet.

Next, in accordance with the results of the aptitude test conducted earlier consisting of the eight fundamental skills discussed in Section 10.3.6.2.3, aptitude and inaptitude will be assessed. Those who are assessed as apt immediately start receiving training at the actual production line. Those who prove to be inapt go through repeated dynamic training, which simulates a similar movement until they reach the skill level designated on the evaluation sheet, after which training at the production line is given.

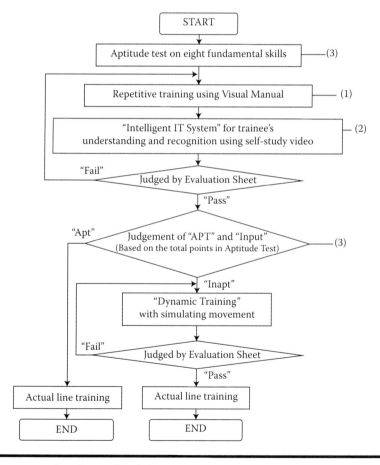

Figure 10.26 Flowchart of training program.

10.3.7 HIA System Application Example

This section explains the specific examples of the implementation of the HIA system and the perceived effectiveness of this system at Toyota, an advanced automotive manufacturer.

10.3.7.1 Application Case of Skill Training for Newly Employed Production Operators

The skill training for newly employed production operators in Japan and abroad was conducted in the production preparation stage using the proposed HIA System for start-up with a stabilized regular production system. Skill training for assembly jobs is explained as an example here. Skill training is conducted for production operators so that they are able to operate correctly at the time of a new plant start-up or model change. Especially in the case of assembly jobs, accurate work completion within the specified time is required for target attainment at an early stage before start-up or changeover.

Since the conventional training used a group of operations for judgment of the training result based on completion in the specified cycle time, it brought about a disparity in the quality of final products.

To be more specific, an example of a trimming process is shown in Figure 10.27. Fundamental skills are divided into eight items: bolt (6 mm) tightening, bolt

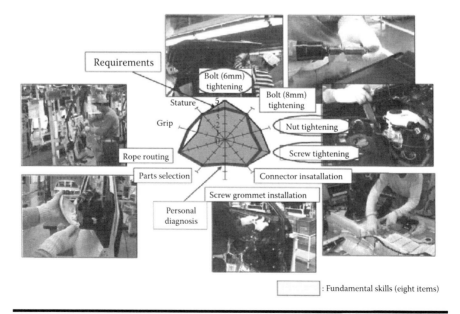

Figure 10.27 Example of skill levels between requirements and personal diagnosis.

(8 mm) tightening, nut tightening, screw tightening, connector installation, screw grommet installation, parts selection, and rope routing.

The training is conducted with the stress placed on bolt (6 mm) tightening and nut/screw tightening. A chart shows the difference between the requirements and the personal diagnosis so as to make every person aware of, and able to overcome, their weak points through training. The training is conducted repeatedly until target level is attained, determined by evaluation using the specified evaluation sheet.

Figure 10.27 shows an example of the visual manual concerning the bolt-feeding operation. First, the operational procedure is broken down into smaller operational movements. Accurate motions are clearly indicated visually using still pictures, movie images, and animation. The visual manual is in a standardized form. The screen consists of a procedure block, description block, and a display image block, which has a main screen and a subscreen (refer to Section 10.3.5).

This manual is read by turning the page using the forwarding button according to the procedure. Particularly in the description block, key points representing the know-how of each intelligence operator are written and visual information not displayed on the main screen is shown in detail in the subscreen. The explanatory text under each image describes why the posture is needed, what role it plays in QA, and other information from the intelligence operator so as to share the best practices in the world.

Figure 10.27 displays a movie showing the nut-feeding procedure for assembling product parts on the main center screen. The subscreen on the right side displays a still picture showing a key point of good posture by maintaining a right angle between the arm and the fingers.

Figure 10.28 shows the learning evaluation conducted for new employees assigned this time to the trimming process. The learning curve for conventional

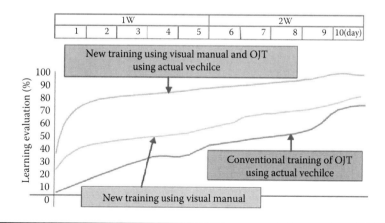

Figure 10.28 Learning evaluation for new employees assigned to trimming process.

training mainly consisting of OJT using the actual vehicle is compared with the new training using the visual manual. The degree of learning indicated in time series for the assigned trimming process job, according to the individual evaluation sheet (details are omitted), shows that it took 4 weeks until satisfaction of the specified accuracy within the specified work time in the case of conventional training. However, this was reduced to one half (2 weeks) with the new method.

The analytical results are as follows:

1. Training with the visual manual to detect primary individual weak points based on personal diagnosis has improved the understanding of the assigned job and achieved faster learning compared to the conventional method.
2. When training with the visual manual is combined with OJT on the actual vehicle, it has been confirmed that the learning speed can be increased through repetition of training that places an emphasis on personal weak points.
3. Efficient training was attained without disparity in the degree of learning by teaching the same contents in the same manner according to the clarified teaching process and because the training procedure was not dependent on the quality of the trainers.

10.3.7.2 Application Case of Shortened Training for New Overseas Production Operators

Below is an application case involving the overseas production operators at a leading automotive manufacturer, Toyota Motor Corporation. The deployment of the proposed HIA system for training newly employed production operators at domestic and overseas manufacturing plants reduced the conventional training period by more than half, from 2 weeks to 5 days, leading the full-scale production to a good start.

Figure 10.29 specifically indicates the skill training curriculum for new overseas production operators. After conducting the aptitude test for fundamental skills, skill training utilizing the visual manual and Intelligent IT System was repeatedly carried out until the standard set out in the evaluation sheet was met.

Those whose results on the aptitude test proved to be inapt went through dynamic training using the simulation training chart to simulate the movement repeatedly until the standard set out in the evaluation sheet was similarly met. Afterward, actual line training was finally given, and the designated skill level was reached in a period less than half compared to before. In this case, the launch of an overseas production plant, the target operating rate was achieved in 4 months after the start-up, as shown in Figure 10.30.

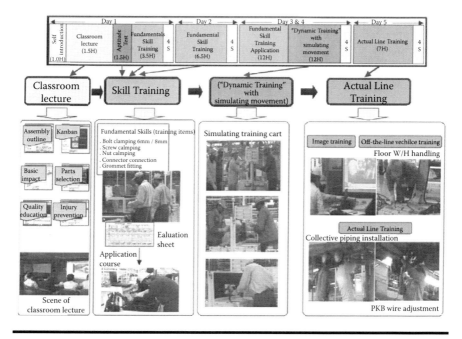

Figure 10.29 Skill training curriculum for new overseas production operators.

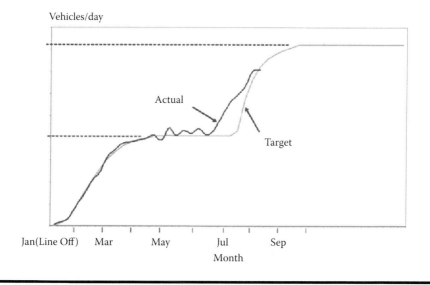

Figure 10.30 Transition of operation rate after launch of production line.

10.3.8 Conclusion

At existing local plants, it was possible to complete the job training for new and seasonal employees earlier, which allowed for a quicker response to the heavy load at the time of new model start-up. In new overseas plants, the fraction of defective products was kept at the same level as in Japan thanks to minimized disparity among intelligence operators in different countries. The authors propose the HIA System as a way forward in the creation of a new, people-centered production system that incorporates rich creativity and the motivation of those involved.

Furthermore, the authors would like to offer the HIA system as a way to combine the following proposed systems to better support production operators' intelligent production activities: V-MICS (Sakai and Amasaka, 2005), which improves production operators' operational technology abilities in regard to high-level equipment, V-IOS, which improves levels of mastered skills, ARIM-BL (Amasaka and Sakai, 1998), which shares intelligent information to maintain or improve the line equipment, and the Human Intelligence Production Operating System (HI-POS) (Sakai and Amasaka, 2006), which improves the level of production operators' technology understanding and skills. In the near future this must become a foundation of global production strategy at advanced car manufacturer A (refer to Section 8.4).

10.4 Comparing Experienced and Inexperienced Machine Workers

10.4.1 Introduction

Today, Japanese manufacturers are aiming for a global production strategy. To achieve this, the author believes it is crucial to improve the intelligence skill level of the production operators, who are the foundation of manufacturing. Amasaka (2002, 2007b) and Amasaka and Sakai (2011) recognized the need for a new production system: Advanced TPS through New JIT (refer to Sections 5.1 and 8.1).

Amasaka and Sakai (2011) and Sakai and Amasaka (2006, 2007, 2008) suggested the HIA System (as a solution and assessed its effectiveness at advanced car manufacturer A (refer to Sections 8.4 and 10.3). So, toward the firm development of Advanced TPS, this research aims to clarify the factors involved in improving the skills of technicians by conducting a comparison of experienced and inexperienced workers engaged in lathe work (Yanagisawa et al., 2013).

With the amount of production from the machining and assembly industries increasing year-on-year, machining technology is becoming increasingly important because machining is the fundamental process underpinning the industry. However, while the global level of technology is improving in line with machine-related companies' expansion of overseas production, the stagnant level of Japanese technology has become a matter for concern. Particularly evident is the sluggish pace at which the technical skills of inexperienced workers improve. This is thought to be due to the implicit style of training, which relies on the intuition and know-how of experienced workers.

Thus, the authors used electroencephalography (EEG) and statistical science to conduct a comparison of experienced and inexperienced workers engaged in lathe work, which is the foundation of machining work, and analyze the skills possessed by experienced workers. The findings were used to analyze the factors involved in improving the skills of inexperienced workers, and the required results have been achieved by utilizing Science SQC (refer to Section 5.4).

10.4.2 Skill Transfer Problem in the Manufacturing Industry

In Japan, occupations in the service industry are by far the most popular, and the number of employees engaged in this kind of work is increasing year-on-year. Employment in the manufacturing industry continued to decline during the 10 year period between fiscal 1992, which was the peak, and fiscal 2002. During this period, the figure fell 22% from 15.69 million people to 12.22 million people, resulting in fewer opportunities to transfer skills. For large companies, such a reduction in the number of employees places a greater burden on individual workers, which itself leads to an increase in the rate of people leaving. For businesses operating on a smaller scale such as small- and medium-sized companies and small factories, this leads to a lack of successors and ultimately to bankruptcy. Thus, Japan's manufacturing industry will begin to shrink.

In addition to the lack of successors, another problem faced by Japan's manufacturing industry is that of skill transfer, or training. Although the manufacturing industry—which relies heavily on specialist technology—is becoming increasingly mechanized, there are still many products where a human touch is required in order to maintain quality. As is the case with traditional crafts, workers need many years of experience to become able to produce quality products, and the lack of established training methods makes it difficult to train new employees. Thus, the two major challenges for Japan's manufacturing industry are a lack of successors and problems with training new employees.

The World Skills Competition (formerly known as the Skill Olympics) is run by World Skills International (WSI). The organization's committee is made up of official and technical delegates from member countries. The purpose of the competition is to promote occupational training in the participating countries and to encourage international exchange and goodwill among young technicians. Japan has been participating since the 11th competition in 1962. Although Japan placed first or second from about the 11th competition to the 20th competition, Korea is recently going from strength to strength, as shown in Table 10.5. In Korea, the large number of technical colleges means that training for machining skills is well organized and therefore technical skills are improving. Japan's manufacturing

Table 10.5 Top Three Ranking Countries in World Skills Competition

Competition Number	Top Three Countries That Won the Gold Medal		
	First Rank	Second Rank	Third Rank
40	South Korea	Switzerland	Japan
39	Japan	South Korea	France
38	Japan, Switzerland, South Tyrol, Italy	Germany, Finland	
37	South Korea	Switzerland	Japan
36	South Korea	Germany	Japan
35	Chinese Taipei	South Korea	Japan
34	South Korea	Chinese Taipei, Switzerland	
33	South Korea		Japan, Germany, Switzerland
32	Chinese Taipei	South Korea	Germany
31	South Korea	Chinese Taipei	Australia
30	South Korea	Chinese Taipei	Australia
29	South Korea	Japan	Chinese Taipei
28	South Korea	Japan	Chinese Taipei
27	South Korea	Chinese Taipei	Australia
26	South Korea	Japan	Germany
25	South Korea	Japan	Switzerland
24	South Korea	Switzerland	Australia
23	South Korea	Germany	Japan
22	Switzerland	South Korea	Spain
21	Germany	South Korea	Japan
20	Japan	Spain	Switzerland
19	Japan	Germany	South Korea
18	Japan	Switzerland	Germany

Table 10.5 (Continued)　Top Three Ranking Countries in World Skills Competition

Competition Number	Top Three Countries That Won the Gold Medal		
	First Rank	Second Rank	Third Rank
17	Switzerland	Japan	South Korea
16	Spain	Japan	West Germany
15	Japan	Netherlands	UK, Italy
14	UK	Japan	Spain
13	Japan	UK	Portugal, Spain
12	Japan	Ireland	West Germany
11	Spain	Japan	

industry, which was once at a global top class level, is currently deteriorating and it is now vital to quickly find solutions to the problems associated with skill transfer within the industry.

10.4.3　Prior Research

Similar research has been conducted on visualizing the knowledge possessed by experienced workers, such as Development of Knowledge Visualization System for Knowledge Transfer (Watanuki and Tanaka, 2003), which visualized knowledge related to metal casting. With the aim of supporting the effective and efficient acquisition of knowledge for inexperienced workers, Watanuki and Tanaka developed a system to systematize metal casting-related knowledge. First, the work was broken down into processes such as planning, mold making, and mold filling. These general work processes were then further broken down into more detailed processes to produce an overview of the tools, terminology, and techniques related to each process.

Furthermore, similar research has also been conducted on visualizing skills, such as Detailed Design and Improvement of a Manual Assembly Task Incorporating How to Efficiently Handle Uncertainties (Murakami and Mizuyama, 2007). This research suggests that efficiency during assembly work relies on skillfully handling uncertainties. Uncertainties in the contact state during work processes are expressed as an aggregate (called a conceptual state) of more than one contact state possessing a degree of probability, not necessarily limited to those perceived by the worker, and the work undertaken by workers is classified into several work elements. Thus, assembly work is expressed as a series of work elements whereby the initial conceptual state is caused to transition into the target conceptual state. Murakami and Mizuyama therefore propose a method of designing assembly work processes that focus on the transition between conceptual states, while taking into account the lack of clarity concerning the conceptual states that relate to the contact states and applying the distance from the target conceptual state, providing a yardstick for deciding which conceptual states a product should pass through.

Research concerning the comparison of experienced workers and inexperienced workers has also been conducted by Miyamoto, "Comparison of Body Movement and Muscle Activity Patterns During Sharpening a Kitchen Knife Between Skilled and Unskilled Subjects" (Miyamoto et al., 2006). This research suggests that knife sharpening, which is fundamentally a grinding process, involves grinding work and decision making that rely heavily on implicit factors such as experience and intuition developed over many years. The research clarifies the distinctive movement of the upper limbs and the muscle activity of the key muscle groups during grinding work performed by experienced workers and inexperienced workers, and studies the corresponding relationships between the movements and muscle activity to determine the stipulated factors affecting the finished state of the blade.

The above clearly demonstrates the fact that there is little research concerning the comparison of experienced workers and inexperienced workers engaged in machining work and that more research in this field is urgently required in order to prevent the current state of the manufacturing industry in Japan from ultimately falling into decline. Thus, based on the example of the application of statistics provided by Amasaka (2004a, 2005, 2007d) and Sakai and Amasaka (2013, 2014), the author has employed statistical science to visualize the unconscious decision-making processes applied by experienced workers in order to enable the experienced workers to train inexperienced workers effectively and efficiently.

10.4.4 Visualizing the Skills of Experienced Workers

10.4.4.1 Comparing the Working Efficiency of Workers

When comparing experienced workers and inexperienced workers, the first thing to compare is working efficiency. Lathe work involves processes such as rough cutting, semifinishing, making adjustments, and finishing. For the comparison, videos were taken of experienced workers and inexperienced workers while working, and the time taken (seconds) for each task was noted. This helped to clarify the differences in working efficiency. Observing the processes separately, the initial setup tasks and actual work tasks were represented with different colors and the work times were noted in seconds, as shown in Table 10.6.

Next, the total times for initial setup and actual work were compared. It was found that, while there was little difference in the amount of time taken by experienced workers and inexperienced workers for the actual work tasks due to the use

Table 10.6 Work Tasks and Work Times for Experienced Workers and Inexperienced Workers

Experienced			Inexperienced		
Work Tasks	Time (s)	Work Times (s)	Work Tasks	Times (s)	Work Times (s)
Scaling	0–3	4	Scaling	0–5	5
Peel off black skin	3–22	20	Peel off black skin	5–24	20
Micrometer	22–27	6	Escape	24–25	2
Escape	27–28	2	Micrometer	28–36	9
Scaling	29–41	13	Scaling	37–48	12
Rough cutting	42–75	34	Rough cutting	49–92	44
Escape	75–76	2	Escape	92–94	3
Scaling	76–78	3	Scaling	94–98	5
Rough cutting	78–111	33	Rough cutting	99–133	35
Escape	111–113	3	Escape	133–136	4
Scaling	113–115	3	Scaling	136–140	5
Rough cutting	115–149	35	Rough cutting	140–175	36
Escape	149–153	4	Escape	175–179	5
Scaling	153–155	3	Scaling	180–188	9
Rough cutting	155–181	26	Rough cutting	188–216	29
Escape	181–183	3	Escape	216–217	2
Scaling	183–185	3	Scaling	217–225	9
Rough cutting	185–210	26	Rough cutting	225–254	30
Escape	211–212	2	Escape	254–255	2
Scaling	213–215	3	Scaling	255–259	5
Rough cutting	216–244	29	Rough cutting	259–285	27
Escape	244–247	4	Escape	285–292	8
Scaling	247–249	3	Scaling	292–300	9

(Continued)

Table 10.6 (Continued) Work Tasks and Work Times for Experienced Workers and Inexperienced Workers

Experienced			Inexperienced		
Work Tasks	Time (s)	Work Times (s)	Work Tasks	Times (s)	Work Times (s)
Rough cutting	250–263	14	Rough cutting	300–316	17
Escape	263–263	1	Escape	317–317	1
Scaling	264–267	4	Scaling	318–322	5
Rough cutting	268–283	16	Rough cutting	322–339	18
Escape	283–286	4	Escape	339–345	7

Table 10.7 Total Times for Initial Setup and Actual Work

	Experienced	Inexperienced
Setup (s)	622	1061
Work (s)	936	962

of automated machines, there was a significant difference for the initial setup tasks, as shown in Table 10.7.

Because the machining conditions differ depending on the work piece, it takes longer for inexperienced workers to make decisions due to their lack of experience and knowledge, which impacts on their working efficiency. The author investigated what forms the basis of efficient and effective working practices for experienced workers and found that their work practices were based on logical cognition, and therefore decisions were influenced more by logic than by physical movements while working, as shown in Figure 10.31.

10.4.4.2 Comparing Experienced and Inexperienced Workers' Cognition Using Electroencephalography

Electroencephalography was used to clarify differences in the cognition of experienced workers and inexperienced workers. The respective workers' beta waves (waves that occur when the brain is particularly active) were monitored while they were performing lathe work, and an X-Rs control chart was used to identify the response points and determine the data that deviate from the control limit lines. The analysis focused on the center line (CL), upper control limit (UCL), and lower control limit (LCL) for the left brain, which performs logical cognition, and

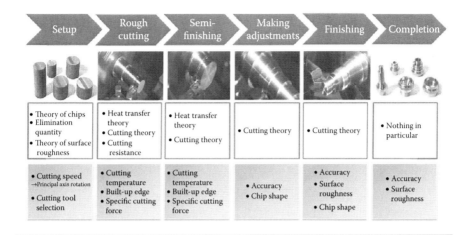

Figure 10.31 Points considered by experienced workers during lathe work.

Table 10.8 Left Brain and Right Brain Control Lines

	Center Line (CL)	Upper Control Limit (UCL)	Lower Control Limit (LCL)
Left brain of experienced	27.6836	31.8045	23.5627
Right brain of experienced	27.3690	30.7490	23.9889
Left brain of inexperienced	26.6430	30.1310	23.1550
Right brain of inexperienced	26.5407	29.5164	23.5650

the right brain, which performs sensible cognition, as shown in Table 10.8 and Figure 10.32.

Points where the left brain and right brain cognition of experienced workers and inexperienced workers deviate from the control limit lines were identified and compared. There are three main types of work: cutting during lathe work, checking measurements using a micrometer, and checking the finish of a completed product.

The results of the comparison reveal that a strong left brain response is common to both experienced workers and inexperienced workers, as shown in Tables 10.9 and 10.10. Therefore, it is clear that both types of workers frequently use logical cognition while working. Much right brain activity is found in experienced workers, while almost no right brain activity is found in inexperienced workers. It was determined from these results that, due to their lack of experience, inexperienced workers concentrate only on knowledge and techniques they have been taught. Thus, it can be seen that the application of logical cognition as well as differences in logical content have a significant influence on completed products in machining work.

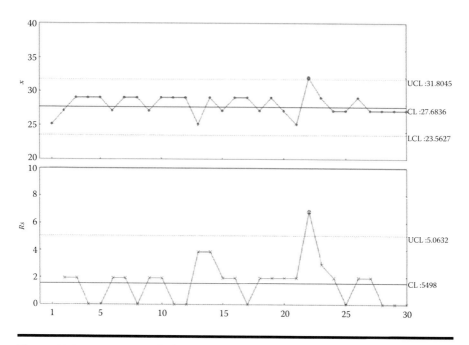

Figure 10.32 X-Rs control chart.

Table 10.9 Correspondence Table for Inexperienced Workers' Left/Right Brain Response Points and Work Tasks

Seconds	Work	Left Brain	Right Brain
3	Cutting	○	–
6	Cutting	○	–
20	Cutting	○	–
111	Micrometer	–	○
152	Micrometer	○	–
238	Cutting	○	–
250	Cutting	○	–
272	Cutting	○	○
310	Micrometer	○	–
331	Checking	○	–
359	Checking	○	–

Table 10.10 Correspondence Table for Experienced Workers' Left/Right Brain Response Points and Work Tasks

Seconds	Work	Left Brain	Right Brain
1	Cutting	–	o
4	Cutting	–	o
22	Cutting	o	–
119	Micrometer	o	o
126	Micrometer	–	o
168	Micrometer	o	–
240	Cutting	o	–
254	Cutting	o	o
256	Cutting	o	–
288	Cutting	o	o
312	Micrometer	o	–
313	Micrometer	o	o
329	Checking	o	–
338	Checking	o	–
360	Checking	o	–
363	Checking	o	o

10.4.4.3 Identifying Decision-Making Criteria of Experienced Workers Using Statistical Analysis

Workers perform the initial setup work, then begin the actual machining work, and finally perform the finishing work to complete the product. There are two types of decisions that experienced workers are required to make during this work flow: decisions that arise during the initial setup stage before the main work commences and decisions that arise after the main work commences.

During the initial setup stage, workers must visually and tactually check the work piece and its characteristics in order to make decisions concerning the most suitable cutting speed and cutting tool for the work piece and also use the visual and other senses to determine whether the cutting work is satisfactory. Thus, statistical analysis was used to clarify the criteria that are applied by experienced workers when making such decisions.

It was confirmed by Dr. Hosoda, a professor at the Institute of Technologists, that experienced workers determine the suitable cutting speed based on the physical and mechanical characteristics of the work piece. The physical and mechanical characteristics of a work piece are its tensile strength, elongation, hardness, thermal conductivity, density, Young's modulus, shear elasticity modulus, and yield strength, and the relative influence of these characteristics is clarified in accordance with the rough cutting conditions, as shown in Table 10.11.

Multiple regression analysis was conducted in order to determine the relationship between the optimum cutting speed and the physical and mechanical factors, as shown in Table 10.12. Values were selected for the variables using the stepwise method, and the partial regression coefficients resulting from the analysis revealed that the shear elasticity modulus, hardness, elongation, and Young's modulus have a significant influence on the rough machining cutting speed (in the order of hardness > shear elasticity modulus > elongation > Young's modulus).

In order to conduct the same multiple regression analysis with respect to cutting work for finishing, which is performed after the rough machining process, the rough machining cutting speed was added as an explanatory variable, as shown in Table 10.13.

Values were again selected for the variables using the stepwise method, and the partial regression coefficients resulting from the analysis revealed that the rough machining cutting speed and elongation had a significant influence (rough machining cutting speed > elongation), as shown in Table 10.14.

Next, the authors investigated the criteria for selecting the cutting tool. The physical and mechanical characteristics were set as the explanatory variables, while the cutting tool materials used during rough machining and finishing were set as the objective variables. Then, four kinds of workpieces (Group 1, aluminum [A5056]; Group 2, industrial pure iron; Group 3, magnesium alloy (HB1), and Group 4, industrial pure titanium) with differing physical and mechanical characteristics were selected. Discriminant analysis of two groups was conducted as follows: Groups 1 vs. 2, Groups 1 vs. 3, Groups 1 vs. 4, Groups 2 vs. 3, Groups 2 vs. 4, and Groups 3 vs. 4. When the discriminant coefficients were converted to absolute values and averages were calculated, the results showed that the influence occurred in the order of density > tensile strength > elongation, as shown in Table 10.15.

When the same analysis was conducted concerning selection of the cutting tool for finishing, the results showed that the influence occurred in the order of thermal conductivity > density > elongation.

Next, the important decision-making factors applied by experienced workers when making decisions during work processes were clarified. It was found that, in order to ensure satisfactory cutting, experienced workers change variable factors such as the tool type and cutting speed in accordance with decision-making factors such as the chips, lathe vibration, and air temperature. For this research, chip color and chip shape—which are the keys to making appropriate decisions—are used as decision-making factors, while the tool type, cutting speed, feed rate, cutting depth, presence/absence of cutting oil, escape angle, and shear angle are the variable factors.

Furthermore, the analysis conducted for this research uses a material called S45C as an example. This material was selected because, as a kind of carbon steel, S45C is the most commonly used material and also the material used for the World

Table 10.11 Work Pieces and Cutting Speeds

Work Pieces	Tensile Strength (N/mm²)	Elongation (%)	Hardness (HV)	Thermal Conductivity (w/m/°C)	Density (kg/m³)	Young's Modulus (GPa)	Shear Elasticity Modulus (G)	Yield Strength (MPa)	Rough Cutting speeds (m/min)
A5056	299	40	95	238	2.6	700	26	195	135
C1100	245	14	50	400	8.9	130	48	200	220
SUS304	520	55	200	16	8	197	74	205	200
SS400	400	20	125	80	7.9	206	79	240	250
Ti–6Al-4V	980	14	318	22	4.4	106	41	920	60
Industrial pure iron	196	60	100	67	7.9	205	81	98	150
SUS410	540	25	230	28	7.8	200	75.8	345	200
SUS430	450	22	90	25.6	7.8	200	76.9	205	300
S45C	828	22	170	44	7.8	205	82	727	250
Nickel	335	28	638	90	8.9	204	81	58	60
Magnesium alloy (MB1)	230	6	52	30.8	1.8	40	17	140	150
Industrial pure titanium (C.P.Ti)	320	27	200	21.9	4.6	106	45	170	80

Table 10.12 Multiple Regression Analysis Results for Rough Machining Cutting Speed Objective Variable

	Objective Variables	Residual Sum of Squares	Multiple Correlation Coefficient	Contribution Ratio R^2	Ajusted Contribution Ratio R^{*2}	
	Cutting speeds	8388.941	0.937	0.877	0.807	
No.	Explanatory variables	Residual sum of squares	Variation	Variance ratio	Partial regression coefficient	Standard partial regression coefficient
1	Constant term	21056.002	12667.06	10.5698	108.931	
2	Tensile strength	8385.841	−3.099	0.0022	−	
3	Elongation	18375.675	9986.735	8.3333	−2.313	−0.478
4	Hardness	49807.393	41418.45	34.5609	−0.401	−0.825
5	Thermal conductivity	8388.933	−0.008	0	−	
6	Density	7970.62	−418.321	0.3149	+	
7	Young's modulus	11366.074	2977.133	2.4842	0.116	0.243
8	Shear elasticity modulus	45024.054	36635.11	30.5695	2.94	0.885
9	Yield strength	7957.383	−431.557	0.3254	−	

Table 10.13 Rough Machining/Finishing Cutting Speeds for Each Work Piece

Work Pieces	Tensile Strength (N/mm²)	Tensile Strength (%)	Hardness (HV)	Thermal Conductivity (w/m/°C)	Density (kg/m³)	Young's Modulus (GPa)	Shear Elasticity Modulus (G)	Yield Strength (MPa)	Cutting Speeds (m/min)	Finishing Cutting Speeds (m/min)
A5056	299	40	95	238	2.6	700	26	195	135	300
C1100	245	14	50	400	8.9	130	48	200	220	300
SUS304	520	55	200	16	8	197	74	205	200	500
SS400	400	20	125	80	7.9	206	79	240	250	300
Ti-6Al-4V	980	14	318	22	4.4	106	41	920	60	100
Industrial pure iron	196	60	100	67	7.9	205	81	98	150	300
SUS410	540	25	230	28	7.8	200	75.8	345	200	500
SUS430	450	22	90	25.6	7.8	200	76.9	205	300	500
S45C	828	22	170	44	7.8	205	82	727	250	500
Nickel	335	28	638	90	8.9	204	81	58	60	200
Magnesium alloy (MB1)	230	6	52	30.8	1.8	40	17	140	150	300
Industrial pure titanium (C.P.Ti)	320	27	200	21.9	4.6	106	45	170	80	150

Table 10.14 Multiple Regression Analysis Results for Finishing Cutting Speed Objective Variable

	Objective Variables	Residual Sum of Squares	Multiple Correlation Coefficient	Contribution Ratio R^2	Ajusted Contribution Ratio R^{*2}	
	Finishing cutting speeds	60238.617	0.854	0.729	0.669	
No.	Explanatory variables	Residual sum of squares	Variation	Variance ratio	Partial regression coefficient	Standard partial regression coefficient
1	Constant Term	60343.498	104.881	0.0157	9.277	
2	Tensile strength	55432.865	−4805.75	0.6936	+	
3	Elongation	75709.123	15470.51	2.3114	2.307	0.264
4	Hardness	53781.995	−6456.62	0.9604	+	
5	Thermal conductivity	52619.435	−7619.18	1.1584	−	
6	Density	60014.957	−223.66	0.0298	+	
7	Young's modulus	60236.063	−2.553	0.0003	+	
8	Shear elasticity modulus	59967.696	−270.921	0.0361	+	
9	Yield strength	59117.164	−1121.45	0.1518	+	
10	Cutting speeds	212160.113	151921.5	22.698	1.494	0.828

Table 10.15 Discriminant Coefficients for Cutting Tool Materials during Rough Cutting

		Discriminant Coefficient					
		Group 1 vs. 2	Group 1 vs. 3	Group 1 vs. 4	Group 2 vs. 3	Group 2 vs. 4	Group 3 vs. 4
	Mahalanobis distance	75.419	40.154	8.69	9.435	38.974	17.092
	Determine the error rate (%)	0.001	0.077	7.025	6.229	0.09	1.936
1	Constant term	86.422	52.761	17.174	−33.661	−69.248	−35.587
2	Tensile strength	4.32	2.94	1.24	1.37	3.08	1.71
3	Elongation	3.97	3.59	0.84	0.37	3.13	2.75
4	Hardness	0.173	0.121	0.048	−0.052	−0.125	−0.073
5	Thermal conductivity	−0.158	−0.097	−0.048	0.061	0.11	0.049
6	Density	−4.558	−3.592	−1.035	0.966	3.524	2.558
7	Young's modulus	0.057	0.034	0.022	−0.023	−0.034	−0.012
8	Shear elasticity modulus	0.031	0.025	0.003	−0.005	−0.027	−0.022
9	Yield strength	0.304	0.203	0.086	−0.101	−0.219	−0.117

Note: Group 1 is aluminum (A5056); Group 2 is industrial pure iron; Group 3 is magnesium alloy (MBI), and Group 4 is industrial pure titanium (CPTi).

Skills Competition. The authors clarified the important criteria used by experienced workers for making decisions based on the chip color and chip shape when cutting S45C.

The color of chips varies depending on the cutting temperature, and the shape of the chips also varies significantly. Covariance structure analysis was used to clarify what kind of influence these elements receive from each factor. This was done separately with respect to cutting work for the rough machining process and cutting work for the finishing process. The results of the covariance structure analysis are as follows. The chip color during rough machining, as shown in Figure 10.33, where surface roughness and subsurface damage are not a consideration, is strongly influenced by the cutting speed.

Due to the emphasis on efficiency, the cutting speed is the only factor to have a significant influence. The chip shape during rough machining as shown in Figure 10.34 is influenced by the cutting speed and the escape angle. This is because chips can become wedged between the work piece and the lathe jaws during cutting. This may cause the chips to melt due to the heat and adhere to the tool blade, resulting in the formation of a built-up edge. Significant factors are an escape angle that allows chips to flow away easily and a cutting speed that causes the chips to be swept away quickly. The chip color during finishing as shown in Figure 10.35 is strongly influenced by the cutting speed and presence/absence of cutting oil.

This is because damaged layers can occur during finishing due to the heat generated as the cutting speed is changed. The chip shape during finishing as shown in Figure 10.35 is strongly influenced by the escape angle and the feed angle. This is because chips can easily become caught during the finishing process, which is required for lowering the surface roughness, and these factors must be adjusted to avoid this. Thus, the above analysis reveals the decision-making factors applied by experienced workers when selecting the cutting speed and selecting the cutting tool.

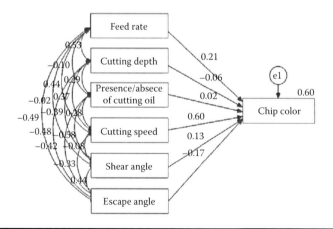

Figure 10.33 Chip color during rough machining.

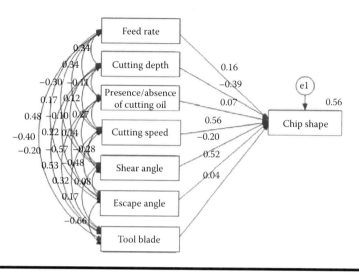

Figure 10.34 Chip shape during rough machining.

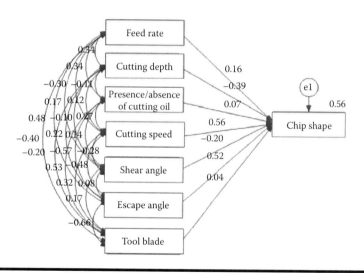

Figure 10.35 Chip color during finishing.

10.4.5 Evaluation of This Research

The decision-making factors and factors involved in improving technical skills that have been made explicit by this research were evaluated by Dr. Hosoda, a professor at the Institute of Technologists, and 10 students (with lathe experience) from the institute's Department of Manufacturing Technologies.

Opinions from an experienced worker's perspective such as the following were received: "The research clarifies points where inexperienced workers need guidance such as cutting tool selection, which is beneficial for both the person giving guidance and the person receiving guidance."; "Comparing workers' cognition using brain waves is an innovative research method, and I rate it highly."

Opinions from an inexperienced worker's perspective such as the following were received: "Logic-based research such as this makes the subject easy to understand and easy to learn about."

10.4.6 Conclusion

For this research, the authors used statistical science and EEG to conduct a comparison of experienced and inexperienced workers engaged in lathe work, which is the fundamental process involved in machining, and analyzed the skills of the experienced workers. They then conducted factorial analysis to clarify the factors involved in improving the skills of technicians. The research received positive evaluations from both experienced and inexperienced workers, and the required results were achieved.

Acknowledgments

Part of this research is supported by Dr. Yasuhiro Hosoda (a professor at the Institute of Technologists, Saitama-ken, Japan).

10.5 New Global Partnering Production Model for Expanding Overseas Strategy

10.5.1 Introduction

In recent years, leading manufacturers in Japan have been deploying a new production strategy called globally consistent levels of quality and simultaneous global launch (production at optimal locations) in order to get ahead in the worldwide quality competition, and high quality assurance in manufacturing—simultaneous achievement of QCD (Amasaka and Sakai, 2010, 2011). This is the key to successful global production and has become a prerequisite for its accomplishment employing New JIT (Amasaka, 2002).

The greatest concern of corporate managers is the success of overseas production strategy—local production as well as to bring overseas manufacturing up to Japanese standards. Therefore, in order to increase the skills of production workers at local manufacturing sites (hereafter referred to as production operators), the key to successful global production is necessary to realize manufacturing suited to the actual situation at production sites of various overseas production bases.

Against this background, Amasaka (2007b) focused on the strategic operation of the evolution of production sites—the New Japanese Global Production Model, which strategically deploys Advanced TPS, Strategic Production Management Model through New JIT strategy. Amasaka (2007c,e, 2008b) proposed New JIT and has verified its effectiveness (refer to Sections 5.1 and 8.3).

Then, Ebioka et al. (2007) proposed a New Global Partnering Production Model (NGP-PM) employing Advanced TPS, which generates a synergetic effect that organically connects and promotes continual evolution of the production plants in Japan and overseas, as well as greater cooperation among production operators. The effectiveness of this model for expanding overseas strategy will be verified at advanced car manufacturer A (the Japanese car top maker of the world extreme big enterprises).

In the context of the rapid progress of globalization, the Republic of Turkey, one of Europe's most prominent newly emerging economies, is expected to achieve significant growth, centering on its manufacturing industries. Given its strong international competitiveness in terms of labor and raw materials, the country's automobile industry has particular potential for growth.

As an opening move in Turkey's new global production strategy, therefore, the authors propose a New Turkish Production System (NTPS) employing strategic New JIT for the growth of the next-generation automobile industry in the Republic of Turkey (Mustafa, 2009, Yeap et al., 2010).

NTPS would represent the integration and evolution of the Advanced TPS through New JIT, which is itself an evolved model of the current leading Japanese Production System, and the Advanced Turkish Production System, which is an evolved model of the Traditional Turkish Production System cultivated to date.

10.5.2 Problems with Achieving Global Production

Recently, leading companies have aimed to succeed in localizing production as a global production strategy; the key to this is success in global production. However, it has been observed that, despite the fact that overseas plants have the relevant production systems, facilities, and materials equivalent to those that have made Japan the world leader in manufacturing, the building up of quality—assurance of process capability (Cp) has not reached a sufficient level due to the lack of skills of the production operators at the manufacturing sites. In this context, many studies are being carried out abroad on globalization (Lagrosen, 2004; Ljungström, 2005) and TQM (Burke et al., 2005; Hoogervorst et al., 2005).

As a countermeasure to this problem, and in order not to lag behind the evolution of digital engineering—the transition to advanced production systems at production plants, Japanese manufacturers expect the production plants in Japan to serve as *mother plants*. They would welcome overseas production operators to these plants and promote a local production program—transplanting the know-how of Japanese manufacturing (Amasaka, 2002, 2007c,e; Amasaka and Sakai, 2010).

However, it is by no means easy to transfer the know-how of Japanese manufacturing directly to overseas production bases as mentioned above. In other words, there is always an obstacle to overcome—a suitable production system for each production base, due to the difference in abilities (level of skill and education) or national characteristics between the local production site and Japan.

Therefore, to cope with this situation, an environment in which the creation of labor values—employee satisfaction (ES), advanced skills, a sense of achievement, and self development—can be realized must be urgently considered (Amasaka, 2007f, 2008b; Yamaji et al., 2007). In order to accomplish this, the authors surmise that it is necessary to develop a type of manufacturing that fits the local circumstances of various overseas production bases and to advance from Japanese mother plants to global mother plants.

10.5.3 Strategic Use of New JIT: the Key to Global Production

10.5.3.1 Current Situation of Quality Management in Japanese Manufacturing

Taking a careful look at the recent repeated cases of market recall in the manufacturing sector, it can be deduced that it is not only an issue of quality of products, but also a severe challenge to the reliability of products—the mode of quality management of Japanese manufacturing. The authors surmise that such a situation points to the brittle fatigue of the function of QA—reliability of products, which once was a strong point of Japanese manufacturing.

Also, it suggests a possible idling of the function of quality management—reliability of human resources (Amasaka, 2008). In this context, there are many studies abroad on globalization (Lagrosen, 2004; Ljungström, 2005) and TQM (Hoogervorst et al, 2005; Burke et al, 2005; Kakkar and Narag, 2007; Prajogn, 2006). Basically, the QA division and TQM promotion division work in harmony as a cornerstone of quality management, which manages the quality of each of the related divisions so as to keep the business cycles for quality management circulating normally.

However, since the burst of the bubble economy, as if epitomizing the advantage of a Japanese production system, (i) super mass production—automation of production and automation of inspection—has been promoted. (ii) As quality defects were reduced in the production process, it was thought that, relatively speaking, (iii) quality assurance and inspection do not produce added value, and therefore, (iv) the organization/operation of the QA division or inspection division was scaled down or mere routine work was introduced therein, and (v) activities for creating new functions for QA or TQM have stagnated.

For the past 10 years or more, efforts for promoting customer-first manufacturing—high quality assurance—that enhances the customer value as

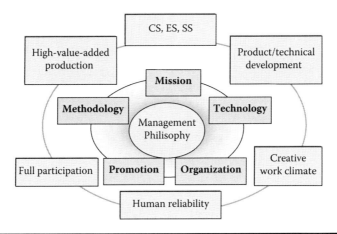

Figure 10.36 Basic principle of quality management.

well as human resource development—high reliability of total business process—have been neglected and therefore are lagging behind (Goto, 1999; Amasaka and Sakai, 2010). Though overseas local production has been advancing recently along with a rapid expansion of global production, the challenge of building up quality that can match that of Japanese manufacturing has not been overcome.

QA-related divisions are busy cleaning up the mess of quality defect cases, and TQM promotion divisions has not been functioning well to cultivate human resources with fighting potential, putting too much emphasis on TQM education for TQM promotion (Evans and Lindsay, 1995).

Having said this, it is deemed necessary to now go back to the basic principle of quality management, as shown in Figure 10.36, and to reform the QA function, as well as the TQM promotion function, which advances the former (Evans and Lindsay, 1995; Yamaji and Amasaka, 2006).

This basic principle involves, as a management philosophy, establishment of mission, methodology, promotion, organization, and technology and realization of customer satisfaction (CS), employee satisfaction (ES), social satisfaction (SS), high-added-value products, product development/technology development, creative corporate environment, human reliability, and full participation. It is vital to strategically deploy this basic principle.

10.5.3.2 Proposal of New Japanese Quality Management Model Employing New JIT

In order to realize the key to global production, globally consistent levels of quality and production at optimal locations, it is absolutely essential that manufacturing via global partnering is precisely developed to suit actual conditions and skill levels at the production sites of the various overseas production bases. In working to resolve these urgent issues, the authors constructed New JIT, New Management Technology Principle (refer to Section 5.1).

Considering the importance of scientific quality management, and on the basis of New JIT, which has verified the effectiveness of the stratified task team in TQM promotion, as well as the universality of the basic principle of quality management presented in the previous chapter, the author (Yamaji et al., 2007, 2008; Amasaka, 2008b) proposes, as indicated in Figure 10.37, a next-generation-type, advanced model of quality management: the New Japanese Quality Management Model (Yamaji et al.,2006; Amasaka, 2007).

The functions of the (i) QA and (ii) TQM promotion as corporate enviroment factors for suceeding in global production are (1) CS,ES,SS, (2) High Quality assurance, (3) simultaneously achieve QCD (4) success in global partnering, and (5) evolution of quality management.

More specifically, the (i) QA division needs to promote highly reliable manufacturing, and cooperative activity across the organization is indispensable to achieve that. In the (ii) TQM promotion division, it is important to cultivate human resources that have even higher skills, knowledge, and creativity. Therefore, the value of human resources must continually be upgraded and the cultivation of intelligent human resources must be promoted in an effort to

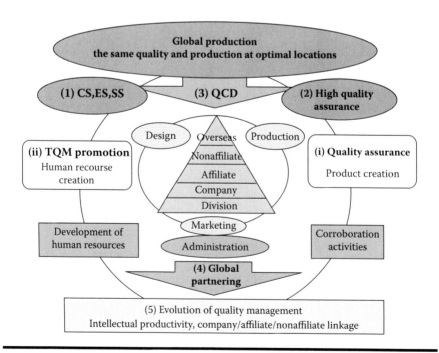

Figure 10.37 Outline of New Japanese Quality Management Model.

improve the productivity of white-collar workers (Yamaji and Amasaka, 2006, 2007; Amasaka, 2009).

In order to realize the above, global partnering that enables strategic cooperation among divisions, such as design, production, marketing, and administration, as well as the entire company, affiliated companies, nonaffiliated companies, and overseas corporations, must be achieved. Improvement of intellectual productivity simultaneously achieves QCD utilizing this model.

10.5.4 Proposal of New Global Partnering Production Model

10.5.4.1 Significance of Global Partnering Research

Sakai and Amasaka (2005) and Ebioka et al. (2007) have defined global partnering as knowledge sharing in order to promote continual evolution of production plants in Japan and overseas, as well as greater cooperation among production operators. In other words, global partnering means that Japanese companies no longer stick to a one-sided promotion of Japanese concepts and systems in overseas plants as they have in the past, but that Japanese and overseas plants exchange opinions, accept each other, and share knowledge in order to continue developing the best-suited manufacturing system for each environment.

Therefore, for expedient achievement of globally consistent levels of quality and simultaneous global launch (production at optimal locations), the authors consider global partnering through cooperation from fresh standpoints to be the key to achieving worldwide quality competitiveness—the simultaneous achievement of QCD (Amasaka, 2007b, 2008; Amasaka and Sakai, 2011).

10.5.4.2 NGP-PM: A Proposal for Bringing Overseas Manufacturing up to Japanese Standards

For advanced Japanese companies in the process of developing global production, the most important issue is how to bring manufacturing up to Japanese standards at overseas plants. Figure 10.38 shows a new idea, NGP-PM, which generates a synergetic effect that organically connects and promotes continual evolution of the production plants in Japan and overseas, as well as greater cooperation among production operators (Ebioka et al., 2007).

The mission of NGP-PM is the simultaneous achievement of QCD in order to realize high quality assurance. The essential strategic policies include the following three items: (1) the establishment of a foundation for global production, realization of global mother plants—advancement of Japanese production sites; (2) achieving the independence of local production sites through the incorporation of the unique characteristics (production systems, facilities, and materials) of both developing countries (Asia) and industrialized countries (US, Europe); and (3) the necessity of structuring

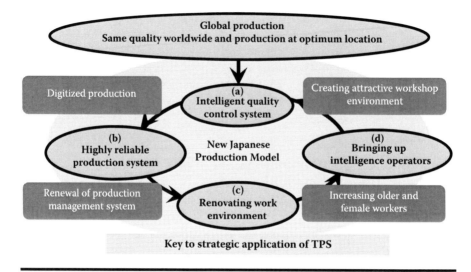

Figure 10.38 New Japanese production model (NJPM).

of a Global Network System for Developing Production Operators (GNS-DPO) to promote knowledge sharing among the production operators in Japan and overseas as well as for the promotion of higher skills and enhanced intelligence.

In order to realize this, with NGP-PM as the foundation, it is essential to create a spiraling increase in the four core elements by increasing their comprehensiveness and high cyclization. Specifically, in realizing global mother plants if Japanese and overseas manufacturing sites are to share knowledge from their respective viewpoints, the core elements must be advanced. To achieve this, a necessary measure is to design separate approaches suited to developing and industrialized countries.

First, in developing countries, the most important issue is increasing the autonomy of local manufacturing sites. At these sites, training for highly skilled operators (focus on manual laborers) that is suited to the manual labor–based manufacturing sites is the key to simultaneous achievement of QCD.

Similarly, in industrialized countries, where manufacturing sites are based on automatization and increasingly high-precision equipment, training of intelligence operators resulting in realizing highly reliable production control systems and ensuring high efficiency is the key to simultaneous achievement of QCD (Sakai and Amasaka, 2006).

Furthermore, production operators trained at global mother plants can cooperate with operators at overseas production bases and, in order to generate synergistic results, can work to localize global mother plants in a way that is suited to the overseas production bases. GNS-DPO can then be effectively utilized to ensure that this contribution continues indefinitely.

GNS-DPO is critical in ensuring the smooth exchange of essential information in order to realize overseas manufacturing at Japan standards. This essential information includes quality control information for each production base as well as facility planning information, *kaizen* information, and information on the level of skill of human resources.

10.5.5 Application: Proposal of the New Turkish Production System (NTPS)

10.5.5.1 Current Status of Japan's Leading Automobile Industry

10.5.5.1.1 New Japanese Production Model Employing Advanced TPS: A Model for Production Site Evolution

Recent years have seen significant changes in manufacturing, including the evolution of digital engineering, the shortening of product lead time, and simultaneous global start-up. In order to prevent a lag in the transformation of production sites, the most critical issue has become creating superior products based on the customer-first policy of QCD.

As mentioned above, in order to achieve this, a model for utilizing Advanced TPS in order to realize New JIT strategy is necessary, as shown in Figure 8.17 (refer to Section 8.4). To this end Amasaka (2007e) and Ebioka et al. (2007) propose the New Japanese Production Model (NJPM), a model for production site evolution, as shown in Figure 10.39.

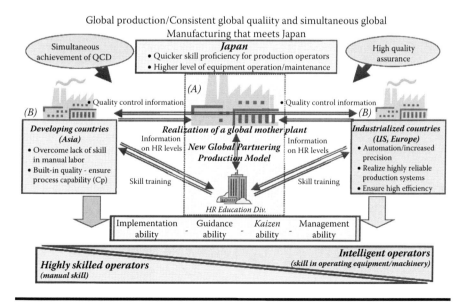

Figure 10.39 New Global Partnering Production Model (NGP-PM).

Current issues related to specific development items include increasingly digitized production, reform of production control systems, and the creation of attractive workshop environments tailored to the increasing numbers of older and female workers. Some ways to address the above issues are (a) intelligence quality control systems, (b) a highly reliable production system, (c) renovation of the work environment, and (d) education for intelligent operators.

The authors therefore consider NJPM for evolution of manufacturing sites to be a critical foundation for bringing overseas manufacturing up to Japanese standards.

10.5.5.2 Current Status of Turkey's Automobile Industry

Turkey currently ranks 16th in the world for production of automobiles, one of the highest rankings among the newly industrialized countries. The strategic advantages of its location in terms of production and distribution have enabled the Turkish automobile industry to establish its own unique production system that reflects local industries, known as the Traditional Turkish Production System, while simultaneously achieving steady growth.

In recent years, in an opening move in the global production strategy of the Turkish automobile industry, leading overseas companies have become increasingly active in local production. For example, since 1994, Toyota Motor Manufacturing Turkey (TMMT) has been conducting training in "the Toyota Way" (Jeffrey, 2003) at Toyota in Japan and has introduced the traditional Toyota Production System called JIT (refer to Chapters 3 and 4).

In the background to this is an undercurrent that Turkey has an environment that is amenable to the introduction of a Japanese-style production system, given the similarities in character between the Turkish and Japanese people and their belief that success will come if one works hard.

10.5.5.3 Proposal of the New Turkish Production System

Having investigated the actual situation in Turkey, the authors propose a NTPS with the objective of the integration and evolution of traditional Turkish production systems and the aforementioned Japanese production system.

10.5.5.3.1 Concept of NTPS

Based on the research of Ramarapu et al. (1995) and Amasaka et al. (Amasaka, 2004c, 2007c, 2007e,f, 2008b,c, 2009; Ebioka et al., 2007; Yamaji et al., 2007, 2008; Amasaka et al., 2008), the author proposes a concept for NTPS, shown in Figure 10.40 (Mustafa, 2009).

The left side of the diagram shows the hierarchy of the Toyota Production System (TPS), which forms the foundation of Japanese-style production systems, and its evolved model, Advanced TPS. Similarly, the right side of the diagram shows

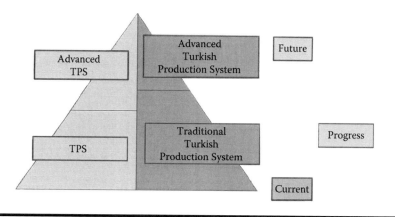

Figure 10.40 NTPS concept.

the Traditional Turkish Production System and its evolved model, the Advanced Turkish Production system.

In order to integrate and evolve the concepts of these two production systems, the common factors (elements) among the various production elements, such as JIT and the Lean system, in Turkey and Japan, as well as those factors (elements) that are unique to Turkey, must be taken into consideration.

10.5.5.3.2 Extraction of Important Keywords for NTPS

In order to extract the common factors (elements) and the Turkey-specific unique factors (elements) required for the NTPS, the authors conducted field surveys of local manufacturers in Japan (Toyota Motor Corporation, Honda, Denso, Central Motors, Nihon Spring, and others) and Turkey (Tofas, Oyak-Renault, Ford Otosan, Toyota Turkey, and Denso Turkey) (Mustafa, 2009).

In these surveys, approximately 500 pieces of language data were obtained from interviews, the findings of manufacturing plant tours, on-site case studies, and reference literature. An affinity diagram of the data thus obtained, based on 5M-E (man, machine, material, manufacturing, measuring, environment), was used to investigate the relationships between the data, based on empirical technologies.

As shown in Figure 10.41, the authors were able to extract as important keywords for the NTPS ten common factors, four unique factors specific to Turkey, and two unique factors specific to Japan.

- i. The factors common to Turkey and Japan are (1) QA, (2) SQC education, (3) QCD activities, (4) *kaizen* activities, (5) creative proposal programs, (6) improvements through environmental regulations, (7) automatic management methods, (8) global partnering, (9) safety, and (10) definition, awareness, and elimination of *muda* (waste).

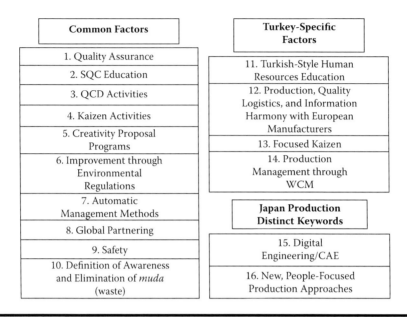

Figure 10.41 Important NTPS keywords obtained from affinity diagram.

ii. The Turkey-specific factors are (11) Turkish-style human resources education, (12) production, quality, logistics, and informational harmony with European manufacturers, (13) focused *kaizen*, and (14) production management through world-class manufacturing (WCM).
iii. The Japan-specific factors are (15) digital engineering/CAE and (16) new, people-focused production approaches.

10.5.5.3.3 Text Mining Analysis

The authors further used text mining analysis to explore in more detail the relationships between the language data obtained in Section 8.2 (Refer to Figure 8.4 and Figure 8.6).

Examples of the results of analysis of the 5M-E keywords for the NTPS are explained below. First, the relationships between TPS and Advanced TPS (necessary keywords that each should possess) are shown clearly in Figure 10.42. Further, in a similar manner, the relationships between the Traditional Turkish Production System and the Advanced Turkish Production System are shown clearly in Figure 10.43.

Figure 10.42 shows that (1) TPS, JIT, workplace environment, Science SQC, and standard work orientation have a high degree of relationship, and that (2) the key elements (factors) that should be present in Advanced TPS include worker education, assurance of high quality, DE, CG, SQC, global production, virtual plant, simulation, and partnering.

New Manufacturing Management Employing New JIT ■ 433

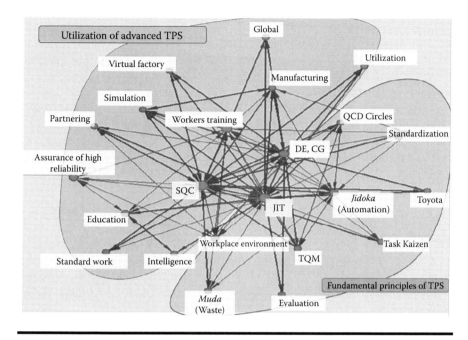

Figure 10.42 Relationships between TPS and advanced TPS.

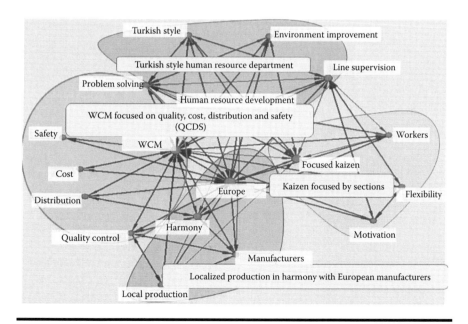

Figure 10.43 Connections of Traditional Turkish Production System and Advanced Turkish Production System.

Similarly, Figure 10.43 shows that (1) in the Traditional Turkish Production System, (1-1) safety, cost, and quality essential to global production and (1-2) line supervision, environmental improvement, and problem-solving methods essential to human resources development and utilization are closely related. The key elements (factors) that should be present in (2) the Advanced Turkish Production System are (2-1) motivated and flexible workers and line, and *kaizen* activities are closely related to focused *kaizen* by sections and (2-2) reinforcement of the influence and collaboration between Turkish and European manufacturers, which is essential to localized production that is in harmony with European manufacturers.

10.5.5.3.4 Outline of New Turkish Production System

Based on the results of analysis of the findings obtained as described, the author has proposed a NTPS, as shown in Figure 10.44. (Mustafa, 2009; Yeap et al., 2010) The NTPS represents the integration and evolution of the Advanced TPS, which is itself an evolved model of the current Toyota Production System, and the Advanced Turkish Production System, which is an evolved model of the Traditional Turkish Production System cultivated to date.

As shown in the diagram, from the perspective of 5M-E, factors that are common to Japan and Turkey, Turkey-specific factors, and Japan-specific factors have been

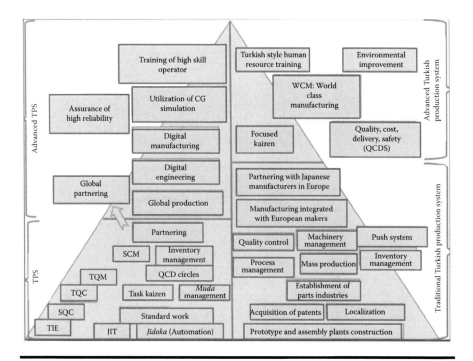

Figure 10.44 Outline of New Turkish Production System.

considered as the technological elements (factors) required for production. Not only will this enable the Turkish automobile industry to enhance its technological capabilities and increase production, but it will also lead to expectations of the production of high-quality products and the development of flexible production systems in the future.

10.5.5.3.5 Application Examples

The measures taken by a leading manufacturing company A to achieve successful global production will be discussed. Company A is taking measures through its Global Production Center (GPC) to realize global production. The GPC offers Japanese plants and its human resources education division to a global mother plant, NGP-PM.

Sakai and Amasaka (2006) propose HI-POS as the creation of a new person-centered production system to increasingly realize improvements in skills and proficiency. An important part of this system is the HIA, which allows human wisdom to be translated into increased skills and oral tradition (Sakai and Amasaka, 2008). Within this, the skill training curriculum for overseas production operators and important tools will be discussed (refer to Section 8.4).

10.5.5.3.5.1 Operator Training Process for Assembly Works — The skill training for newly hired domestic and overseas production operators, which used to take more than 2 weeks, has been shortened to just 5 days—less than half the time—and has ensured a stable mass production system. Specific content of the skill training curriculum for overseas production operators is shown in Figure 10.45. This figure shows an example of operator training processes for automobile assembly works.

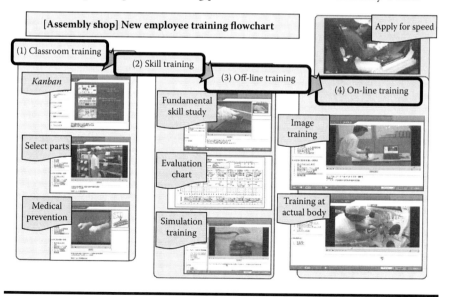

Figure 10.45   Example of operator training processes for automobile assembly works.

436 ■ *New JIT, New Management Technology Principle*

After everything from basic skills training to a competency evaluation test is completed, skills training (utilizing tools such as the visual manual and the HIA-Intelligent IT system) is conducted repeatedly until a certain standard is achieved on an evaluation sheet. Finally, actual training on the line is conducted, and achievement of a set skill level was shown to be achieved in half the time previously needed (refer to Section 8.4).

Furthermore, Figure 10.46 is an example of a hyper visual manual concerning bolt trading operations. This Figure shows the screen setup of the visual manual, a communication tool that allows for increased exchange of knowledge and information regarding production processes.

With this tool, production operators can be guided using shared content worldwide, and operators can also engage in self-study between actual skill training sessions. Through repeated image training, skill level can drastically improve at the initial stages, contributing significantly to proficiency improvement in highly skilled operators.

10.5.5.3.5.2 Bringing Up Intelligent Operators: V-IOS Construction —

As mentioned earlier, Sakai and Amasaka (2005) proposed V-MICS (refer to Section 8.4), which improves production operators' operational technology abilities in regard to integrated equipment, in preparation for global production. This system works to support improvements in production operators' skills and techniques, such as equipment availability administration, defect analysis (Amasaka and Sakai, 1996, 1998), and other intelligence-based functions.

The authors, therefore, proposed the intelligent operator educational system Virtual Intelligence Operator System (V-IOS) (Sakai and Amasaka, 2002) by utilizing a visual manual format, exemplified in Figure 10.47, shows an example of the visual manual format using V-IOS. This figure improves levels of mastered skills

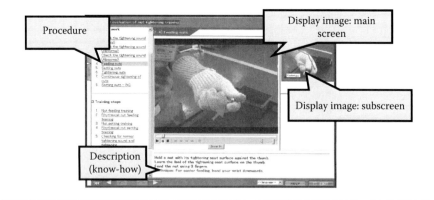

Figure 10.46 Example of hyper visual manual concerning bolt feeding operation.

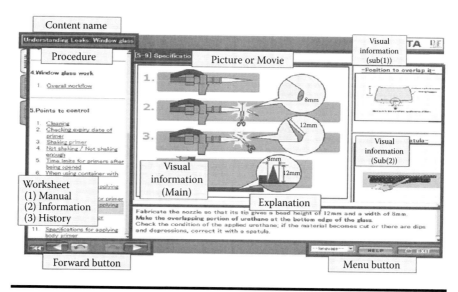

Figure 10.47 Example of the visual manual format using V-IOS.

among production operators. Manuals describing the operation procedure have so far been kept by each operator or shop station, and the know-how obtained through use is a personal asset. To make manufacturing operators strong, the equipment (manufacturing) manual itself should allow simple revision by each handling person.

The screen consists of paging block, procedure explanation block, visual information display block, and explanation display block. The manual is to be read by turning the pages using the forwarding button according to the procedure. Especially in the explanation display block, the key point representing the know-how of each operator is written and the visual information not indicated on the main screen is shown in detail in the subscreen.

The contribution V-10s makes to the evolution and dissemination of production operators' mastered skills leads to improved productivity among production operators when setting up a new overseas plant, and the given benefits have already been acknowledged. As a result of execution using the engine block line at K plant as a model line, the target of zero failures was attained after a lapse of two years since the start of the activity in fiscal 2000, thereby establishing a new autonomous production line. Deployment to production lines in Japan and abroad has been promoted with satisfactory results.

10.5.6 Conclusion

The author proposed a model and conducted verification for the strategic development of New JIT through the strategic use of NGP-PM with the aim of establishing basic principles and realizing a Japanese standard of quality in overseas manufacturing. As a concrete example of deployment, the author proposed a an NTPS for the growth of the Turkish automobile industry.

Hereafter, the author will conduct verification research into the effectiveness of the proposed NTPS. Furthermore, critics have recommended construction of the New Malaysia Production System, the New China Production System, and the New Vietnam Production System based on the results of this research (Shan et al., 2011; Shan, 2012; Miyashita and Amasaka, 2013).

References

Abegglen, J. C. 1958. *The Japanese Factory: Aspects of Its Social Organization*, The MIT Press, New York.
Alba, E. 2005. *Parallel Metaheuristics: A New Class of Algorithms*, Addison Wiley, Boston.
Amasaka, K. 2000. Partnering chains as the platform for quality management in Toyota, in *Proceedings of the 1st World Conference on Production and Operations Management*, pp. 1–13, August 30, 2000, Seville, Spain, (CD-ROM).
Amasaka, K. 2002. New JIT, a new management technology principle at Toyota, *International Journal of Production Economics*, 80: 135–144.
Amasaka, K. (ed.). 2003. *Manufacturing Fundamentals: The Application of Intelligence Control Charts—Digital Engineering for Superior Quality Assurance*, Japan Standards Association, Tokyo, Japan [in Japanese].
Amasaka, K. 2004a. Customer Science: Studying customer values, *Japan Journal of Behavior Metrics Society*, 32(1): 196–199.
Amasaka, K. 2004b. *Science SQC, New Quality Control Principle: The Quality Strategy of Toyota*, Springer-Verlag, Tokyo.
Amasaka, K. 2004c. Development of "Science TQM", a new principle of quality management: Effectiveness of strategic stratified task team at Toyota, *International Journal of Production Research*, 42(17): 3691–3706.
Amasaka, K. 2005. Constructing a customer science application system CS-CIANS—Development of a global strategic vehicle Lexus utilizing New JIT, *WSEAS Transactions on Business and Economics*, 3(2): 135–142.
Amasaka, K. (ed.). 2007a. Establishment of a needed design quality assurance: Framework for numerical simulation in automobile production, Working Group No. 4 studies in Japanese Society Quality Control, Tokyo, Japan [in Japanese].
Amasaka, K. 2007b. High linkage model "Advanced TDS, TPS & TMS": Strategic development of "New JIT" at Toyota, *International Journal of Operations and Quantitative Management*, 13(3): 101–121.
Amasaka, K. 2007c. Highly reliable CAE model, the key to strategic development of Advanced TDS, *Journal of Advanced Manufacturing Systems*, 6(2): 159–176.
Amasaka, K. 2007d. The validity of TDS-DTM: A strategic methodology of merchandise development of New JIT—Key to the excellence design LEXUS, *International Business and Economics Research Journal*, 6(11): 105–116.
Amasaka, K. 2007e. New Japan Production Model, An advanced production management principle: Key to strategic implementation of New JIT, *International Business and Economics Research Journal*, 6(7): 67–79.

Amasaka, K. 2007f. Applying New JIT—Toyota's global production strategy: Epoch-making innovation in the work environment, *Robotics and Computer-Integrated Manufacturing*, 23(3): 285–293.

Amasaka, K. 2008a. An integrated intelligence development design CAE model utilizing New JIT, *Journal of Advanced Manufacturing Systems*, 7(2): 221–241.

Amasaka, K. 2008b. Strategic QCD studies with affiliated and non-affiliated suppliers utilizing New JIT, *Encyclopedia of Networked and Virtual Organizations*, III(PU-Z): 1516–1527.

Amasaka, K. 2008c. Science TQM, a new quality management principle: The quality management strategy of Toyota, *The Journal of Management and Engineering Integration*, 1(1): 7–22.

Amasaka, K. 2009. The foundation for advancing the Toyota Production System utilizing New JIT, *Journal of Advanced Manufacturing Systems*, 8(1): 5–26.

Amasaka, K. 2010a. Proposal and effectiveness of a high quality assurance CAE analysis model, *Current Development in Theory and Applications of Computer Science, Engineering and Technology*, 2(1/2): 23–48.

Amasaka, K. 2010b. Chapter 4: Product Design, *Quality Assurance Guidebook,* New Edition. *Quality Assurance Guidebook*, Japanese Society for Quality Control, pp. 87–101, Tokyo, Japan [in Japanese].

Amasaka, K. 2012. Constructing optimal design approach model: Application on the advanced TDS, *Journal of Communication and Computer*, 9(7): 774–786.

Amasaka, K. 2013. Developing a Higher-Cycled Product Design CAE Model: The evolution of automotive product design and CAE (plenary lecture), *The 1st International Conference on Intelligent Control, Modelling and Systems Engineering*, pp. 26–38, August 6, 2013, Valencia, Spain.

Amasaka, K., Ito, T., and Nozawa, Y. 2012. A new development design CAE employment model, *Journal of Japanese Operations Management and Strategy*, 3(1): 18–37.

Amasaka, K., Kurosu, S., and Morita, M. 2008. *New Manufacturing Theory: Surpassing JIT—Evolution of JIT*, Morikita Shuppan, Tokyo [in Japanese].

Amasaka, K., Onodera, T., and Kozaki, T. 2014. Developing a Higher-Cycled Product Design CAE Model: The evolution of automotive product design and CAE, *International Journal of Technical Research and Applications*, 2(1), 1728.

Amasaka, K. and Sakai, H. 1998. Availability and Reliability Information Administration System ARIM-BL by methodology in inline-online SQC, *International Journal of Reliability, Quality and Safety Engineering*, 5(1): 55–63.

Amasaka, K. and Sakai, H. 2010. Evolution of TPS fundamentals utilizing New JIT strategy—Proposal and validity of advanced TPS at Toyota, *Journal of Advanced Manufacturing Systems*, 9(2): 85–99.

Amasaka, K. and Sakai, H. 2011. The New Japan global production model "NJ-GPM": Strategic development of Advanced TPS, *Journal of Japanese Operations Management and Strategy*, 2(1): 1–15.

Aragóna, A., Alegrea, J., and Gutiérrez-Solana, F. 2005. Effect of clamping force on the fatigue behavior of punched plates subjected to axial loading, *Engineering Failure Analysis*, 13(2): 271–281.

Aspy, P. P. and Wai, K. M. 2001. An online evaluation of the compete: Online decision entry system (CODES), *Developments in Business Simulation and Experimental Learning*, 28: 188–191.

Aspy, P. P., Wai, K. M., and Dean, S. R. 2000. Facilitating learning in the new millennium with the complete online decision entry system (CODES), *Developments in Business Simulation and Experimental Learning*, 27: 250–251.

Burke, R. J. et al. 2005. Effects of reengineering on the employee satisfaction-customer satisfaction relationship, *The TQM Magazine*, 17(4): 358–363.

Chawla, R. 2002. An automated 3D facilities planning and operations model: Generator for synthesizing generic manufacturing operations in virtual reality. *Journal of Advanced Manufacturing Systems*, 1(1): 5–17.

Ebioka, K., Sakai, H., and Amasaka, K. 2007. A New Global Partnering Production Model "NGP-PM" utilizing "Advanced TPS", *Journal of Business and Economics Research*, 5(9): 1–8.

Evans, J. R. and Lindsay, W. M. 1995. *The Management and Control of Quality*, South-Western College Publishing, Cincinnati.

Faraj, S. and Spoull, L. 2000. Coordinating expertise in software development teams, *Management Science*, 46(12): 1554–1568.

Gabor, A. 1990. *The Man Who Discovered Quality: How Deming W. E. Brought the Quality Revolution to America*, Random House, New York.

Gallivan, J. M. and Keil, M. 2003. The user-developer communication process: A critical case study, *Information Systems Journal*, 13(1): 37–68.

Goto, T. 1999. *Forgotten the Origin of Management*, Seisansei-syuppan, Tokyo [in Japanese].

Hoogervorst, J. A. P., Koopman, P. L., and van der Flier, H. 2005. Total quality management: The need for an employee-centred, coherent approach, *The TQM Magazine*, 17(1): 92–106.

Hounshell, D. 1984. *From the American System to Mass Production*, Johns Hopkins University Press, Baltimore MD.

Hunt, V. D. 1992. *Quality in America: How to Implement a Competitive Quality Program*, Technology Research Corporation, Homewood, Illinois.

Iiyoshi, T. and Hannafin, M. 2002. Cognitive tools and user-centered learning environments: Rethinking tools, functions, and applications, in P. Barker and S. Rebelsky (eds), *World Conference on Educational Multimedia, Hypermedia and Telecommunications*, vol. 1, pp. 831–836, June 26, 2002, Denver, Colorodo, USA, AACE, Chesapeake, VA.

ISO 4161/15701/15702/10663. 1991. BS (British Standards)/ES (European Standards).

Izumi, S., Yokoyama, T., Iwasaki, A., and Sakai, S. 2005. Three-dimensional finite element analysis on tightening and loosening mechanism of bolt joint, *Transaction of the Japan Society of Mechanical Engineering, Series A*, 71(702): 204–212.

Jeffrey, K. L. 2003. *The Toyota Way: 14 Management Principles from the World's Greatest Manufacturer*, McGraw Hill, New York.

Kakkar, S. and Narag, A. S. 2007. Recommending a TQM model for Indian organizations, *The TQM Magazine*, 19(4): 328–353.

Kawai, K. 2005. The issue and solution for information sharing in software development, in *Project Management Gakkai Kenkyu Happyo Taikai Yokoshu 2005 Shuuki*, pp. 209–211, September 27, 2005, Toyo University, Tokyo, Japan [in Japanese].

Keil, M. and Carmel, E. 1995. Customer-developer links in software development, *Communications of the ACM*, 38(5): 33–44.

Kirsch, J. M., Sambamurthy, V., Dong-Gil, K., and Purvis, L. R. 2002. Controlling information systems development projects: The view from the client, *Management Science*, 48(4): 484–498.

Kozaki, T., Yamada, H., and Amasaka, K. 2011. A highly reliable development design CAE analysis model: A precision approach to automotive bolt tightening, in *Proceedings of 2011 Asian Conference of Management Science and Applications*, pp. 1–8, December 22, 2011, Sanya, Hainan, China, (CD-ROM).

Lagrosen, S. 2004. Quality management in global firms, *TQM Magazine*, 16(6): 396–402.

Leonard-Barton, D.A. 2003. *The Factory as Learning Laboratory*, In Operations: Management Critical Perspectives on Business and Management. vol.2, pp. 456.

Ljungström, M. 2005. A model for starting up and implementing continuous improvements and work development in practice, *TQM Magazine*, 17(5): 385–405.

Lu, Y., Xiang, C., Wang, B., and Wang, X. 2011. What affects information systems development team performance? An exploratory study from the perspective of combined socio-technical theory and coordination theory, *Computers in Human Behavior*, 27(2): 811–822.

Mihara, R., Nakamura, M., Yamaji, M., and Amasaka, K. 2010. Study on business process navigation system "A-BPNS", *International Journal of Management and Information Systems*, 14(2): 51–58.

Ministry of Economy, Trade and Industry. 2010. The Field Survey of Service Industries, Tokyo, Japan [in Japanese].

Miyamoto, N. et al. 2006. Comparison between expert and non-expert in sharpening a kitchen knife, in *Symposium on Sports Engineering: Symposium on Human Dynamics*, pp. 187–192, November 8, 2006, Japan Society of Mechanical Engineers Kanazawa Ishikawa-ken, Japan [in Japanese].

Miyashita, S. and Amasaka, K. 2013. The proposal of the improved quality model "NVPS" of the Vietnam Automotive Manufacture, in *5th Annual Technical Conference of Japanese Operations Management and Strategy Association*, pp. 48–57, June 2, 2013, Osaka City University, Oosaka, Japan [in Japanese].

Murakami, Y. and Mizuyama, H. 2007. Detailed design and improvement of a manual assembly task incorporating how to efficiently handle uncertainties, *Journal of the Society of Plant Engineers Japan*, 19(3): 177–188.

Murata, D., Yutaka Kudo, Y., Hirai, C., Inoya, Y., and Mitomi, A. 2001. Development and evaluation of information sharing infrastructure for software development projects, *Joho Shori Gakkai Kenkyu Hokoku (GN)*, 48: 35–40 [in Japanese].

Mustafa, M. S. 2009. The proposal of New Turkish Production System utilizing Advanced TPS, Master's Thesis, Aoyama Gakuin University [in Japanese].

Nakamura, M., Enta, Y., and Amasaka, K. 2011. Establishment of a model to assess the success of information sharing between customers and vendors in software development, *Journal of Management Science*, 2: 165–173.

Nihon Keizai Shinbun. 2000. Worst record: 40% increase of vehicle recalls (July 6, 2000), Tokyo, Japan [in Japanese].

Nihon Keizai Shinbun. 2005. The manufacturing industry: Skill tradition tools uneasy (May 3, 2005), Tokyo, Japan [in Japanese].

Nihon Keizai Shinbun. 2012. Toyota recalls 880,000 trucks, (August 2, 2012), Tokyo, Japan [in Japanese].

Nomiya, H. and Uehara, K. 2004. Visual learning by set covering machine with efficient feature selection, in *Proceedings of 16th European Conference on Artificial Intelligence (ECAI2004)*, vol. 110, pp. 525–529, August 24, 2004, Valencia Spain.

Nozawa, Y., Yamashita, R., and Amasaka, K. 2012. Analyzing cavitation caused by metal particles in the transaxle: Application of high quality assurance CAE analysis model, *Journal of Energy and Power Engineering*, 6(12): 2054–2062.

Onodera, T. and Amasaka, K. 2012. Automotive bolts tightening analysis using contact stress simulation: Application to high quality CAE analysis approach model, *Journal of Business and Economics Research*, 10(7): 435–442.

Prajogo, D. 2006. Progress of quality management practices in Australian manufacturing firms, *TQM Magazine*, 18(5): 501–513.

Ramarapu, N. K., Mehra, S., and Frolick, M. N. 1995. A comparative analysis and review of JIT implementation research, *International Journal of Operations and Production Management*, 15(1): 38–49.

Roos, D., Womack, J. P., and Jones, D. T. 1990. *The Machine That Changed the World*, Rawson Associates, New York.

Sakai, H. and Amasaka, K. 2002. Construction of "*V-IOS*" for promoting intelligence operator, in *Proceedings of the 18th Annual Technical Conference of Japan Society for Production Management*, pp. 173–176. September 21, 2003, Nagasaki Institute of Applied Science, Nagasaki, Japan [in Japanese].

Sakai, H. and Amasaka, K. 2005. V-MICS, advanced TPS for strategic production administration: Innovative maintenance combining DB and CG, *Journal of Advanced Manufacturing Systems*, 4(6): 5–20.

Sakai, H. and Amasaka, K. 2006. Strategic HI-POS, intelligence production operating system: Applying Advanced TPS to Toyota's global production strategy, *WSEAS (World Scientific and Engineering and Society) Transactions on Advances in Engineering Education*, 3(3): 223–230.

Sakai, H. and Amasaka, K. 2007. Human intelligence diagnosis method utilizing advanced TPS, *Journal of Advanced Manufacturing Systems*, 6(1): 77–95.

Sakai, H. and Amasaka, K. 2008. Human-integrated assist system for intelligence operators, *Encyclopedia of Networked and Virtual Organization*, II(G-Pr): 678–687.

Sakai, H. and Amasaka, K. 2012. Creating a business monitoring system "A-IOMS" for software development, *Chinese Business Review*, 11(6): 588–595.

Sakai, H. and Amasaka, K. 2013. Establishment of Body Auto Fitting Model "BAFM" using "NJ-GPM" at Toyota, *Journal of Japanese Operations Management and Strategy*, 4(1): 38–54.

Sakai, H. and Amasaka, K. 2014. How to build a linkage between high quality assurance production system and production support automated system, *Journal of Japanese Operations Management and Strategy*, 4(2): 20–30.

Sakai, H., Waji, Y., Nakamura, M., and Amasaka, K. 2011. Establishment of "A-PPNS", a navigation system for regenerating the software development business, *Journal of Industrial Engineering and Management Systems*, 10(1): 44–54.

Shan, H. 2012. The proposal of new China production system utilizing New JIT, Master's Thesis, Aoyama Gakuin University [in Japanese].

Shan, H., Yeap, Y. S., and Amasaka, K. 2011. Proposal of a new Malaysia production model "NMPM", a new integrated production system of Japan and Malaysia, in *Proceedings of International Conference on Business Management 2011*, pp. 235–246. October 29, 2001, Miyazaki Sangyo-Keiei University, Miyazaki, Japan.

Suzuki, M. (ed.). 2005. *Encyclopedia of Physics*, Asakura-syoten [in Japanese].

Takahashi, T., Ueno, T., Yamaji, M., and Amasaka, K. 2010. Establishment of highly precise CAE analysis model—An example the analysis of automotive bolts, *The International Business and Economics Research Journal*, 9(5): 103–113.

Tanaka, M., Miyagawa, H., Asaba, E., and Hongo, K. 1981. Application of the finite element method to bolt-nut joints: Fundamental studies on analysis of bolt-nut joints using the finite element method, *Bulletin of the Japan Society of Mechanical Engineers*, 24(192): 1064–1071.

Ueno, T., Yamaji, M., Tsubaki, H., and Amasaka, K. 2009. Establishment of bolt tightening simulation system for automotive industry: Application of the highly reliable CAE model, *International Business and Economics Research Journal*, 8(5): 57–67.

Watanuki, K. and Tanaka, S. 2003. Development of knowledge visualization system for knowledge transfer, *The Japan Society of Mechanical Engineers*, 23: 111–112.

Whaley, R. C., Petitet, A., and Dongarra, J. J. 2000. Automated empirical optimization of software and the ATLAS project, Technical Report, University of Tennessee, Department of Computer Science, Knoxville, TN.

Yamaji, M. and Amasaka, K. 2006. New Japan quality management model, Hyper-cycle Model "QA & TQM Dual System"—Implementation of New JIT for Strategic management technology, in *Proceedings of the International Manufacturing Leaders Forum*, pp. 1–6, October, 24, 2006 (CD-ROM), Taipei, Taiwan.

Yamaji, M. and Amasaka, K. 2007. New Japan quality management model—Implementation of New JIT for strategic management technology, *International Business and Economics Research Journal*, 7(3): 107–114.

Yamaji, M., Sakai, H., and Amasaka, K. 2007. Evolution of technology and skill in production workplaces utilizing advanced TPS, *Journal of Business and Economics Research*, 5(6): 61–68.

Yamaji, M., Sakatoku, T., and Amasaka, K. 2008. Partnering performance measurement "PPM-AS" to strengthen corporate management of Japanese automobile assembly makers and suppliers, *International Journal of Electronic Business Management*, 6(3): 139–145.

Yamaji, M., Suzuki, K., Shimizu, K., Ishizuka, T., and Amasaka, K. 2006. Revolutions in QA & TQM, key to the success of globalized quality management—Implementation of new Japan model Science TQM, in *Proceedings of the 23rd Annual Conference of Japanese Society for Production Management*, pp. 85–88, March 19, 2006, Osaka, Japan [in Japanese].

Yanagisawa, K., Yamasaki, M., Yoshioka, K., and Amasaka, K. 2013. Comparison of experienced and inexperienced machine workers, *International Journal of Operations and Quantitative Management*, 19(4): 259–274.

Yeap, Y. S., Mustafa, M. S., and Amasaka, K. 2010. Proposal of "new Turkish production system, NTPS": Integration and evolution of Japanese and Turkish production system, *Journal of Business Case Study*, 6(6): 69–76.

Zhang, M. and Jiang, Y. 2004. Finite element modeling of self-loosening of bolt joints: Analysis of bolted joints, *ASME PVP*, 478: 19–27.

Chapter 11
Conclusion

The Japanese administrative management technology that contributed the most to the world in the latter half of the twentieth century is typified by the Japanese production system represented by the Toyota Production System. This system was kept at a high quality level by a manufacturing quality management system generally called just in time (JIT). However, a close look at recent corporate management activities reveals various situations where advanced manufacturers that are industry leaders are having difficulty due to unexpected quality-related problems. Some companies have slowed down their production engineering development and are thus facing a crisis of survival as manufacturers. Against this background, improvement of Japanese management technology is sorely needed at this time.

In the remarkable technologically innovative competition seen today, in order to realize manufacturing that ensures customer-first QCD, it is indispensable to first create a core technology capable of reforming the business process used for the technological development of divisions related to engineering design. Equally important, even for production-related divisions, is to develop new production technologies and establish new process management that, when combined, enable global production.

In addition, even the product promotion, sales, and service divisions are expected to carry out rationalized marketing activities that are not merely based on past experiences, so that they can strengthen ties with their customers. It is believed that the foundation of corporate survival is to establish a new quality-centered management technology that can link the management of the activities carried out by the divisions above with a view to enhancing the quality of their business processes.

Given this context and by predicting the form of next-generation manufacturing, the author established the New JIT, New Management Technology Principle in response to the urgent need to create a next-generation management technology. This world-leading, uniquely Japanese quality management principle is vital in ensuring that Japanese manufacturing does not fall behind in the advancement of technology. New JIT aims to contribute to producing globally consistent levels of quality and simultaneous production worldwide, which is the key to success in expanding global production.

New JIT consists of the following core technologies: Total Development System (TDS), Total Production System (TPS), Total Marketing System (TMS) employing, TQM-S (TQM Utilizing Science SQC, New Quality Control Principle, called Science TQM, New Quality Management Principle). These systems are then effectively linked to rationally achieve strategic management using the driving force in developing New JIT: Strategic Stratified Task Team Model, Epoch-Making Innovation in the Work Environment Model, Partnering Performance Measurement Model, and Strategic Employment of the Patent Appraisal Method.

Future successful global marketers must develop an advanced management system that impresses users and continuously provides high-quality products in a timely manner through corporate management. Since providing what customers desire before they are aware of that desire will become a more essential part of any successful manufacturing business, the author has constructed a customer science that utilizes New JIT.

More specifically, the author proposed a high-linkage model that is composed of a "triple management technology system." This system combines advanced TDS, advanced TPS, and advanced TMS for expanding uniform quality worldwide and production at optimum locations.

New JIT established by the author is now being subjected to verification of its validity at a number of Japanese manufacturers with a view to it being further developed and established as the New Japan Management Technology Model.

In recent years, New JIT has been applied at many leading Japanese companies where its effectiveness has been verified, and it is now known as a strategic global management technology model, surpassing JIT. It is hoped that this book will contribute to the evolution of management technology for expanding global production.

Index

A

Advanced production system, 24–25
   digital engineering and information technology, 25
   issues to be addressed, 24–25
   quality assurance, 25
   top management and manager, concerns, 24–25
Advanced TDS, 227–231
   design model, 227–229
   effectiveness, 231
   by Science SQC, 227, 230–231
   Strategic Development Design Model, 250–251, 303–304
Advanced TMS, 230–233
   core elements, 230
   effectiveness, 232–233
   SCCM, 276
   Strategic Development Marketing Model, 276
Advanced TPS, 42–46, 260–266
   foundation, 42–46
      demand in management technology, 43
      high-linkage cycle, business processes improvement, 45
      New JIT, strategic implementation, 44–45
      QCD achievement, manufacturing technology, 46
      strategic QCD studies, affiliated/nonaffiliated suppliers, 43–44
      Typified Japanese Production System, JIT, 42–43
   HIA, 393–396
      easy creation and modification, 393–394
      for intelligence operators, strategic implementation, 391
      simple know-how accumulation and easy access, 394–395
      system construction, 393–396
   and TPS, 433
   utilization of CAD and CAE data, 395–396
   Strategic Production Management Model, 260–266
   demand for new management technologies, 260
   HDP, 265
   high-cycle system, 262–264
   highly reliable production system, 265
   intelligent operators development system, 264
   intelligent quality control system, 264
   IPOS, 265
   New Japanese Production Model, 260–261
   NJ-GPM, 264
   principle, 260–261
   Renovated Work Environment Model, 264
   requirements, 229
   six core technologies, 264
   TPS-LAS, 264
   TPS-QAS, 265
   upgrading intelligence of production, 263
   V-MICS, 266
AECD-DM, *see* Automobile Exterior Color Design Development Model (AECD-DM)
Aging and Work Development Comfortable Operating System (AWD-COS), 268
AGTCA, *see* Attention-Grabbing Train Car Advertisements (AGTCA)
AIDA Model, *see* Attention-Interest-Desire-Action (AIDA) Model

APD-CM, *see* Automobile Profile Design Concept Method
ARIM, *see* Availability and Reliability Information Monitor System (ARIM)
ARIM-BL, *see* Availability and Reliability Information Administration Monitor System in the Body Line (ARIM-BL)
Attention-Grabbing Train Car Advertisements (AGTCA), 194, 200–201
    CAID analysis, short-term passenger group, 209
    cluster analysis, riding conditions, 208
    posters advertising events, 209
    train car ads, 210
Attention-Interest-Desire-Action (AIDA) Model, 345
Automobile Exterior Color Design Development Model (AECD-DM), 233–247
    application example, 239–243
        analysis of customers in 40s, 239
        Automobile Body Color Development Approach Model, 239–240
        Automobile Exterior Color Design Approach Model, 241
        categorizing preferences using cluster analysis, 241
        painted panels used for surveys, 242
        path from color elements to exterior color, 242
        textual expressions of exterior colors, 240–243
    business approach model, 238–239
    CS-CIANS, 233
    extraction example of text mining, 238
    fact-finding research at manufacturers, 236
    group into clusters, 235
    group into the color development process, 235
    indexing customer values, 238–239
    on-site investigations, 234–235
    problem areas and work processes, 235
    proposal, 243–248
        aesthetic evaluation data (preference), 245
        color development (step 4), 246–247
        effectiveness, 247–248
        elements of color, each panel, 245
        exterior color design, self-expression, 247–248
        goodness-of-fit test, 246
        market research (step 1), 243
        path from color elements to exterior color, 246–247
        preference survey (step 2), 243–244
        principle component analysis, 244
        survey on textural expressions (step 3), 244–245
    success factors and elements, 236–238
        both manufacturers, 237
        different manufacturers, 236–237
        identification, 236–237
        at paint manufacturers, 238
        visualizing, 236
Automobile Form and Color Design Approach Model, 293–300
    color and subjective customer impressions, 295–296
    covariance structure analysis and standardized estimates, 296
    eye camera and EEG, 293
    form/color and subjective color customer impressions, 296–298
    line-of-sight heat map, 300
    numerical representation of form, CAD, 293
    parameters, 297
    priority areas, 296
    research on design support method, 293–294
    self-expressive group, 295
    target selection, 295
    vehicle model for inspection, 300
    verification, 297–300
Automobile Profile Design Concept Method (APD-CM), 147–148
    CS-CIANS, 143–146
    customer science: consumer value, 142–143
    Lexus, application: development of customer science, 147–155
    new management technology principle, 142
Automotive product design and CAE, 372–388
    automotive bolt-tightening analysis simulation, 377
    bolt-loosening mechanism, 385–388
    bolt-tightening experiment, 377–380
    bolt-tightening simulator, 380–384
    higher-cycled product design, 372, 374–375, 388
    highly reliable analysis technology component model, 375–377
    product design process, 373–374
    product design reform, problems, 373
    status and issues, 374

Automotive transaxle oil seal leakage, 161–169
  CG navigation and intelligence CAE software, 167–169
  design changes and process control, 169–171
  effectiveness of market claim rate reduction, 170
  fault and factor analyses, 164–166
  influential effect of each factor, 167
  mechanism through visualization, 161–162
  oil leakage mechanism, 163–164
  oil seal function, 161
  oil seal visualization equipment, 162–163
  Weibull analysis, result, 165
Availability and buffer number, relation model, 178–179
  changing of buffer number, 178
  flowchart, optimized buffer number, 179
  line processes, 178
  optimized buffer number, 178
  production number, 179
Availability and Reliability Information Administration Monitor System in the Body Line (ARIM-BL), 40
Availability and Reliability Information Monitor System (ARIM), 271
AWD-COS, *see* Aging and Work Development Comfortable Operating System (AWD-COS)

B

Body Auto Fitting Model (BAFM), 328–344
  advanced robot illustration, 342
  advanced TPS, 329
  algorithm, 339
  body assembly plant and automation level, 331
  bolt design influences, 340
  bolt position measurement study, 341
  bolt recognition program, 341
  conventional production system, 328
  coordinate conversion algorithm, 340
  effect of floor space reduction, 343
  equipment specifications, 341
  establishment, 329–331
  exterior component installation, 332
  fitting accuracy, 333
  fitting adjustment method, 334
  floor space-efficient system, 342
  hole center detection, 338
  ideas for global production, 328
  linkage among production processes, 329–330
  luggage compartment panel, 333
  luggage-fitting standards, 334
  measurement system, 336
  measuring fitting position, 338
  preliminary study and initial problems, 335
  problem solutions, 337
  standards for door fitting, 335
  statistical processing of image recognition data, 338
  strategic project target, 331–333
  system evaluation, 344
  thrust force control method, 342
  tightening reliability cause-and-effect diagram, 341
  two-dimensional CCD camera, 337
Bolt-loosening mechanism, 385–388
  CAE analysis results, 388
  3-D analysis of pitch differences, 387
  highly precise CAE analysis, 387
  results of axis symmetry 2-D model, 385–386
  stress distribution, nut bearing surface, 385–386
Bolt-tightening experiment, 377–381
  angle and torque, relationship, 378
  bolt and nut measurements, 379
  bolt vibration test, 380
  CAE, 377
  friction coefficients, 377–378, 380
  transition of bolt load, 380–381
Bolt-tightening simulator, 377–385
  automotive, using experiments, 377
  axis symmetry 2-D model, 382
  CAE results, 384–385
  creation of CAE software, issues, 380–381
  3-D finite element model, 382–383
  finite element method (FEM), 381
  highly reliable CAE analysis technology component model, 381–382
  stress of surface of substrate, 383–384
  stress of the thread contact section, 384–385

C

CAD, *see* Computer-aided design (CAD)
CAE, *see* Computer-aided engineering (CAE)
CAM, *see* Computer-aided manufacturing (CAM)
CD, *see* Customer delight (CD)
CE, *see* Concurrent engineering (CE)

450 ■ *Index*

CIM, *see* Construction information modeling (CIM)
CMP-FDM, *see* Customer Motion Picture-Flyer Design Method (CMP-FDM)
Computer-aided design (CAD), 6
Computer-aided engineering (CAE), 6, 155
  advanced TDS, 303–304
  aligning prototype testing and CAE, 306–307
  automotive product design (*see* Automotive product design and CAE)
  bolt-loosening mechanism, 387
  current status and issues, 303
  define the problem, 304
  development design, 303
  drivetrain oil seal leaks, 308–314
    basic oil seal lip structure, 311–312
    CAE analysis, 313
    cavitation analysis, 313–314
    defining problem and visualization device, 309–310
    fluid speed analysis, 313–314
    oil leakage mechanism, 309–310
    oil seal function, 308
    oil seal visualization device, 310
    optimal design approach model, 310–311
    pressure analysis, 313–314
    three-dimensional analysis, 313–314
    two-dimensional analysis, 312
    visualization, oil seal leakage mechanism, 308
  faulty mechanism, 307
  highly precise CAE analysis, 307
  Highly Reliable CAE Analysis Technology Component Model, 306
  predict and evaluate, 307–308
  product design: application and issues, 302–303
  requirements, 303
  TPS-LAS, 172–177
    computer simulation manufacture, 172–173
    defect and of defect recurrence prevention, 176
    defect data collection method, 177
    digital engineering, 172–173, 177
    digital factory simulation, 174
    functions and requirements, 175
    Inline-Online SQC methodology, 176
    LIS, 174
    process layout CAE system, 173
    TVAL, 174
    WIS, 174
  visualization experiment, 306
Computer-aided manufacturing (CAM), 6
Concurrent engineering (CE), 6
Construction information modeling (CIM), 6
CR, *see* Customer retention (CR)
CS, *see* Customer satisfaction (CS)
CS-CIANS, *see* Customer science utilizing the Customer Information Analysis and Navigation System (CS-CIANS)
Customer delight (CD), 39
Customer Motion Picture-Flyer Design Method (CMP-FDM), 194, 199–200, 205–208
Customer Purchasing Behavior Model of Advertisement Effect (CPBM-AE) for automotive sales, 278–280
Customer retention (CR), 39
Customer satisfaction (CS), 39
Customer science utilizing the Customer Information Analysis and Navigation System (CS-CIANS), 77, 143–146, 233
  consumer value utilizing New JIT, 142–143
  customer information, 144
  generating customer value, 233
  networking, 144–145
  proposal, 142
  schematic drawing, 143
  strategic implementation of New JIT, 143
  strategic product development, conception, 143–144
Customer value (CS), 88

D

DFS, *see* Digital factory simulation (DFS)
Digital engineering, 7, 25, 44, 259
Digital factory simulation (DFS), 171
Driving force in developing New JIT, 86–112
  Epoch-Making Innovation, Work Environment Model, 98–112
  Partnering Performance Measurement Model, 112–134
  Strategic Stratified Task Team Model, 86–98

E

Electroencephalogram (EEG), 293
Employee satisfaction (ES), 57

Epoch-Making Innovation in Work
  Environment Model
  activity examples, 106–112
    break pattern comparison, 108
    changed break time and effect, 107
    comfortable work clothes and test on
      model lines, 109
    effect on productivity, 108
    fatigue, types and evaluation, 106–112
    line stop time, work delays, 109
    outline of projects, 111
    perceived effect and free opinions, 108
    principal component analysis, awareness
      survey results, 110
    study work standards to reduce fatigue, 106
    temperature conditions, minimizing
      fatigue, 109
  definite plan for concept actualization,
    104–106
    AWD6P/J structures, 104–105
    problem solving for projects, diagram,
      105
    Science SQC approach, 106
    total task management team activity,
      104–105
  new management technology principle,
    99–100
    establishment of, New JIT, 99–100
    TPS, 100
  prerequisites, quality, 101–104
    considerations for older workers, 101–103
    new concept for assembly, 102
    relation diagram, objectives of projects
      and team, 104
    variation of aging workers, 101
Experienced workers, skills, 408–420
  comparing the working efficiency, 408, 410
  decision-making criteria, statistical analysis,
    413–414, 420
    brain response points and work tasks,
      413
    discriminant coefficients for cutting tool
      material, 414, 419
    multiple regression analysis results, 414,
      416, 418
    rough machining/finishing cutting
      speeds, 414, 417
    work pieces and cutting speeds,
      414–415
    X-Rs control chart, 412
  and inexperienced workers' cognition,
    electroencephalography, 410–412
  visualizing the skills, 408–421
  work tasks and work times, 409

F

Failure mode and effect analysis (FMEA), 55
Fault tree analysis (FTA), 55

G

Global Network System for Developing
  Production Operators (GNS-DPO),
  428

H

HDP system, *see* Human digital pipeline
  (HDP) system
HIA system, *see* Human-Integrated Assistance
  (HIA) System
Higher-Cycled Product Design CAE Model,
  374–375, 388
  application to similar problems, 388
  automotive product design and production,
    372
  proposal, 374–375
Highly Reliable Development Design CAE
  Model, 249–259
  advanced TDS, 250–251, 255
  collective strengths, 256
  development targets, 254
  evolution of CAE, 252
  intelligence and high precision
    implementation, 258
  issues, 253
  need for new design model, 249–250
  prediction-based design process, 254
  problematic phenomena, 256–257
  technical element analysis, 257
  theoretical analysis model, 257–258
  urethane foam molding simulator
    development, 257, 259
  utilization and education, 252
High-probability customers (HPCs), 282
HI-POS, *see* Human Intelligence Production
  Operating System (HI-POS)
HPCs, *see* High-probability customers
  (HPCs)
Human digital pipeline (HDP) system, 260,
  271–272
  accumulated work operations, 326
  advance setting for avoidance points, 322

assembly procedure manual using design
data, 324
calculation of coordinate locations, 322
contribution to Toyota's global production,
326
element instruction sheet, 324
hardware configuration, 320–321
HI-POS, intelligence production system,
317–318
human resources compatible with digital
production, 316–317
operator interference, 322
productivity/quality evaluation, 327
simulation algorithm, 321–322
simulation result, 324–325
software configuration, 321
structuring and components, 318–319
verification of operators' interference, 325
work operation combination, 323, 325
Human-Integrated Assistance (HIA) System,
390–403
advanced TPS, 391, 393–396
developing intelligence operators for
production, 390–393
shortened training methods, 396–399,
402–403
skill training for newly employed
production operators, application
case, 400–402
Human Intelligence Production Operating
System (HI-POS), 260, 268

I

Inexperienced workers, 410–412
Intelligent customer information marketing
system, 274–281
business and sales activities innovation, 275
CPBM-AE, 278–280
effective mixed-media model, establishment,
281
HPCs, 282
innovating dealers' sales activities, 275
JSMS, 281
LPCs, 282
market/customer domain, 278
marketing strategy domain, 278
MPCs, 282
need for marketing strategy, 274–275
product, price, place, and promotion (4P), 274
SCCM, 274, 276–277
SMDM, 274, 276

Intelligent Production Operating System
(IPOS), 265, 268

J

Japanese manufacturing, 25–26
administrative management technology, 1–2
new product, launch of, 25–26
QC circle activities, 25
quality technology, 25–26
Japanese Sales Marketing System (JSMS), 281
Japanese TQM, 34
Japan supply system, 112
JSMS, *see* Japanese Sales Marketing System (JSMS)

K

Kiichiro Toyoda (Toyota Motor's founder), 11

L

Lean System, 7, 13–14, 34
machining process, 14
one-by-one (single part) production, 13–14
standard work, 14
Lexus, application: customer science
development, 147–155
APD-CM, 147–148
collage panel image analysis, 149–150
customer satisfaction and vehicle appearance
assessment, 152
design preferences, 151–153
preferences by generation, 149–151
psychographics, 153–155
questionnaire for image survey by
generation, 148
researching design images, collage panels, 148
sense of values, collages, 148
TDS-DTM, 155
vehicle model classification, class degree/
year model, 154
Logistics investigation simulation (LIS), 171, 174
Low-probability customers (LPCs), 282

M

Machine workers, comparing experienced and
inexperienced, 404–422
prior research, 407–408
research evaluation, 421–422
skill transfer problem, 405, 407
visualizing skills, 408–421

Man, machine, material, method, and environment (4M-E), 14, 27
Management tasks, manufacturing companies, 5–7
    Japanese TQM (TQC), 7
    progress of production control, 5–7
    Toyota Production System and TQM, 7
Management technology issues, 7–8
    Lean System, 7
    New JIT, 8
    new marketing methods, 8
    next-generation production system, 8
    QCD research, 8
    Quantification Class III, 7
    TQM, 7
Manufacturing fundamentals surpassing JIT, 24–30
    advanced production system, technologies and skills, 24–25
    Japanese manufacturing: quality technology, 25–26
    overseas production plants, 27–29
    reinforcement, production management function, 29–30
    technological capability, Japanese manufacturers, 30
Medium-probability customers (MPCs), 282

N

New Global Partnering Production Model (NGP-PM)
    global production, problems, 423–424
    NTPS (*see* New Turkish Production System (NTPS))
    overseas manufacturing up to Japanese standards, 427–429
    significances, 427
    strategic use of New JIT, 424–427
New Japanese Global Production Model (NJ-GPM), 260–261, 264
New Japanese Production Model (NJPM), 428
New JIT, New Management Technology Principle, 2, 8
New Software Development Model (NSDM)
    current status, 360
    information sharing, 360–361
    successful information sharing between customers and vendors, 361–370
        classifying, 362, 365
        cluster analysis, result of, 365
        data collection, 361, 366
        diagnostic system, 370
        in software development (*see* Software development information sharing)
        success-boosting factors, 363–369
New Toyota Production System, 40–41
New Turkish Production System (NTPS)
    and advanced TPS (*see* Advanced TPS)
    automobile industry, current status, 430
    bringing up intelligent operators, 436–437
    concept, 430–431
    extraction of important keywords, 431–432
    Japan's leading automobile industry, current status, 429–430
    New Japanese Production Model, 429–430
    operator training process for assembly works, 435–436
    outline, 434–435
    text mining analysis, 432, 434
Next-generation production control system, 6, 8
NGP-PM, *see* New Global Partnering Production Model (NGP-PM)
NJ-GPM, *see* New Japanese Global Production Model (NJ-GPM)
NJPM, *see* New Japanese Production Model (NJPM)
NSDM, *see* New Software Development Model (NSDM)
NTPS, *see* New Turkish Production System (NTPS)

O

Optimal selection using a model formula, 214–217
    model formula, 215–216
    recipient selection process, 216–217
Overseas production plants, 27–29
    production system issues, 28–29
        on-site capabilities, 28
        uniform quality and simultaneous launch, 29
    style of manufacturing, 27–28
        man, machine, material, method, and environment (4M-E), 27
        reinforced inspection, process management, 27–28
        same quality and time as Japan, 27
        statistical quality control (SQC), 27

P

Partnering Performance Measurement for Assembly and Suppliers (PPM-AS), 114–119
  assembly makers evaluators, 115
  cluster analysis, grouping, 116
  formulation model of PPM-A, PPM-S, and PPM-AS, 116
  General Motors, 122–123
  GM and Jatco, radar chart of, 123
  NHK evaluation sheet, 121
  Nissan and Jatco, radar chart of, 122
  Nissan Motor Co., Ltd., 121–122
  and PPM-A, PPM-S, evaluation sheets/formulation model, 114–116
  PPM-A derivation, 116–117
  PPM-A radar chart, 118
  PPM-S derivation, 117
  PPM-S radar chart, 118
  principle component analysis, grouping, 116
  questionnaire result of Toyota and NHK, 121
  radar chart, 119
  suppliers evaluators, 115
  Toyota and NHK, radar chart, 122
  Toyota evaluation sheet, 121
  Toyota Motor Corporation, 120
  verification, 119–123
Partnering Performance Measurement Model, 114–119
Patent value appraisal method (PVAM), 125–134
  intellectual property department, 125–126
    elements, 125
    value (cash flow) of intellectual property, 126
  proposal, 127–133
    chartered patent agents, 128–129
    cluster analysis, 129
    development engineers, 129
    equation, 130–132
    intellectual property department staff, 129
    potential factors, 127
    PVAM software, 132–133
    radar chart, potential factors, 132
    strategic patents, 128
    variables, 127
  significance of patent acquisition, 124–125
  corporate reform and patent acquisition, 124–125
  patent applications in Japan, 124
  strategic patents, 126–127
    engineers' values, 126
    potential structures, 126
  verification of effectiveness, 133–134
PM, *see* production management (PM)
PMOS-DM, *see* Practical Method Using Optimization and Statistics for Direct Mail (PMOS-DM)
PPM-AS, *see* Partnering Performance Measurement for Assembly and Suppliers (PPM-AS)
Practical Method Using Optimization and Statistics for Direct Mail (PMOS-DM), 194, 202
  effectiveness, 212–213
  putting to work, 210–211
  response likelihood by attribute, 211
  verification results, 213–214
Process control and process improvement, 14–21
  condition controls, 17
  daily control activities, production site, 17
  daily improvement activities, production site, 17–18
  innovation, QCD research, 18
  manufacturing methods and costs, 15
  process control, poor cases, 16
  process improvements, examples, 20–21
  production layout, example, 19
  QCD improvement measures, results, 20
Process layout CAE system, intelligence TPS, 171–182
  DFS, 171
  LIS, 171
  relation model, availability and buffer number, 177–179
  TPS-LAS application, 171–177, 179–182
  WIS, 171
Production control, manufacturing industry, 5–7
  methodologies, 6
  next-generation production control system, 6
  supply chain management (SCM), 7
Production management (PM), 76
  cooperation, 29
  experience-based implicit knowledge, visualization, 29–30
  quality control, 30
  PVAM, *see* Patent value appraisal method (PVAM)

Q

QCD studies, *see* Quality, cost, and delivery (QCD) studies
QCIS, *see* Quality Control Information System (QCIS)
Quality, cost, and delivery (QCD) studies, 47–50
    assembly process for rear axle units, 50
    customer-first QCD, 1–2
    douki-ka production, 47
    high quality assurance, shock absorbers, 19
    improvement measures, 20
    machining process for rear axle shaft, 49
    parts assembly, 19–20
    principle of manufacturing, 18
    process improvement, 19
    production process of rear axle units, 47–48
    rear axle unit processes, 48
    TQM, 34
    worker's standard operation time, 19
    work-piece types, 20
Quality Control Information System (QCIS), 271

R

Rear axle unit
    assembly process
        achievement of QCD, 51–54
        bolt-tightening tools, 54
        improvements, 53
        of paint, 52
        styrene modified alkyd resin solvent coating, 52
    shaft processing
        automated stress relief equipment, 51
        extra coating operations, 50
        midfrequency tempering equipment, 50–51
Renovated Work Environment Model, 264
Robot Reliability Design and Improvement Method (RRD-IM), 186–193
    actual line evaluation, 192–193
    body assembly line, 189–193
    line availability tendency, 192
    practical robot failures, 189
    random failure, 190–191
    reliability achieved situation, 192
    TPS, 183–188
        accelerated life via high temperature, 186
        Arrhenius plot between temperature and holding time, 186
        body assembly model line, 187–188
        checking wear under no lubrication conditions, 188
        concept and structural elements, 185
        evolution, 183–184
        in global production, 184
        random failure, 187
        wear failure, 188
    wear failure, 191
    Weibull analysis results, 190

S

Sakichi Toyoda (Toyota Group founder), 11
SCCM, *see* Scientific Customer Creative Model (SCCM)
Science SQC, 62–64
    advanced course, curriculum, 72
    beginner and business courses, 72
    as demonstrative scientific methodology, 65
    discriminate analysis result, 75
    division of seminarian and human resources training, 71
    DOS-Q5, achievement, 77
    drive system unit, quality improvement, 72–74
    engineering problem, 73
    establishment, technical methods, 67
    estimated oil leakage mechanism, 76
    hierarchical seminar and human resources growth, 71
    implementation of management, 74
    Integrated SQC Network TTIS, construction, 67
    jidoka (automation), 64
    lectures, 72
    Management SQC, schematic drawing, 70
    manufacturers' management technology, improvement, 64
    5M-E, 64
    need for new SQC, 64–66
    next-generation TQM management, 66
    next-generation-type management technology, 65
    practical application, 66
    principles, 36
    problem solving, 75
    promotion cycle, consistency/schematic drawing, 69–70
    quality improvement, conventional approaches, 73
    recommendation, 69

relation chart method of activities, 74
schematic drawing, 66
technical methods, schematic drawing, 68
total task management team activities, 73–74
TQA-NM development using PM, 76
TTIS, schematic drawing, 68
visualization of oil leakage mechanism using TM, 76
Science TQM, 2
Scientific Customer Creative Model (SCCM), 274, 276
Scientific Mixed Media Model (SMMM), 194–213
    AGTCA, 200–201, 208–210
    boosting automobile dealer visits, 198
    CMP-FDM, 199–200, 205–208
    customer purchase behavior, 202–204
    customer science, 196
    evolution of market creation employing TMS, 196
    flyers for the purchase group, 207
    marketing strategy, market trends, 194–195
    new attractive flyer design, 207
    PMOS-DM, 202, 210–213
    survey questions, 203–204
    VUCMIN, 199, 204–206
SCM, *see* Supply chain management (SCM)
Shortened training methods, 397–403
    assessment of aptitude/inaptitude by aptitude test, 398
    intelligent IT system, 397–398
    issues, 396
    for new overseas production operators, application case, 402–403
    optimization of training steps, 399
    repetitive training utilizing the visual manual, 397
Simultaneous engineering (SE) activities, 156
SMDM, *see* Strategic Marketing Development Model (SMDM)
SMMM, *see* Scientific Mixed Media Model (SMMM)
Social satisfaction (SS), 57
Software development information sharing, 365–366
    diagnostic system, 369–370
    information-sharing factors, 362
    problem and solution factors, 361
SQC, *see* Statistical Quality Control (SQC)
Statistical Quality Control (SQC), 2, 6, 12
    integrated network, 36
    management, 36
    scientific (*see* Science SQC)
    technical methods, 36
Strategic Development Design CAE Model, TDS, 155–171
    automotive development production and simulation technology, 156–157
    CAE and SE activities, 155–156
    CG navigation function, 156
    Intelligence CAE Management Approach System, 159–161
    proposal, 157
    Stratified Intelligence CAE Development Design System, 158–159
    Total QA High Cyclization Business Process Model, 157
    transaxle oil seal leak mechanism, 155–171
Strategic Marketing Development Model (SMDM), 274
Strategic Stratified Task Team Model, developing New JIT, 86–97
    cooperative creation team, 92
    creative corporate climate, business management and development, 86
    practice and effectiveness, 92–97
        changes in team activities, advanced car manufacturer, 93
        joint total task management team activity, 97
        outline of management SQC seminar, 95
        progress of team activities, 92
        task management team activities, 95
        task team activity, 94
        total task management team activity, 96–97
    QC theory, 90
    Science SQC, 91
    SQC promotion cycle, 89
    SQC renaissance, 89, 91
    strategic cooperative creation team, construction, 91–92
    Japan supply system, 91
    methodology, 91
    promotion, 91
    strategy, 91–92
    technology, 91
    structured model, 90
    team activities (*see* Team activities and requirements)
    TQM-S, 91
    TTIS, 90
Supply chain management (SCM), 88

T

TDMM, *see* Total Direct Mail Model (TDMM)
TDS, *see* Total Development System (TDS)
TDS-DTM, *see* Total Development System-Design Technical Methods (TDS-DTM)
Team activities and requirements, 86–91
    creativity and strategy, 89–91
        cross-functional collaboration, 89
        customer value improvement, 88
        elements and requirements for promotion, 88
        implementation stage, 89
        QCD activity, improvement, 88
        social satisfaction (SS), 88
        supply chain management (SCM), 88
    promotion, 88
        special knowledge, 88
        thinking capability, 88
        value of job, 88
        vision and mission, 88
    reliability of job and importance, 86–87
        corporate organization chart, 87
        incorporating team activities, 86
        job quality, 86
Text mining analysis, 432, 434
TMS, *see* Total Marketing System (TMS)
Total Development System (TDS), 2, 36–40
Total Development System-Design Technical Methods (TDS-DTM), 155
Total Direct Mail Model (TDMM), 344–356
    advertising method, 344
    AIDA Model, 345–346
    core systems, 345
    customer purchase data, 347–348, 352
    and customer retention, 345
    effectiveness, 351
    likelihood of dealership visit, 349
    necessity, 345
    optimizing direct mail content, 349–350
    premier customer segments, 347
    promotion system for sales staff, 351
    research on direct mail activities, 346
    static and dynamic customer, 351
    strategically determining direct mail recipients, 350–351
Total Link System Chart (TLSC) for HI-POS, 269
Total Marketing System (TMS), 2, 36, 38–39
Total Production System (TPS), 2, 36, 38

QAS, 40
and TQM, JIT, 34
Total Quality Assurance Networking Model (TQA-NM)
    chart creation, 58–59, 63
    deploying TQA-NM, 62
    ES and SS, 57
    experience and skill, technical personnel, 55
    FMEA, 55
    FTA, 55
    guidelines, 60
    level-based application chart, 59–60
    manufacturing in Japan today, 55
    necessity of preventing defects, 55–56
    occurrence and outflow prevention rankings, 56, 59–60
    outline, 58
    product management, 62
    quality assurance departments, issues faced by, 56–57
    for strategic distribution, 60–61
    technical management, 62
    technological components, 55
Total quality control (TQC), 7, 12
Total quality management (TQM), 7, 12
Total Quality Management System (TQM-S), 2
Total SQC Technical Intelligence System (TTIS), 90
Toyota Production System, 11–21
    JIT, core concept, 11–13
        definition, 1
        leaning process, 12
        Lean System, 12–13
        QCD research, 12
        Sakichi Toyoda mottos, 11
        SQC, 12
        TQC and TQM, 12
    manufacturing, basic principle, 13–21
        lot production, 13–14
        one-by-one production, 13–14
        process control and improvement, 14–21
        production engineering and process design, 14
        quality and productivity via Lean System, 13–14
    and total quality management, relation, 12
Toyota Sales Marketing System, TMS, 41
Toyota Verification Assembly Load (TVAL), 174
TPS, *see* Total Production System (TPS)

TPS Intelligent Production Operating System (TPS-IPOS), 260
TPS Layout Analysis System (TPS-LAS), 260, 267–268
    advance verification, 180
    effectiveness, 181–182
    process layout analysis simulation, 171
    proposal, process layout CAE system using, 172–177
    TPS-LAS-DFS, 180
    TPS-LAS-LIS, 181
    TPS-LAS-RCS, 180
TPS Quality Assurance System (TPS-QAS), 260, 265, 271
TQA-NM, *see* Total Quality Assurance Networking Model (TQA-NM)
TQC, *see* Total quality control (TQC)
TQM, *see* Total quality management (TQM)
TQM-S (TQM utilizing Science SQC), 35
TTIS, *see* Total SQC Technical Intelligence System (TTIS)
TVAL, *see* Toyota Verification Assembly Load (TVAL)
Twenty-First Century Manufacturing Vision: New Wheel-Type Manufacturing Model, 30

V

Validity of advanced TDS, TPS, and TMS, 231–233
    effectiveness, 231–233
    market creation, 232
    SMMM, 232
    TMMM, 232
Videos that Unite Customer Behavior and Manufacturer Design Intentions (VUCMIN), 199, 204–206
Virtual Intelligent Operator System (V-IOS), 270
Virtual Maintenance Innovated Computer System (V-MICS), 260, 266, 273
Visual manual creation process, 393–394
V-MICS, *see* Virtual Maintenance Innovated Computer System (V-MICS)
VUCMIN, see Videos that Unite Customer Behavior and Manufacturer Design Intentions

W

Weibull analysis, 165
Workability investigation simulation (WIS), 171, 174